T0318637

STRUCTURE AND TECTONICS OF THE INDIAN CONTINENTAL CRUST AND ITS ADJOINING REGION

Deep Seismic Studies

STRUCTURE AND TECTONICS OF THE INDIAN CONTINENTAL CRUST AND ITS ADJOINING REGION

Deep Seismic Studies

HARISH C. TEWARI
B. RAJENDRA PRASAD
PRAKASH KUMAR

Elsevier
Radarweg 29, PO Box 211, 1000 AE Amsterdam, Netherlands
The Boulevard, Langford Lane, Kidlington, Oxford OX5 1GB, United Kingdom
50 Hampshire Street, 5th Floor, Cambridge, MA 02139, United States

Library of Congress Cataloging-in-Publication Data
A catalog record for this book is available from the Library of Congress

British Library Cataloguing-in-Publication Data
A catalogue record for this book is available from the British Library

ISBN: 978-0-12-813685-0 (print)
ISBN: 978-0-12-813686-7 (online)

For information on all Elsevier publications
visit our website at https://www.elsevier.com/books-and-journals

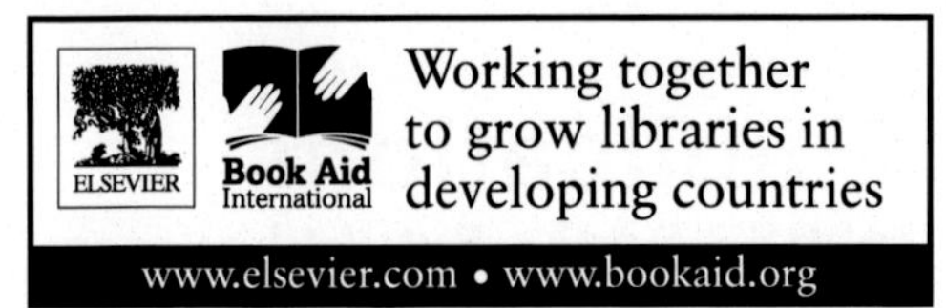

Acquisition editor: LaFleur, Marisa (ELS-CMA)
Editorial Project Manager: Hilary Carr
Production PM: Maria Bernard
Publisher: Candice Janco
Cover Designer: Christian Bilbow

Typeset by SPi Global, India

CONTENTS

CHAPTER ONE

Overview of Crust and Introduction to Seismic Observations on Indian Plate

Five billion years ago the planet Earth was formed as a large conglomerate. The immense amount of heat energy released by massive high-velocity bombardment of meteorites and comets melted the entire planet. Since then, the planet has been cooling off and the process continues even today. During the cooling process, denser materials, such as iron from meteorites, sank into the core of the Earth while lighter material, e.g., silicates, oxygen compounds and water from comets, rose near to the surface.

Studies of the Earth's structure over the last 100 years have proven that the Earth consists of several layers. Characteristic properties of each layer are different in terms of physical and chemical parameters. The chemical parameters, e.g., alkalinity, acidity, salinity, etc., and the physical parameters, e.g., pressure, temperature, density and elasticity, vary from layer to layer. The parameters of elasticity and density determine the seismic wave velocity, which normally is different for each of these layers.

From the study of the seismic wave velocity and density, the Earth has been subdivided into four main units (Fig. 1.1): the inner core, outer core, mantle and crust (Ritter Michael, 2006). The equatorial radius of the Earth is 6378 km, out of which the inner core is about 1250 km, the outer core 2200 km and the mantle 2900 km thick, respectively. The core is composed mostly of iron and is so hot that its outer part (outer core) is molten, with about 10% sulfur. The inner core is under extreme pressure and remains solid. Most of the Earth's mass is in the mantle, which is composed of iron, magnesium, aluminum, silicon and oxygen compounds. At over 1000°C, the mantle is solid but can deform slowly in a plastic manner. The crust is composed of the least dense calcium and sodium/aluminum-silicate minerals. Being relatively cold, the crust is rocky and brittle and therefore is easier to fracture.

Structure and Tectonics of the Indian Continental Crust and Its Adjoining Region
https://doi.org/10.1016/B978-0-12-813685-0.00001-7

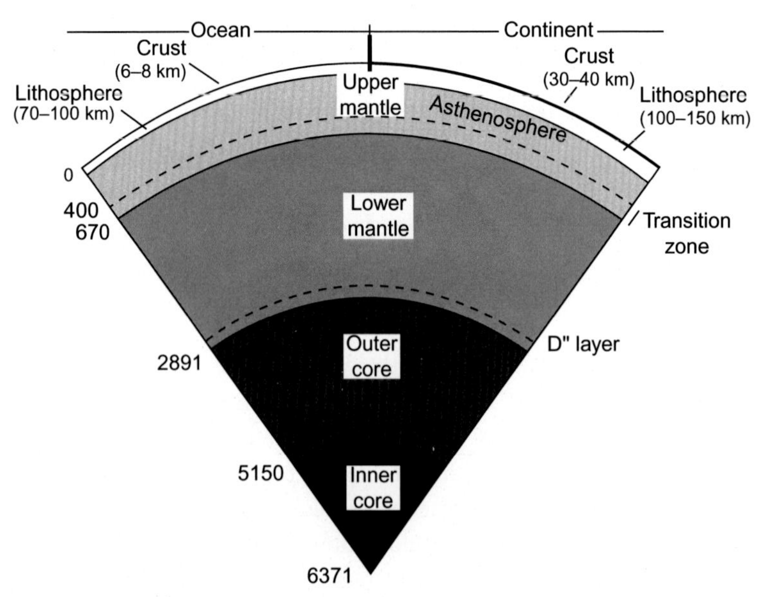

Fig. 1.1 Generalized structure of the Earth. Depths to major boundaries are given.

1.1 LITHOSPHERE AND ASTHENOSPHERE

The crust and the uppermost mantle, down to a depth of about 70–100 km under the deep ocean basins and 100–200 km under the continents, is called the lithosphere. It is rigid and forms a hard outer shell that deforms in an essentially elastic manner. The lithosphere is composed of various plates that float on partially molten asthenosphere. Delineation of an unambiguous boundary that separates the lithosphere from the underlying asthenosphere has not yet become possible, most likely because the asthenosphere under old continental platforms is imaged as a broad zone in the seismic velocities. Here, instead of a single low-velocity zone, a series of high- and low-velocity layers are intermingled (Fuchs et al., 1987).

The upper mantle plays a crucial role in structural development of the Earth's crust. Critical levels of thermodynamic conditions prevail in individual zones of the upper mantle, under the influence of differential and thermoelastic stresses. A discontinuous increase or decrease of volume takes place due to polymorphic, phase and chemical transformation of the inhomogeneous mantle substance. Distribution of structural forms within the Earth's crust and mineral deposits on the surface have a close dependence on the processes occurring in the upper mantle.

The asthenosphere is a rheologically weak, semiviscous layer in the upper mantle. In this layer the velocities often decrease, suggesting lower rigidity. This weaker layer is thought to be partially molten; the melt may be able to flow over long periods of time like a viscous liquid or plastic solid, in a way that depends on its temperature and composition. The asthenosphere plays an important role in plate tectonics, because its viscous state allows relative motion of the overlying rigid lithospheric plates. Some of the researchers suggest that the asthenosphere should be defined not as a weak upper mantle layer but as a zone of partial melting (Pavlenkova, 1988).

The lithosphere is divided into several plates, of which the crustal component could be either continental or oceanic. Very little progress has been achieved so far in understanding the evolution of the continental lithosphere, due to the inaccessibility of its subcrustal part for direct studies. Explosion seismology studies in different tectonic settings (viz. old Precambrian shields, young continental platforms and the oceans) show that several velocity layers exist in the upper mantle (Mooney and Meissner, 1992). The most important findings are: (1) occurrence of the low velocity layers at shallow depths in the continental upper mantle, with large velocity contrasts at their boundaries; and (2) observation of unexpectedly high-compressional (P) wave velocities, up to 8.6–8.9 km s^{-1}, and high-velocity gradients of 0.02–0.04 km s^{-1} at depths of 10–30 km below the crust-mantle boundary (Bean and Jacob, 1990). These findings provide indirect evidence that the elastic anisotropy continues within the uppermost mantle.

1.1.1 The Crust

The crust covers the mantle and is the Earth's hard outer shell, the surface on which we are living. Compared to other layers of the Earth, the crust is much thinner, like a stamp on a football. It generally consists of solid material, but this material is not the same everywhere and is less dense and more rigid than the material of the Earth's mantle. The crust over the oceans is different in nature as compared to the rocks of the continental crust. The oceanic crust is about 6–11 km thick and the rocks in it are very young, not older than 200 million years, compared to the rocks of the continental crust. Its igneous basement consists of a thin (about 500 m thick) upper layer of superposed basaltic lava flows underlain by basaltic intrusion, the sheeted dike complex and the gabbroic layer. A greater part of the oceanic crust consists of the tholeiitic basalt (basalt without olivine), which has a dark, fine and gritty volcanic structure. It is formed out of liquid lava, which cools off quickly. The grains are so small that they are only visible under a microscope. The average density of the oceanic crust is 3000 kg m^{-3}.

The crust under the continents and areas of shallow seabed close to their shores (continental shelves) is called the continental crust. It covers more than one-third of the Earth's surface. It is thicker than the oceanic crust, 35–40 km thick under the stable areas and 50–80 km under the young mountain ranges, and mainly consists of igneous rocks. It is divided into two layers. The upper crust mainly consists of sediment, gneiss, granite and granodiorite rocks, while the lower crust consists of basalt, gabbros, amphibolites and granulites. The average density of the upper crust is 2700 kg m^{-3} while that of the lower crust is 2850 kg m^{-3}. It is older than the oceanic crust; some rocks are as old as 3800 million years. When active margins of the continental crust meet the oceanic crust in subduction zones, the oceanic crust is subducted due to relatively lower density of the former. The lower density does not allow the continental crust to be subducted or recycled back into the mantle. For this reason, the oldest rocks on the Earth are within the cratons or cores of the continents, rather than in the repeatedly recycled oceanic crust. Increasingly younger units surround the older cores in the center of the continents.

Six principal types of crust are identified based on the sedimentary thickness, crustal thickness and mean seismic velocities (Beloussov and Pavlenkova, 1985). Type I, in the regions of the most recent mountains where high relief is accompanied by mountain roots, is 50–70 km thick. It has generally high-heat flow values and mean velocities are in the range of 6.4–6.5 km s^{-1}, but in some subtypes velocities as high as 7.0 km s^{-1} are also seen. Type II crust covers almost half the area of the continents and is common to areas with thin (<3 km) sedimentary cover and also crystalline shields. It is about 40-km thick, mean seismic velocity in the consolidated part is around 6.5 km s^{-1}, and has low heat flow values. Type III is an attenuated, low-velocity crust in exterior parts of the continents, e.g., West European platform. It is 25–30 km thick, has an inconsistent heat flow and a mean seismic velocity of 6.1–6.3 km s^{-1}. Type IV is the transitional crust between the continents and oceans and is similar to Type III but with higher mean seismic velocity of about 6.6 km s^{-1}. It is observed in continental margins. Type V crust varies in thickness between 15 and 40 km and is associated with deep basins. It has large thickness of the sediments (5–15 km). The heat flow values are generally high and determination of its mean velocity is not possible. A typical feature of this type is the presence of the high-velocity lower crustal layers (>6.8 km s^{-1}) and a lower thickness to the crust-mantle boundary. Type VI is the crust of oceanic basins where the crustal thickness is small and the upper part of the crust (granitic crust of velocity 6.0–6.3 km s^{-1}), whose thickness in the continental crust is in excess of 10 km, is absent.

The crust itself has no influence on the Earth but constant movement of the lithospheric plates (crust + uppermost part of the mantle), better known as the plate movement caused by influence of the convection current in the asthenosphere, does. To be more precise, the convection currents actually cause the Earth's plates to move and sometimes collide with each other. These movements cause earthquakes and at weak zones of the Earth's crust volcanoes can erupt. Because of ongoing plate movement in the last several millions of years, mountains and valleys have been formed, and that is why the surface of the Earth looks as it does now. Knowledge of characteristic features of the continental crust, therefore, is very important to an understanding of the relationship between the processes in the Earth's mantle and the geological and geomorphological phenomena observed on the surface, as the present-day configuration of the continental crust is mostly an outcome of lithospheric evolution and crust-mantle interaction. This knowledge, together with that of the physical and chemical properties of the crust, is also vital for understanding the mechanism of crustal evolution and tectono-thermal processes in the Earth's interior. It is also important for understanding characteristics of earthquakes and other natural hazards, formation/distribution of the natural resources, and evolution of various structural features present on the Earth's surface, as deep-seated structural variations in the crust are manifested in near-surface geological patterns of direct human socioeconomic interest.

1.1.2 The Mechanisms of Crustal Evolution

The mechanisms through which the continental crust evolved are debatable. However, in general, a majority of geologists agree that the continental crusts were formed in the oceanic arcs (Taylor, 1967; Kusky and Polet, 1999; Stern, 2008; Xiao et al., 2010), which is evidenced by the similarity in bulk composition of continents to that of the oceanic arcs (Taylor and McLennan, 1985; Rudnick and Gao, 2003; DeBari and Sleep, 1991; Davidson and Arculus, 2006). Fig. 1.2 explains the mechanisms for the formation of the continental crust, which shows the turbulent vs. stable convection patterns. In Archean time, temperature was high and convection was prevalent in the mantle (Olson, 1989), which resulted in thinner crust, possibly due to the lack of supply of mantle materials. However, in the Proterozoic, when the temperature was low, mantle convection reached

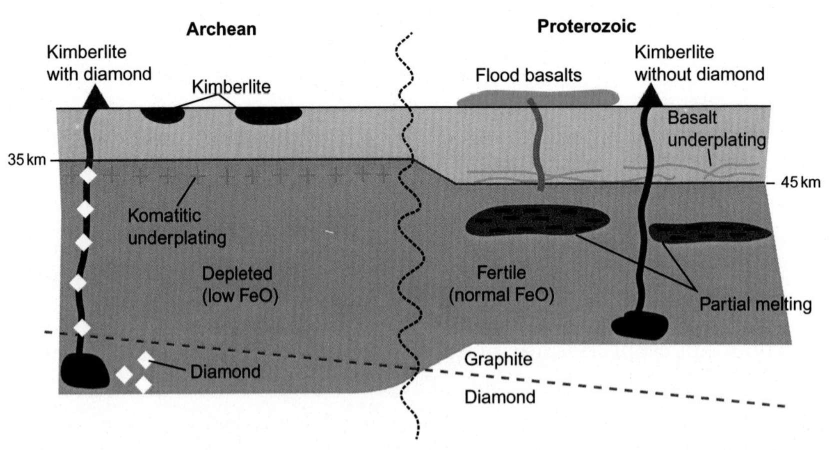

Fig. 1.2 A schematic diagram showing the Archean and Proterozoic crustal evolution. *(After Durrheim, R.J., Mooney, W.D., 1991. Archcan and Proterozoic crustal evolution: evidence from crustal seismology. Geology 19, 606–609.)*

a stable condition. As a result, the basaltic underplating played an important role in crustal thickening and also in altering the composition from felsic to mafic (Fig. 1.2). This also explains the high-mantle temperature in Archean reflected in the eruption of komatiitic lavas and the formation of a refractory lithosphere depleted in FeO (Jordan, 1978; Hawkesworth et al., 1990). Pavlenkova (1987) suggests that the crustal thickening resulted due to transformation of the mantle material during crustal cooling.

With the advent of more sophisticated seismological tools, our knowledge of the crust has increased tremendously. The continental crust was formed at about 4.2 Ga (Compston and Pidgeon, 1986; Bowring et al., 1989). It is suggested that the crust was stabilized before ~2.9 Ga, when it was thinner with a flat and sharp Moho (Durrheim and Mooney, 1991; Abbott et al., 2013). Numerous geochronological studies propose that the age of the Archean crust may range between 2.5 and 3.4 Ga and that of the Proterozoic crust between 0.5 and 2.5 Ga (De Wit et al., 1992; Rudnick and Fountain, 1995; Condie, 2005; Hawkesworth and Kemp, 2006; Van Kranendonk et al., 2011; Dhuime et al., 2012; Lowe, 1994; Kusky and Polet, 1999; Choukroune et al., 1997; Bastow et al., 2011; Arndt, 2013).

The continental crust is formed by the solidification of lava or magma through successive crystallization of erupted molten material extracted from the Earth's interior. The formation rate of the crust has, therefore, decreased with time because the heat generation rate decreased inside the Earth due to the progressive decay of radioactive elements. Two models (Fig. 1.3) have

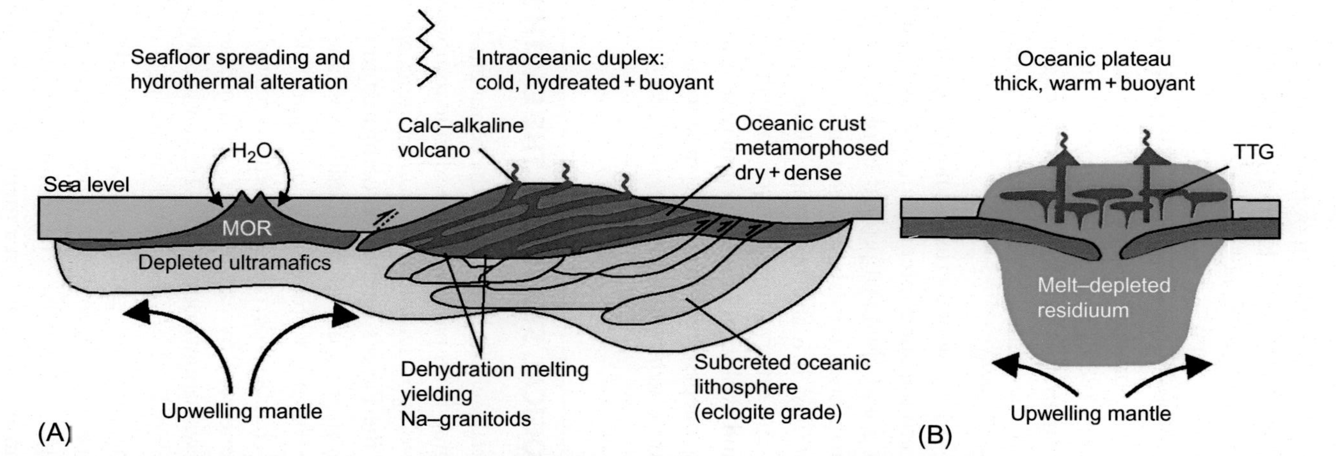

Fig. 1.3 Models for Paleoarchean crust formation: (A) horizontal accretion (subduction) tectonics; (B) vertical (plume accretion) tectonics (Van Kranendonk et al., 2014). *((A) Modified from De Wit, M.J., Roering, C., Hart, R.J., Armstrong, R.A., de Ronde, C.E.J., Green, R.W.E., et al., 1992. Formation of an Archaean continent. Nature 357, 553–562; Helmstaedt, H., Gurney, J.J., 1995. Geotectonic controls of primary diamond deposits: implications for area selection. J. Geochem. Explor. 53, 125–144.)*

been advocated for the crustal formation: horizontal accretion of island arcs or vertical accretion due to differentiation of magmatic material above hotspots (Hoffman, 1988; Percival, 1989; Kroner, 1984). It implies that the composition of the crust may either be andesitic (Taylor and McLennan, 1985; Rudnick and Gao, 2003) or basaltic (DeBari and Sleep, 1991; Davidson and Arculus, 2006) and is different from the single dominant crust-forming process, i.e., basaltic volcanism throughout Earth's history (Ashwal, 1989; Turcotte, 1989). As the fundamental process is different for Proterozoic and Phanerozoic crusts, the seismic characteristics of the crust are expected to be different at different geologic provinces (Drummond and Collins, 1986).

The configuration of continental crust is mostly an outcome of lithospheric evolution and crust-mantle interaction. This knowledge, together with that of the physical and chemical properties of the crust, is also vital to understand the mechanism of crustal evolution and tectono-thermal processes in the Earth's interior. It is also important for understanding characteristics of earthquakes and other natural hazards, formation/distribution of the natural resources, and evolution of various structural features present on the Earth's surface, as deep-seated structural variations in the crust are manifested in near-surface geological patterns of direct human socioeconomic interest.

1.2 SEISMOLOGICAL STUDIES OF THE EARTH'S CRUST

Due to the essentially elastic behavior of the continental crust, all types of seismic studies, ranging from passive source seismic tomography to controlled source, high-resolution seismic reflection imaging, are preferred for gaining knowledge of the Earth's interior, as compared to other methods. Strong lateral variation in structure of the Earth's crust and mantle makes their study a complex problem. Virtually all of our direct information about the interior of the Earth has been derived from studies related to propagation of the elastic waves generated by earthquakes. Earthquakes generate two types of waves: surface waves and body waves. The surface waves are guided by the density and velocity layering, particularly at and near the surface, and are important in elucidation of the crustal and upper mantle structures. But of greater general interest and more easily understood are the body waves that propagate through the Earth. The compressional (P) and shear (S) waves that travel through the body of a medium are known as the body waves. A body wave in which the particle motion is in the direction of propagation is called a P-wave and a wave in which the particle motion is perpendicular to the direction of propagation is known as an S-wave. These conform to the laws of geometrical optics, being reflected and refracted at interfaces where the velocities change.

Historically, seismological methods have provided the first information about the crust. Among the oldest and most fundamental problems in seismology are: (a) determining the velocity-depth relation accurately, and (b) asserting the nature of the discontinuities within the Earth and translating this into knowledge about the interior of the Earth. Early in the 20th century, study of the seismic waves generated through earthquakes showed that the interior of the Earth has a radially layered constitution and the boundaries between the layers are marked by abrupt changes in seismic velocities/velocity gradient. In 1906 Oldham noted that at large epicenter distances, travel times of the seismic compression waves that traversed through the body of the Earth were greater than expected; the delay was attributed to a fluid outer core. Support for this idea came in 1914, when Guttenberg described a shadow zone for the seismic waves at distances of greater than 105 degrees (1 degree = about 110 km at the equator). In 1936, Lehmann observed weak arrivals of the compression waves in the gap between 110 and 143 degrees and interpreted it as evidence for a solid inner core. An anomalous layer, 150–200 km thick and termed the D layer, was identified just above the core-mantle boundary. In this layer the body wave gradients are very small and may even be negative. During 1940–42 Bullen developed a model, consisting of concentric shells, of the internal structure of the Earth. Later a parameterized model PREM (Preliminary Reference Earth Model) based upon the inversion of body waves, surface waves and free-oscillation data was prepared (Dziewonski and Anderson, 1981). This is the current standard model of the Earth's internal structure.

Mohoroviĉiĉ obtained first results of the crustal velocities in the year 1909. He identified a boundary at 30–35 km depth within the continents, where the seismic wave velocities in the Earth's interior showed a sudden increase, and termed this the crust-mantle boundary. This discontinuity, since then known as the Moho, is generally represented by either a large velocity jump or a steep velocity gradient (transition zone) in the lowermost part of the crust and indicates a change from mafic to ultramafic rocks. In seismology the crust is defined by a P-wave velocity (V_P) of less than 7.8 km s^{-1} or S-wave velocity (Vs) of less than 4.3 km s^{-1}, overlying an upper mantle with a P-wave velocity of about 8.1 km s^{-1}. The earthquake studies identified three main P-wave phases in the crust: the Pg, PmP and Pn.

The Pg wave follows the top of granitic crystalline basement below the sediments, the PmP wave appears as a critical or postcritical reflection from the Moho and the Pn wave is supposed to run along the top of mantle. In addition to these, other P-wave phases were also identified from various layers within the crust.

Earlier, the detection of Moho and its nature were determined by the explosion seismic and travel time observations from earthquake data. However, since the 1980s, converted wave techniques in passive source seismology have become robust tools to map the crust-mantle discontinuities and thereafter an enormous amount of work has been published about the measurements and nature of the Moho. Nevertheless, its genesis has remained a subject of controversy. Based on seismic wave propagation, the Moho has been defined as the change in P-wave velocity ($V_P > 7.6$ km s^{-1}) and density (Oliver, 1982; Prodehl et al., 2013; Rabbel et al., 2013; Jarchow and Thompson, 1989; Steinhart, 1967; Hammer and Clowes, 1997). Sometimes changes in anisotropic direction (Jones et al., 1996) and also the scale lengths of heterogeneity may occur across this boundary (Enderle et al., 1997). In seismic sections, the depth of the Moho typically coincides with the boundary between the reflector-rich crust and the reflector-poor uppermost mantle (Cook, 2002; Cook et al., 2010; Mooney and Brocher, 1987; Carbonell et al., 2013). It is unquestionably a geophysical boundary where a change in elastic properties takes place, and this leads to the conversion of seismic waves.

The Moho can thus be characterized as noted here:

(1) It is an interface that separates the Earth's crust from the underlying mantle.
(2) It is a first-order discontinuity based on velocity and composition.
(3) Compressional-wave velocity increases to 7.6 km s^{-1} at the Moho.
(4) Petrologically, the Moho is the boundary between homogeneous layers of mafic (above) and ultramafic (below) rocks.
(5) The Moho subsists worldwide with laterally variable depth on broad wavelengths.
(6) Its sharpness and reflectivity vary from region to region. Here, by sharpness we mean that the crust-mantle transition occurs over a vertical distance of less than 2 km (James et al., 2003).

Although the Moho is a seismic definition, petrologically it is a transition from spinel lherzolite to garnet lherzolite (Griffin and O'Reilly, 1987; O'Reilly and Griffin, 1985), where a marked increase in seismic velocity exists in the lower crust. The seismic Moho is a phase change within the upper mantle (Griffin et al., 1987; Griffin and O'Reilly, 1986; O'Reilly, 1989; O'Reilly and Griffin, 1991) and is related to the pressure and temperature histories. Therefore, it is associated with a boundary, which separates the mafic lower crust (gabbros for the oceanic crust and mafic granulites for the continental crust) from the ultramafic upper mantle (e.g., Christensen, 1996; Christensen and Mooney, 1995; Gao et al., 1998; Mengel and Kern, 1992; Rudnick and Fountain, 1995; Jarchow and Thompson, 1989). An electrical Moho has also been defined as the bottom of the relatively conductive continental lower crust (10^{-4}–10^{-1} S m^{-1}) and is often not coincident with the seismic Moho (Wang et al., 2013; Mengel and Kern, 1992).

The seismic and petrological Moho are generally noncoincident, as the temperature regime plays an important role in determining the depth of the petrological Moho. Griffin and O'Reilly (1987) argue that the Moho in a lithologic column is determined by the temperature regime, as the tectonics and the temperature regime modify the lower crust and/or the upper mantle. These processes act on the rocks overprinting a series of additional factors that affect the seismic signature: for example, rock fabrics and anisotropy. In this sense, the seismic and the petrological Moho can be placed at different levels. However, Brown et al. (2009) suggest that the petrological Moho is somewhat deeper than the seismic Moho. In the petrological Moho, the transition from granulites to ultramafic rock (considered to be mantle lithologies) would be located slightly deeper in the column. The seismic Moho has been considered to correspond to the lithological contact between gabbros and rocks with ultramafic compositions (harzburgites), and the petrological Moho has been correlated with the deeper structures/fabrics and lithological contrasts within the ultramafics (e.g., peridotites, dunites, etc.).

1.3 CONTROLLED SOURCE SEISMIC STUDY

The controlled source seismic study [also known as the deep seismic sounding (DSS) study] is a definitive geophysical technique for exploring the structure of the Earth's crust and uppermost mantle. It is a highly sophisticated technique involving use of controlled explosions, of known energy release, to generate elastic waves in the Earth's crust. This method uses the technique of transmitting into the Earth the seismic waves that are generated by exploding charges and recording them back on the surface. The large success achieved by the seismic studies in hydrocarbon exploration has prompted their use in studying the Earth's crust. Through these studies the continental crust has been characterized by three broad parameters: its velocity, reflectivity and thickness. These parameters are also related to the heat flow, temperature and viscosity at various depths in the crust. The controlled source seismic study (5–30 Hz) effectively bridges the gap (Fig. 1.4) between the high-frequency conventional exploration for hydrocarbon (30–100 Hz) and low-frequency passive source seismological studies (0.1–5 Hz). For finding out the crustal structure it can be divided into two broad groups: (1) the refraction/postcritical (wide-angle) reflection also known as refraction seismology, and (2) the deep continental reflection study.

The most reliable information on velocity variation in the continental crust comes from the seismic refraction profiles. Though the variation of seismic velocity with depth can be fitted into five broad ranges (Meissner

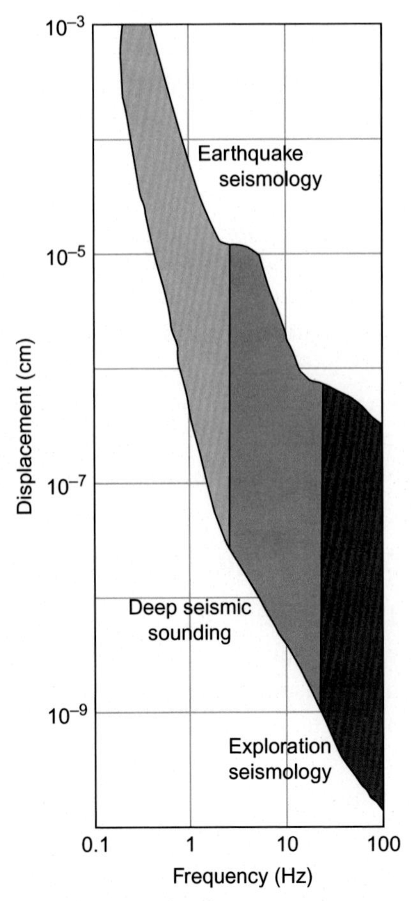

Fig.1.4 Schematic diagram showing the frequency range used in different seismic techniques.

Table 1.1 Velocity Ranges in the Crust

Part of the Crust	Velocity (V_p) Range (km s^{-1})
Sediments or near-surface rocks	<5.7
Upper crust	5.7–6.4
Lower crust	6.4–7.1
Lowermost crust (in rifts, shields and platforms)	7.1–7.8
Uppermost mantle	>7.8

After Meissner, R., Weaver T.H., 1989. Continental crustal structure. In: James, D.E. (Ed.), The Encyclopaedia of Solid Earth Geophysics. Vam Nostrand Reinhold Co., New York, pp. 75–89.

and Weaver, 1989), it differs according to the geology and tectonic configuration of a region (Table 1.1). Studies of the continental crust and subcrustal lithosphere using seismic refraction/postcritical reflection measurements are of first-order importance for determining reliable models of the seismic velocity structure and physical properties of the lithosphere. The Eastern block of the European countries, led by the erstwhile USSR, made major contributions to our understanding of the chemical and physical properties of the Moho through large-scale refraction/postcritical reflection studies. The crust-mantle boundary (referred to as the Moho in forthcoming text) is regarded as the most important velocity boundary in crustal seismic observations. The general model of the continental Moho is a variable thickness transition zone, composed of layers progressing from mafic to ultramafic rocks with increasing depth (Jarchow and Thompson, 1989).

The deep reflections (near-vertical incidence) from the Moho boundary were reported as early as the late 1950s but it took almost two decades to substantiate them through deep continental reflection studies. The COCORP (Consortium for Continental Reflection Profiling, United States) studies paved the way in recognizing the velocity layers as reflecting horizons and hence identifying the reflectivity patterns that determine structural variations in the crust of different geological terranes. Major deep seismic reflection study groups and consortia, e.g., BIRPS (United Kingdom), DEKORP (Germany), ECORS (France) and LITHOPROBE (Canada), followed it. High-resolution seismic reflection and refraction programs were also launched over the subduction zones (ANCORP Working Group, 1999), the arc-continent collisions along the continental margins (Lundberg et al., 1997), the convergent plate boundaries (Davey and Stern, 1990), and the orogenic belts (Snyder et al., 1997).

The seismic studies are capable of resolving shallow as well as deep structures in the crust by acquiring suitable refraction and postcritical reflection data sets and also near-vertical reflections. Advances over the past 20 years, both in the areas of seismic data acquisition as well as processing and modeling techniques, offer wide-ranging possibilities for exploring complex subsurface structures that may be heterogeneous and anisotropic. At present it is believed that combined refraction and deep reflection experiments (also known as coincident reflection studies) supported by the other geophysical, geological and geochronological studies on selected geo-transects provide the most reliable seismic images of the continental crust and uppermost mantle. The two seismic techniques are complementary to each other and, when used together, are capable of resolving the structural and physical property variations in the Earth. The refraction/postcritical reflection data sets provide viable models of the velocity distribution required to infer the petrologic composition, grade of metamorphism and material properties such as brittle/ductile regimes. These lead to consistent interpretation of the reflectivity patterns obtained through deep continental reflection studies and provide necessary clues for understanding the complex geodynamic processes that might be operative during the evolution of various geological provinces. In the tectonically active regions accurate mapping of the intracrustal boundaries, including the Moho, and delineation of the deep penetrating steep/low-angle faults reveal various blocks in the crust that may have been relatively displaced due to movements along these faults.

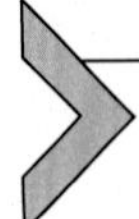

1.4 CONTROLLED SOURCE SEISMIC STUDIES OVER THE INDIAN PLATE

India has a geologically unique position, as rocks ranging from 3800 million years (Archean) to the present (Quaternary) are exposed here. Geologists are of the opinion that several episodes of continental collision and extension have taken place during this time. However, the history of the earlier episodes (from Archean to Triassic) is not well documented. Geological history from the Triassic to the Quaternary time is much better understood. During the Triassic (about 210 Ma), a major rifting episode split the then existing Gondwanaland into two parts: East and West Gondwanaland. The Indian continent was in East Gondwanaland together with Antarctica and Australia. East and West Gondwanaland and the oceanic crust between them were together till the Jurassic time (about 160–155 Ma). The Indian subcontinent then broke off from Antarctica and Australia in the early Cretaceous (about 130–125 Ma) and the Indian Ocean opened up. During the upper Cretaceous time (about 84 Ma) the Indian plate started its very rapid northward drift at an average speed of 16 cm $year^{-1}$, covering a distance of about 6000 km and a rotation of 33 degrees in an anticlockwise direction, until the northwestern part of the Indian passive margin collided with Eurasia in the Early Eocene (about 50 Ma). The Indian plate is still continuing its northward drift at a slower but still surprisingly fast rate of about 5 cm $year^{-1}$. The tectonic map of India (Fig. 1.5) shows that the mobile belts separate various cratons from each other in its shield region (Vijaya Rao and Reddy, 2002).

Prior to availability of the results from crustal seismic studies, very little was known about the character of the Indian continental crust though a few studies based on earthquakes provided velocity-depth models for the crust and upper mantle of the Indian plate, particularly the Himalaya, western part of the Indian shield and north eastern and central Indian region. These models based on the P- and S-wave structure, and also surface wave dispersion studies, have proved very useful in determining the earthquake parameters and seismicity pattern of the Indian subcontinent. In recent times teleseismic and tomographic studies have revealed the crustal and lithospheric thickness in various parts of the Indian plate (Rai et al., 2003, 2005).

Though the seismological studies provide a gross picture of the crust, they are unable to detect the layering and also the velocity structure within it because of their low frequency range. Controlled source seismic studies were started in India in the year 1972 for a better understanding of the geological history of the Indian plate through the velocity and structural configuration of the continental crust. The National Geophysical Research Institute (NGRI), Hyderabad, India, has recorded a number of seismic profiles in different geological and tectonic provinces within the country. Between 1972 and 1985 all the crustal seismic data in India were acquired through analog seismic equipment. A major improvement in data acquisition took place when digital units replaced the analog equipment in the year 1985.

Between 1972 and 1991 only seismic refraction/postcritical (also known as wide-angle) reflection studies were carried out to understand the velocity models of the Indian crust. During this period, major progress was made in processing of crustal seismic data sets due to their availability in digital form, leading to better data interpretation in other parts of the world. Therefore, another program of digitizing the earlier acquired data sets and reinterpreting them through the newer software packages was initiated. Acquisition of deep reflection data was initiated in the year 1992. Another milestone in data

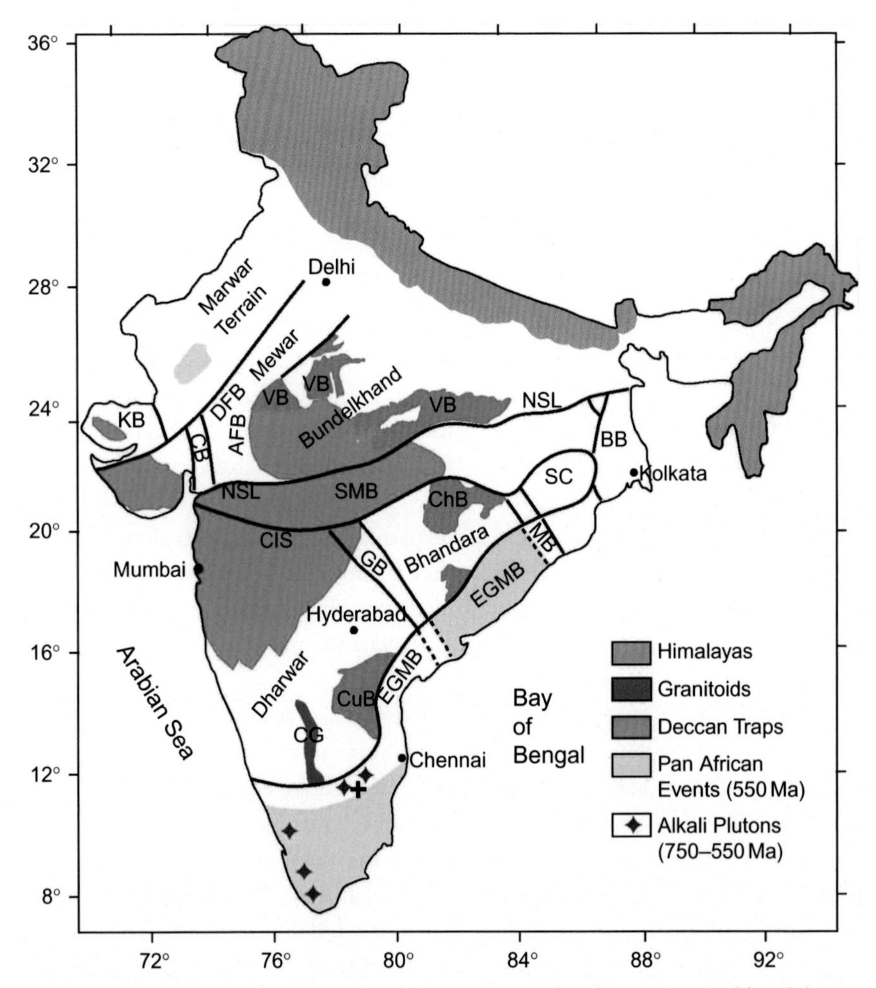

Fig. 1.5 Tectonic Map of India. *KB*, Kutch basin; *CB*, Cambay basin; *CuB*, Cuddapah basin; *ChB*, Chattisgarh basin; *DFB*, Delhi fold belt; *AFB*, Aravalli fold belt; *VB*, Vindhyan basin; *NSL*, Narmada-Son lineament; *SMB*, Satpura Mobile belt; *CIS*, central Indian suture; *BB*, Bengal basin; *MB*, Mahanadi basin; *GB*, Godavari basin; *EGMB*, Eastern ghat Mobile belt; *BPMB*, Bhavani-Palghat Mobile belt; *CG*, Closepet Granite. *(Map source Vijaya Rao, V., Reddy P.R., 2002. A mesoproterozoic supercontinent: evidence from the Indian shield. In: Rogers, J.J.W., Santosh, M. (Eds.), Special Volume on Mesoproterozoic Supercontinent. Gondwana Res. 5, 63–74.)*

acquisition was achieved when wireless telemetry systems replaced the digital units in the year 1998. This made acquisition of crustal seismic data possible even in areas with difficult topography (e.g., the Himalayan region). Since the data acquisition process for all types of seismic data is more or less similar and can be found in several textbooks, it is not discussed here.

Under the program to study the continental crust of India, seismic refraction and postcritical reflection data sets were acquired along various profiles in different geological and tectonic provinces in the Indian shield (peninsular shield, Southern Granulite Terrane, Aravalli-Delhi fold belt in the northwest), Narmada-Son Lineament in central India, Deccan Trap covered regions in the west, the sedimentary basins (Cambay, West Bengal, Mahanadi, Godavari) and the Himalaya (Fig. 1.6). In some of the terranes

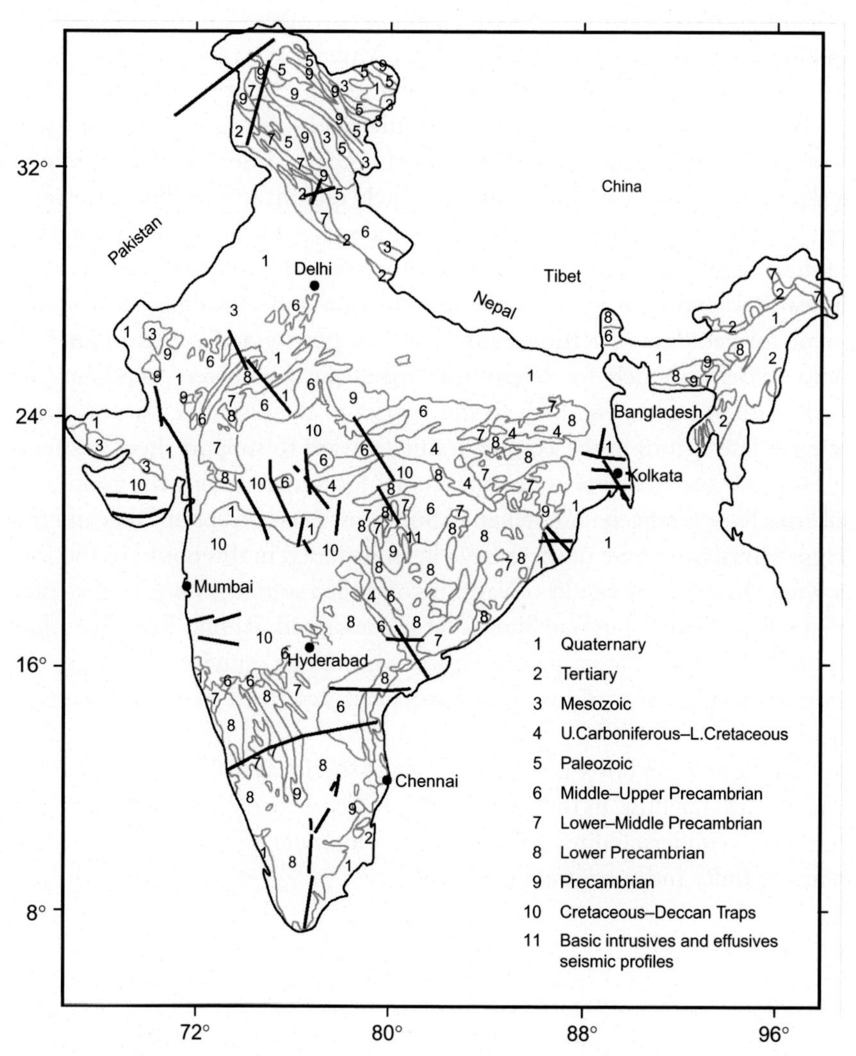

Fig. 1.6 Seismic profiles recorded to study the crust in the Indian subcontinent plotted on Geological Map of India. *(Basic Map source: Geological Map of India, Scale 1:5,000,000, 1993.)*

(Aravalli-Delhi fold belt, central India, Southern Granulite Terrane, Himalayan foothills) deep seismic reflection data were also acquired. Interpretation of these data has brought out the crustal velocity configuration and structure down to the Moho boundary, and occasionally in the uppermost mantle, leading to better understanding of the evolutionary processes involved in the formation of various terranes. Since acquisition of the crustal seismic data in India was started with support of the then USSR, it is natural that most of the earlier results were inspired by them and consist of only the velocity models, while the data acquired at later stages consist of the structural patterns as well as velocity models. Most of the earlier models have, however, been revised by digitizing them to the extent possible.

Before the availability of crustal seismic results, it was difficult for earth scientists to understand the evolution and history of various Indian geological terranes, particularly the peninsular shield region. Availability of the seismic velocity models, and in some cases structure, made better understanding of these regions possible. The results from these data have been widely used by earth scientists to provide new insight into understanding various elements of geologic structure, explanation of gravity and other anomalies, new velocity models to determine earthquake parameters, modeling of hydrocarbon basins, etc. Due to the impact of these studies, various earth science institutions in the country came forward to support these studies.

Outside the Indian plate seismic studies for understanding the continental crust have not been undertaken in Southeast Asia, except in Tibet and the High Himalaya. Some of these have been included in this book. In the following chapters, the results of various controlled source seismic studies carried out in India and its neighborhood are described. Results from the other geophysical studies (e.g., passive source seismology, gravity and magnetic, magnetotelluric, heat flow, etc.) have been described in brief wherever necessary.

The aim of this book is to make all these studies available in one place to enable earth scientists in India and abroad to be aware of them and their contribution in understanding the tectonic development of various geologic terranes in India and surrounding regions.

Indian Peninsular Shield

The Indian peninsular shield (Fig. 2.1), a mosaic of crustal blocks of independent evolutionary history, was juxtaposed together during different geological periods (Radhakrishna, 1989). It is one of the oldest cratonic blocks of the world where rocks as old as 3800 Ma are exposed. It comprises the low-grade granite-greenstone of the Dharwar craton in the north and the high-grade granulite terrane in the south (Ramakrishnan, 2003). Seismic studies to determine configuration of the crust in the Indian peninsular shield can be divided into two phases. In the first phase refraction and post-critical reflection studies were carried out, during 1972–80, along two east-west profiles, one of which crosses the Proterozoic-Archean Cuddapah basin as well as the Dharwar craton, while the other crosses only the Cuddapah basin. In the second phase the refraction, postcritical reflection and also deep reflection studies along a profile crossing the southern granulite terrane in a north-south direction were carried out from the year 2001 onwards. The data acquired in the first phase were in analog form, while in the second phase they were in digital form. Since the two regions are geologically different from each other and the data quality is also very different, the results of the east-west and north-south profiles are discussed independently.

The seismic profiles in the Precambrian terranes reveal several structural features in the crust that were probably developed during the Precambrian time and have persisted since then in several regions. This implies that this crust, and probably also the upper mantle, has remained coherent through subsequent time. Analysis of the velocity models for several Proterozoic and Archean terranes in different parts of the world has shown that the velocity distribution in the crust of these two terranes is significantly different. The Proterozoic crust has a greater thickness (40–45 km) as compared to a 27–40 km thick crust (except at collision boundaries) in the Archean terrane (Durheim and Mooney, 1994). The former has a thick high-velocity (>7.0 km s^{-1}) layer at its base that is absent in most of the latter terranes. Even within the Proterozoic terranes thickness of the high-velocity layer at the base of the crust is a major factor of difference. In most of the models for the

Structure and Tectonics of the Indian Continental Crust and Its Adjoining Region
https://doi.org/10.1016/B978-0-12-813685-0.00002-9

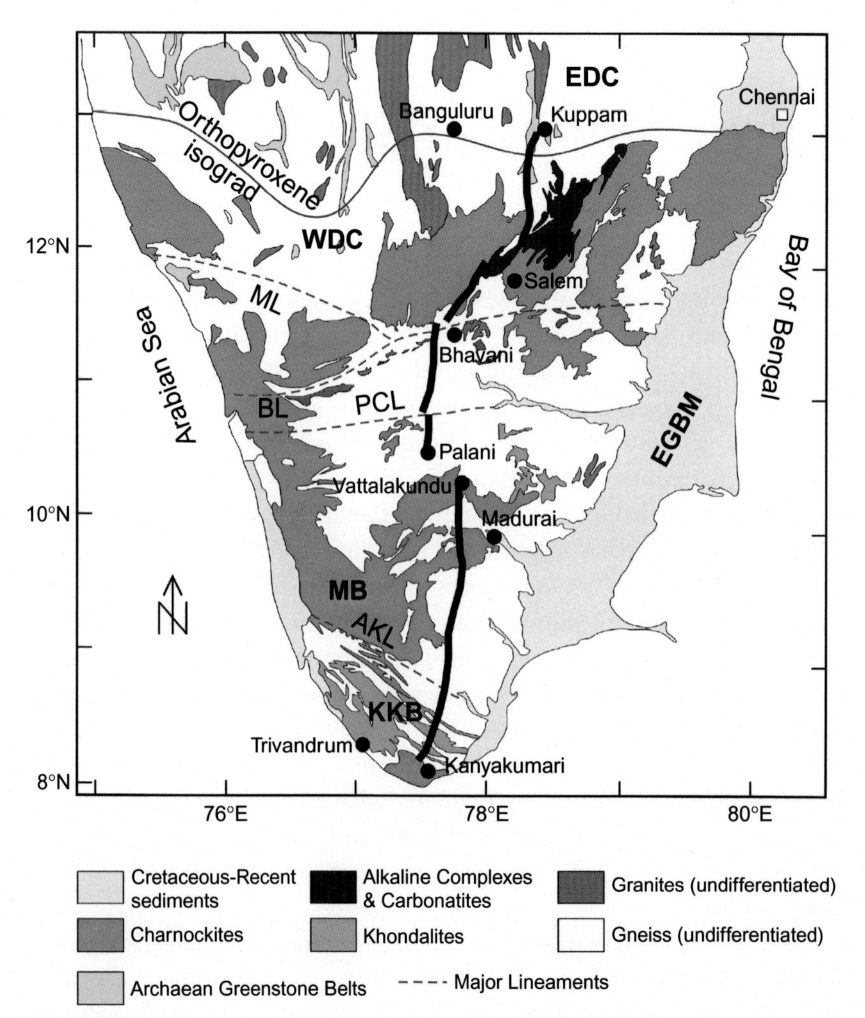

Fig. 2.1 Simplified tectonic map of south Indian shield. *AKL*, Achankovil shear zone; *PCL*, Palghat-Cauvery shear zone; *EDC*, Eastern Dharwar Craton; *WDC*, Western Dharwar Craton; *EGMB*, Eastern Ghat Mobile Belt; *MB*, Madurai block. *(Modified from Chetty, T.R.K., Bhaskar Rao, Y.J., Narayana, B.L., 2003. A structural cross section along Krishnagiri-Palani corridor, Southern Granulite Terrene of India. In: Ramakrishnan, M. (Ed.), Tectonics of Southern Granulite Terrene, Kuppam-Palani Geotransect. Geol. Soc. India, Mem. 50, 255–277; Bhaskar Rao, Y.J., Janardhan, A.S., Vijay Kumar, T., Narayana, B.L., Dayal, A.M., Taylor, P.N., Chetty, T.R. K., 2003. Sm-Nd model ages and Rb-Sr isotopic systematics of charnockite and gneisses across the Cauvery Shear Zone, southern India: implications for the Archaean-Neoproterozoic terrane boundary in Southern Granulite Terrene. In: Ramakrishnan, M. (Ed.), Tectonics of Southern Granulite Terrene, Kuppam-Palani Geotransect. Geol. Soc. India, Mem. 50, 297–317.)*

Archean terranes velocity in the lower crust is less than 7.0 km s^{-1}; they are all sharp, distinctive crust-mantle boundaries at about 35–40 km depth.

The crust-building processes during the Proterozoic are different as compared to those of the Archean. The following two models explain these differences.

(1) The first model considers a change in composition of the upper mantle. In this model, higher temperature during the Archean resulted in an ultradepleted lithosphere that was unable to produce significant volumes of basaltic melt. Proterozoic crust developed above fertile mantle, and subsequent partial melting resulted in basaltic underplating and crustal inflation.

(2) In the second model convection in the hot Archean mantle is considered too turbulent to sustain stable long-lived subduction zones. By the Proterozoic the mantle had cooled sufficiently so that substantial island and continental arcs were constructed, and basaltic underplating formed the high-velocity layer at the base of the crust. Some of the workers, however, find no significant difference in either the crustal thickness (average 43 km for both) or the velocity structure between the crusts of Archean and Proterozoic terranes, although they agree to considerable variation within each type of crust (Rudnick and Fountain, 1995).

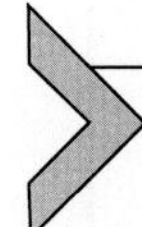

2.1 EAST-WEST PROFILES ACROSS THE EASTERN AND WESTERN DHARWAR CRATONS

The seismic refraction/postcritical reflection studies along the 600-km long Kavali-Udipi profile (Profile I, Fig. 2.2), during the years 1972–75, were path breaking as these were the first of their kind in India. These were initiated under a collaborative program between the erstwhile USSR and India. This profile, located between the east and west coasts, cuts across several Proterozoic and Archean geological terranes of eastern ghat, eastern and western Dharwar cratons. Some of these terranes are associated with known mineralized zones of the Indian peninsular shield. Major geological formations crossed by this profile are: (from east to west) alluvium and laterites of the east coast, peninsular gneiss, granites, Nellore schist belt, southern portion of the Cuddapah basin, Closepet granite, peninsular gneiss (second time), Chitradurga and Shimoga schist belts and alluvium of the west coast. A major geological feature in the eastern Dharwar craton is the Cuddapah basin. During the years 1979–80 similar studies were carried out along

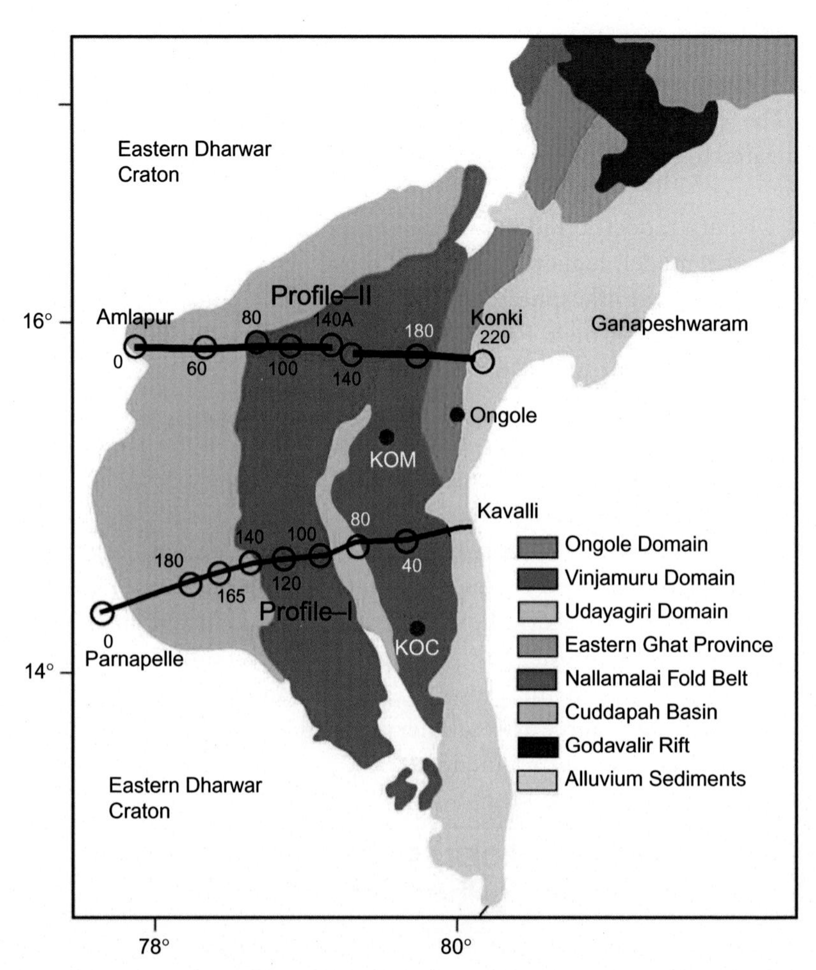

Fig. 2.2 Geological map of the Cuddapah basin and area to its east with deep seismic profiles plotted on it. *(Modified from Kaila, K.L., Tewari, H.C., Roy Choudhury, K., Rao, V.K., Sridhar, A.R., Mall, D.M., 1987. Crustal structure of the Northern part of the Proterozoic Cuddapah basin of India from deep seismic soundings and gravity data. Tectonophysics 140, 1–12).*

another 300-km long, east-west profile (Alampur-Konoki-Ganapeswaram, Profile II of Fig. 2.2) which crosses the northern part of the Cuddapah basin, eastern ghat and alluvium of the east coast.

In view of differences between the Proterozoic and Archean crust, results of seismic studies in the Proterozoic Cuddapah basin and the Archean western Dharwar craton are described separately here.

2.1.1 Cuddapah Basin

The Cuddapah basin, located in the southeastern part of the Indian peninsular shield (Fig. 2.2), is a unique tectonic and orogenic belt of unfossiliferous Proterozoic rocks and comprises unmetamorphosed sediments and a number of intrusive/extrusive bodies of basic igneous rocks. This crescent-shaped basin has a maximum length of 340 km in the north-south direction and a maximum width of 140 km in the middle.

2.1.1.1 Geology and Tectonics

The axis of the Cuddapah basin is northeast-southwest in its northern part but swerves to northwest-southeast in the south. The main structural elements of the basin are: (i) a north-plunging, low-amplitude, asymmetrical syncline made up of gently dipping, practically unfolded western limb that consists of unmetamorphosed Paleoproterozoic sediments of lower Cuddapah and Neoproterozoic Kurnool systems, and (ii) an intensely folded and thrusted eastern limb that contains Neoproterozoic upper Cuddapah sediments and several dome-shaped upwarps (Narayanaswamy, 1966). A fault, where the upper Cuddapah formation thrusts over the much younger Kurnool formation, separates the western and eastern limbs of this basin. The volcanic rocks are exposed in the western part of the basin, and this activity is marked by several occurrences of mainly dolerite trap and compact basalt. These generally contemporaneous flows are partially intrusive as sheets and sills. A large number of dikes cut across the geological formation, connecting some of the sills and sheets. Beyond the eastern margin of the basin the Cuddapah group of rocks comes into contact with the Dharwar group (Archean-Proterozoic).

2.1.1.2 Seismic Results

Several velocity models, which differ from each other in minor detail, have been proposed for the Cuddapah basin. The generalized seismic velocity model for the Indian crust (Kaila et al., 1979) shows three layers for the entire Indian peninsular shield, including the Cuddapah basin (Fig. 2.3).

Main features of this model are: an initial velocity of 5.70 km s^{-1} at a depth of 2 km that gradually increases to 6.5 km s^{-1} at 15 km depth in the upper crust and remains constant till 23-km depth, where it increases to 6.6 km s^{-1}, 6.6–7.15 km s^{-1} between 23- and 40-km depth in the lower crust and a jump to 8.13 km s^{-1} at 40-km depth in the upper mantle. A specific velocity model for the Cuddapah basin (Kaila et al., 1987) shows three

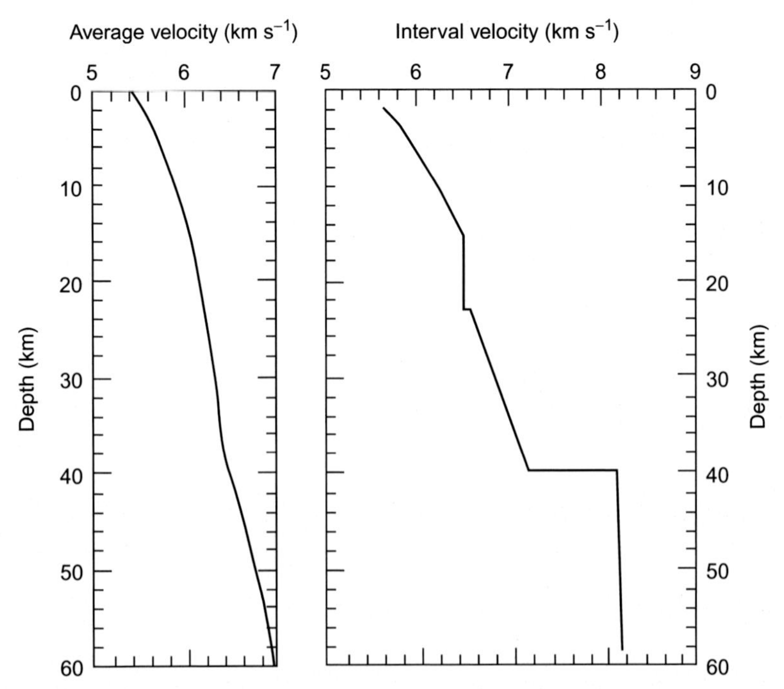

Fig. 2.3 Generalized velocity model for the Indian peninsular shield. *(After Kaila, K.L., Roy Choudhury, K., Reddy, P.R., Krishna, V.G., Narain, H., Subbotin, S.I., Sollogub, V.B., Chekunov, A.V., Kharetchko, G.E., Lazarenko, M.A., Ilchenko, T.V., 1979. Crustal structure along Kavali-Udipi profile in the Indian Peninsular Shield from the deep seismic soundings. J. Geol. Soc. India, 20, 307–333).*

velocities of 5.1, 5.85 and 6.1–6.3 km s^{-1} respectively in the Kurnool sub-basin. The first belongs to the Kurnool sediments, the second to the lower Cuddapah sediments and the third to their Archean basement. In other parts of the basin where the Cuddapah sediments are exposed, the first velocity is 5.25–5.35 km s^{-1} and belongs to the upper Cuddapah sediments, while the other velocities in the lower layers remain unchanged. The velocity in the upper mantle varies between 7.91 and 8.54 km s^{-1}. Some of the models identify a layer of velocity 7.2 km s^{-1} at the bottom of the lower crust in the eastern part of the basin (Kaila and Bhatia, 1981). To the east of the basin (eastern ghat region) a high-velocity gradient is observed, between 3- and 15-km depth (Tewari and Rao, 1990). A high-velocity gradient is also observed between 32- and 45-km depth in most of the Cuddapah basin and the eastern ghat regions, respectively.

In the southwest Cuddapah basin the velocity variation with depth is somewhat complex. Several models show a velocity inversion in the upper crust. In one of the models the inversion is shown at a depth of about 10 km (Chekunov et al., 1984) where a body, with velocity increasing from 6.45 to 6.9 km s^{-1}, overlies a layer with velocity of 6.7 km s^{-1}. Another model shows three layers with velocities of 5.60, 6.05 and 6.90 km s^{-1} respectively in the top part of the upper crust, the 6.90 km s^{-1} velocity layer being limited to the western part of the basin at depths varying between 1.6 and 6.0 km (Tewari and Rao, 1987).

The seismic depth section for the Cuddapah basin (Kaila et al., 1979), along Profile I, shows several blocks separated by faults (Fig. 2.4). A strong reflection, between 7- and 10-km depth, represents the basement of the Proterozoic Cuddapah sediments. A low angle thrust fault at the eastern margin of the basin, along which the Dharwar formation is thrust over the upper Cuddapah formation, terminates the basin. A refractor (velocity 6.68–6.91 km s^{-1}) at depth of 18–25 km in the western part of the basin, and a reflection between 18- and 23-km depth, represent the transition boundary from the upper to the lower crust. The general dip of the reflections is from west to east except in the middle part of the basin, which is uplifted. The Moho is at a depth of 34 km in the western part and at 38–40 km in the eastern part. Another reflection at a depth of 43–48 km, also identified as the Moho, is seen in the eastern part of the basin. Beyond the eastern margin of the basin, reflections show an updip towards the east.

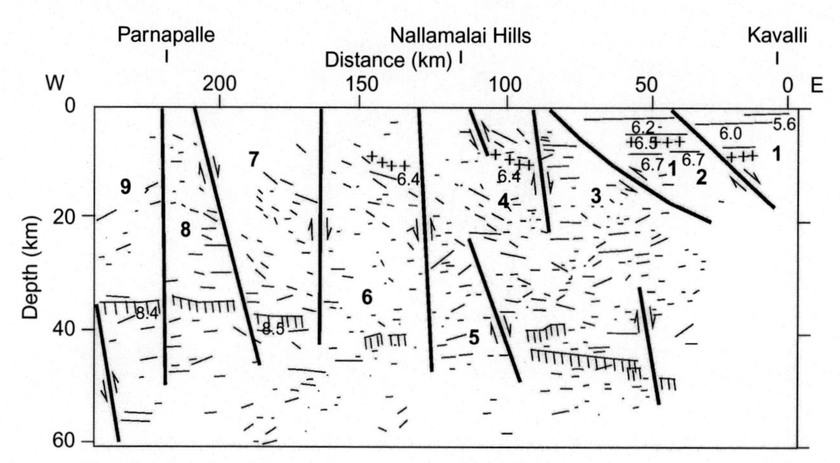

Fig. 2.4 Crustal depth section at the Cuddapah basin and area to its east along Profile I. Velocity values are in km s^{-1}. *(Modified from Kaila, K.L., Roy Choudhury, K., Reddy, P.R., Krishna, V.G., Narain, H., Subbotin, S.I., Sollogub, V.B., Chekunov, A.V., Kharetchko, G.E., Lazarenko, M.A., Ilchenko, T.V., 1979. Crustal structure along Kavali-Udipi profile in the Indian Peninsular Shield from the deep seismic soundings. J. Geol. Soc. India, 20, 307–333).*

On Profile II (Kaila et al., 1987) the Kurnool sediments (velocity 5.1 km s^{-1}) directly overlay the basement (velocity 6.1 km s^{-1}) in the western part and are 200–500 m thick (Fig. 2.5). In the middle of the basin, where the eastward dipping basement attains a maximum depth of about 4.5 km, another layer with velocity of 5.85–5.90 km s^{-1} (Cuddapah sediments) is seen. In the eastern part, the basement of the Cuddapah sediments is at a depth varying between 3.5 and 6.5 km. The general dip of the basement remains from west to east except for the basement uplift in the middle.

The basement configuration of the Cuddapah basin reveals that a fault boundary separates its eastern and western parts (Kaila and Tewari, 1985). The basement of the Cuddapah sediments, in the western part of the basin, dips to the southeast to a maximum depth of about 9000 m on Profile I, near the fault. To the east of this fault, the depth to the basement is 3500 m on Profile II and 8000 m on Profile I, indicating basement uplift. Further east, the depth to the basement increases to a maximum of about 10,000 m at Profile I and 6500 m at Profile II at the eastern end of the basin. Here the basement is upthrown to the east by 5000–6000 m. According to some workers, the basement structure appears to have controlled the convex to the west, arc-shaped thrust fronts in the Cuddapah basin (Venkatakrishnan and Dotiwalla, 1987).

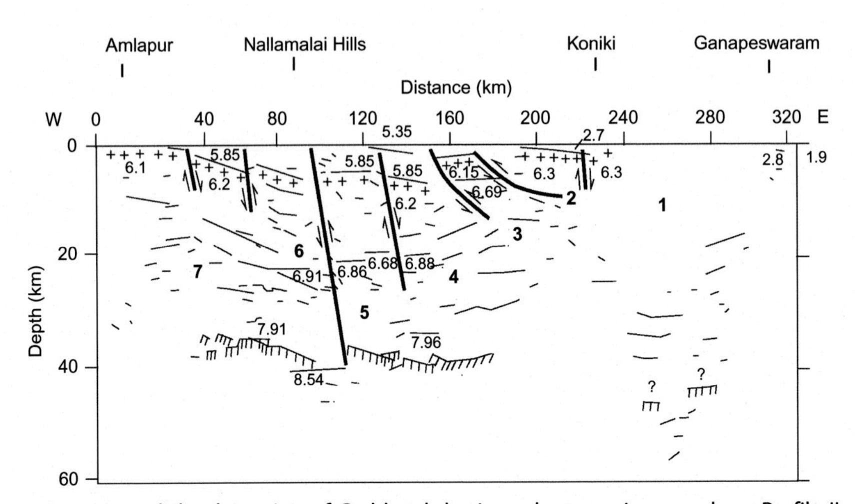

Fig. 2.5 Crustal depth section of Cuddapah basin and area to its east along Profile II. Velocity values are in km s^{-1}. *(Modified from Kaila, K.L., Tewari, H.C., Roy Choudhury, K., Rao, V.K., Sridhar, A.R., Mall, D.M., 1987. Crustal structure of the Northern part of the Proterozoic Cuddapah basin of India from deep seismic soundings and gravity data. Tectonophysics 140, 1–12).*

To the east of the basin, the basement depth pattern along the two profiles is quite different. The thickness of the Dharwar rocks (Archean) seems to be increasing southwards. The granites are exposed at a few places on Profile II beyond which the charnockite and khondalite rocks (maximum thickness 500 m) overlie them. The granite, exposed to the east of the basin, dips under the exposed charnockite rocks to the further east. A 1300-m thick sedimentary layer of velocity 2.7 km s^{-1} overlies the basement near the east coast of India.

Combined results of the two profiles in the Cuddapah basin reveal that the general dip direction of the Moho reflection is from northwest to southeast. The Moho, at a depth of 35 km in the western part of Profile II, has a maximum depth of 38 km in the middle of this profile and 41 km in the middle of Profile I. East of the uplift, in the middle of the basin, the Moho is at a depth of about 35 km. Close to the eastern end of the basin the Moho depth increases to 39 km at Profile II and 40 km at Profile I. Beyond the eastern margin of the basin its depth decreases to 37 km.

Gradually eastward plunging reflections to a depth of 20 km are seen to the east of the eastern margin of the basin (Figs. 2.4 and 2.5), where multiple eastward dipping over thrusts occurs at the ground surface. These indicate that the eastern margin of the basin is a gently sloping thrust where the ancient granite gneiss, overlain in the east by the early Proterozoic Dharwar system, is overthrust westward for a distance of 20–30 km (Chekunov et al., 1984). It is a typical feature for this region and appears to have occurred after the basin formation (Kaila et al., 1979, 1987). Towards the further east, in the region of the eastern ghat mobile belt (EGMB of Fig. 1.3), a post-Dharwar upthrust brought the granitic gneiss, charnockite and khondalite against the exposed Dharwar formation. Parts of the charnockite and khondalite probably were eroded, leaving the granite as exposed rock (Kaila and Tewari, 1985). Modeling of the travel times and relative amplitudes of the digitized seismic refraction/postcritical reflection data around the thrust faults to the east of the basin has shown that these thrusts extend deep into the crust. High velocity of 6.7–7.2 km s^{-1} at the upper-middle crustal depths and 7.0–7.2 km s^{-1} at the lower crustal depths is seen along these thrust faults (Krishna et al., 1990).

Chandrakala et al. (2010) digitized and reinterpreted the seismograms at the western edge of the Cuddapah basin and identified four reflection phases towards the basin compared to three to the west of the basin, indicating that there is one more layer present below the western part of the Cuddapah basin. This additional layer of velocity 7.10–7.35 km s^{-1}, at the base of the crust, is 15- to 20-km thick in the western Cuddapah basin and thins down to about 5 km to the west of the basin. This high-velocity layer has high density (3070–3160 kg m^{-3}) as well and is interpreted as an underplated magmatic layer related to infusion of mantle material. The massive underplating led to thickening of the crust to about 40–44 km below the southwestern part of the Cuddapah basin, compared to about 34 ± 2 km in the surrounding regions, indicating thermal restructuring of the crust/mantle boundary.

The Bouguer gravity picture (Fig. 2.6) in the Cuddapah basin (NGRI, 1975) shows a high positive gradient at the eastern margin of the basin and a large gravity high in its southwest part. Before the availability of the seismic model of the crust, the positive gradient at the eastern margin was attributed to accumulation of lower density sediments whose thickness was estimated as 12 km within the basin in the downthrown western block (Kailasam, 1976). Combined interpretation of the seismic and gravity data (Kaila and Bhatia, 1981, Tewari et al., 1986) shows that coming up of an anomalous high-velocity (7.2 km s^{-1})/high-density (3050 kg m^{-3}) layer at the bottom of the lower crust in the eastern part of the Cuddapah basin, updip of Moho towards the east coast, a 1.5-km thick layer of density

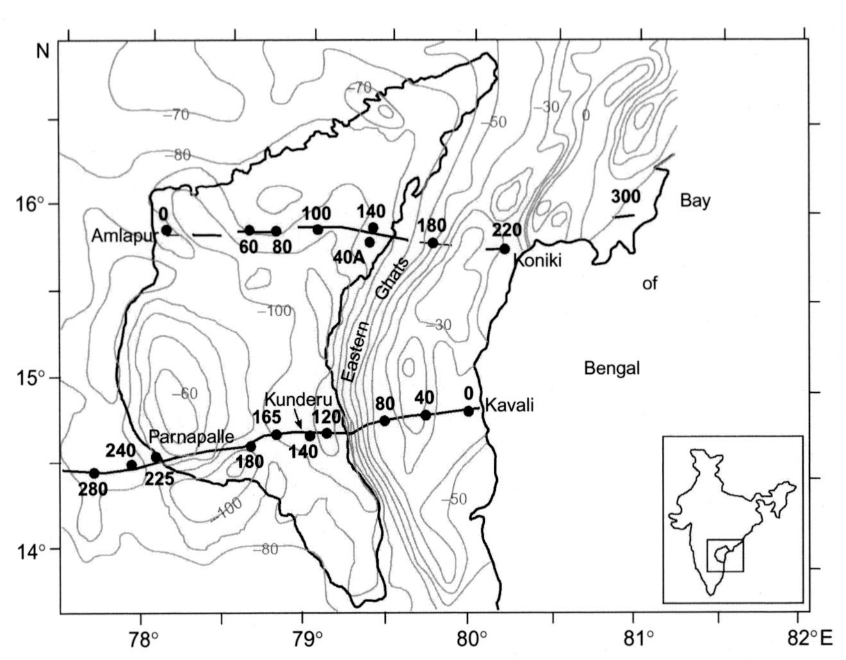

Fig. 2.6 Bouguer gravity anomaly map of Cuddapah basin and area to its east (NGRI, 1975).

2600 kg m^{-3} at the surface, and about 2-km thick sediments on the east coast together produce the two-dimensional gravity picture that agrees well with the observed values (Fig. 2.7).

The basaltic intrusion in the southwest Cuddapah basin is accompanied by a large eastward dipping fault at the western border of the basin, increased crustal thickness to the east of the fault, a smaller number of seismic boundaries and a velocity inversion. This magmatic intrusion is generated by the fault at western border of the basin and the outcrops of basic and ultrabasic rocks belong to the apical parts of the intrusion (Chekunov et al., 1984). The high velocity of 6.90 km s^{-1} (Fig. 2.8), uncharacteristic at shallow

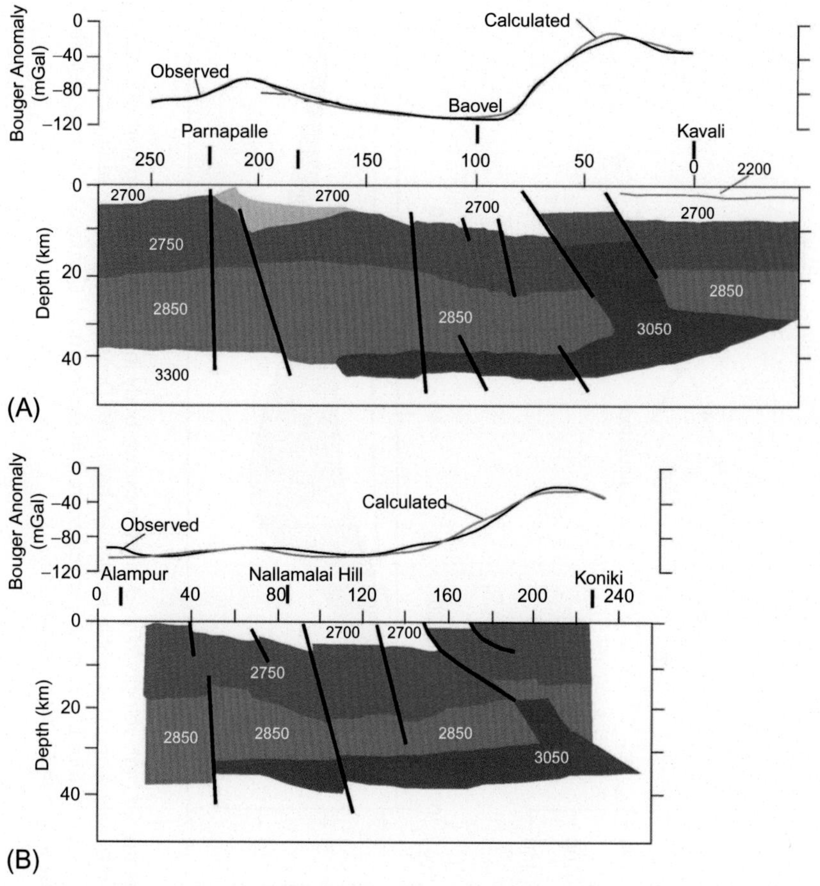

Fig. 2.7 Gravity models along (A) Profile I and (B) Profile II in the region of Cuddapah basin and eastern ghat. Density values are in kg m^{-3}. *(Modified from Tewari, H.C., Rao, V.K., 1995. Seismic studies in the Cuddapah basin: a review. In: Tirupati '95, Seminar on Cuddapah basin, Geol. Soc. of India, Sri Venkateswara University, pp. 67–79).*

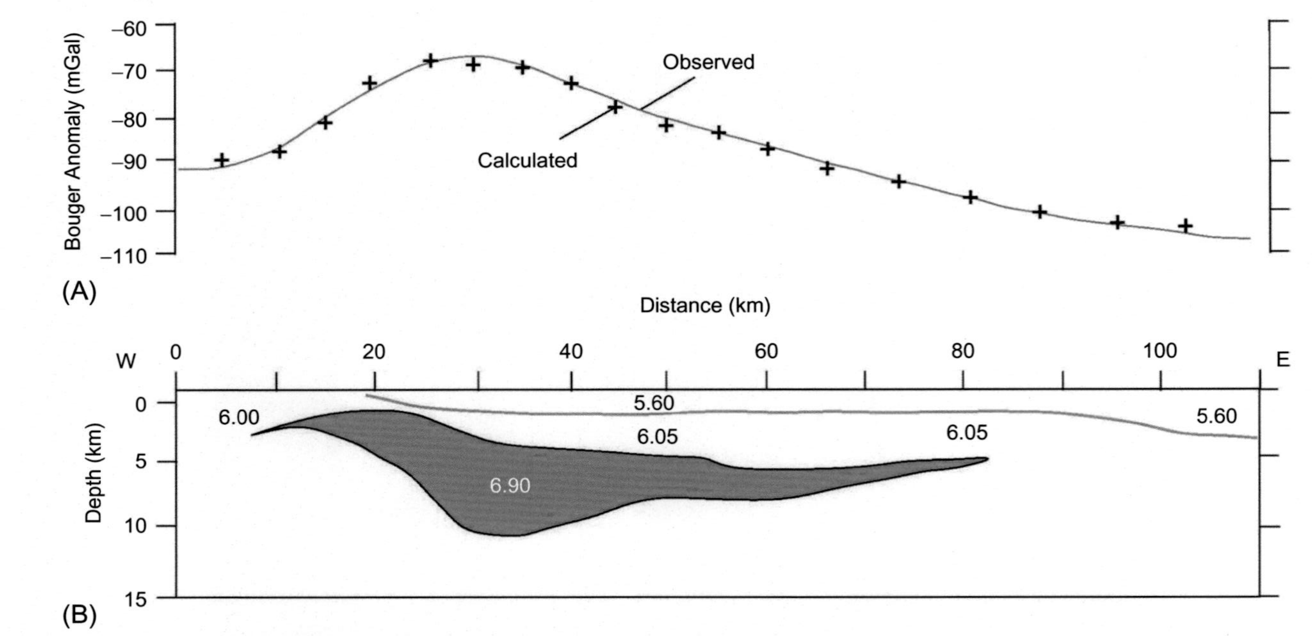

Fig. 2.8 Interpreted high-velocity/high-density intrusive body in the southwestern part of Cuddapah basin. Velocity values are in km s^{-1}. (A) Bouger anomaly along the profile in Marwar basin. (B) Depth model of the sedimentary basin. *(Modified from Tewari, H.C., Rao, V.K., 1987. A high velocity intrusive body in the upper crust in southeastern Cuddapah basin as delineated by deep seismic sounding and gravity modelling. Geol. Soc. India, Mem. 6, 349–356).*

crustal depths, identifies this body as an intrusion from lower crust/upper mantle. This body of higher density, with a maximum thickness of about 7 km, is also responsible for the high gravity values in the southwest Cuddapah basin (Tewari and Rao, 1987).

The total magnetic intensity anomaly in the western part of the Cuddapah basin has an amplitude of more than 500 nT and covers the entire high-density mafic dike body within the basement. Combined interpretation of seismic results with the gravity and magnetic values suggests mantle perturbation and a volcanic source in the southwest Cuddapah basin—maximum depth to top of the source is about 8 km. This model in the southwest Cuddapah basin and to its west (west of Parnapalle in Fig. 2.9) suggests a three-layered crust with velocities of 6.4, 6.6–6.8, 7.4 km s^{-1} and two Moho layers with upper mantle velocities of 7.8 km s^{-1} at 37-km depth and 8.1–8.4 km s^{-1} at 45-km depth. A low-velocity (6.5 km s^{-1}) layer (Fig. 2.9) separates the two Moho layers (Reddy et al., 2004).

Chandrakala et al. (2010) opine that the 15- to 20-km thick magmatic layer at the base of the crust has resulted in high conductivity (Naganjaneyulu and Harinarayana, 2004) as well as a high gravity anomaly. Considering the shape, size and lateral extent of such an anomalously high-density/high-velocity

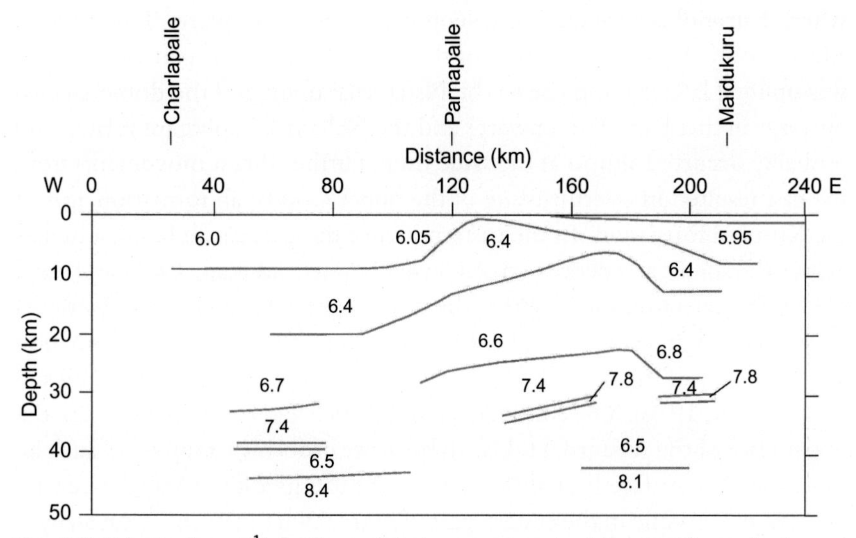

Fig. 2.9 Velocity (km s^{-1})-depth model in western part of the Cuddapah basin and adjoining eastern Dharwar Craton. Velocity values are in km s^{-1}. *(Modified from Reddy, P.R., Chandrakala, K., Prasad, A.S.S.S.R.S., Rama Rao, Ch., 2004. Lateral and vertical crustal velocity and density variation in the Cuddapah basin and adjoining Eastern Dharwar craton. Curr. Sci. 87, 1607–1614).*

feature, they are of the opinion that it may be an imprint of a mantle plume, as the Moho is quite warm and the asthenosphere seems to be quite shallow, at around 100 km, beneath this region.

2.1.1.3 Evolutionary Model

Proterozoic terranes across the world show significant differences in accretion models. A subduction-related horizontal accretion is suggested for the crust of North America and the Baltic shield, but a vertical accretion is favored for the Australian crust. In northeast Africa arc micro plate accretion, with extensive ophiolite obduction in the Neoproterozoic, is suggested. In contrast, a model of intracrustal orogeny is preferred to explain the evolution of some other Neoproterozoic African terranes. Proterozoic fore deeps have been recognized adjacent to some major thrust belts in North Africa.

A seismic study-based evolutionary model proposed for the Cuddapah basin (Kaila and Tewari, 1985) shows that the basin was created in its western part by downfaulting of that crustal block during the Paleoproterozoic and lower Cuddapah sediments were deposited there. Subsequently, the eastern half of the basin was down faulted in the Neoproterozoic and upper Cuddapah sedimentation took place. After close of this sedimentation the western part of the basin was downfaulted, creating a shallow subbasin, where Kurnool sedimentation (Neoproterozoic) took place. This faulting affected only the shallower parts of the basin. The eastern part of the basin was uplifted later, giving rise to the Nallamalai uplift and the dome-shaped upwarps in that part. The upwarps and the Nallamalai uplift are related and probably occurred almost at the same time. Further thrust movements from the east resulted in overthrusting of the upper Cuddapah formation against the Kurnool formation. In the eastern part of the Cuddapah basin, a higher velocity gradient at a deeper crustal level (Tewari and Rao, 1987) and double Moho reflection indicate that the crust-mantle boundary here is transitional. The high-velocity gradient in the lower crust indicates deformation of a large part of the crust due to intrusion from the upper mantle (Glikson and Lambert, 1976). The eastern ghat mobile belt and intrusive bodies in the upper crust of the western Cuddapah basin are probably examples of such an intrusion. The surface heat-flow values in the southwest Cuddapah basin are 27 mW m^{-2}, while in the eastern part they are 63–64 mW m^{-2}. An analysis of heat-flow values and upper mantle (P_n) velocity distribution suggests that lateral inhomogeneity together with Moho temperature is responsible for the variation in P_n velocity (7.91–8.59 km s^{-1}) beneath the Cuddapah basin (Sharma et al., 1991).

Different levels of Moho in the Cuddapah basin indicate horizontal displacement of separate blocks (Sollogub et al., 1977). The geological evidence suggests that collision of different blocks has taken place at the eastern boundary of the basin (Radhakrishna and Naqvi, 1986). The accretion process suggested for the Cuddapah basin has to take into account the easterly dips of the Moho, the double Moho, the high-velocity gradient in the lower crust, the thrust fault at the eastern margin and high heat-flow values (63–64 mW m^{-2}) in the eastern part. These features, seen together with discordance in the crustal reflections at various levels, indicate a horizontal accretion process for the basin where delamination at the crust-mantle boundary and subsequent thermal subsidence/orogenic collapse has possibly taken place due to collision at its eastern boundary (Tewari and Rao, 1995).

2.1.2 Western Dharwar Craton

A large number of geological classifications and sequences have been proposed for the Dharwar craton (Fig. 2.10). It is divided into two main tectonic units, the western Dharwar craton and the eastern Dharwar craton (Ramakrishnan et al., 1976). The dividing line between the two units approximately coincides with the western margin of the exposed granitic belt. Mature sediment dominated greenstone belts with subordinate volcanism and intermediate pressure metamorphism characterize the western craton. In contrast, greenstone belts dominated by volcanic rocks characterized by low-pressure metamorphism typify the eastern craton.

2.1.2.1 Geology and Tectonics

The western part of the Kavali-Udipi deep seismic profile cuts across two important geological features of this craton, the Chitradurga and Shimoga schist belts. Both these belts were formed during the Paleoproterozoic. The Chitradurga schist belt is interbedded with metasediments and overlain by the Mesoproterozoic sediments. On its eastern side, the Neoproterozoic rocks are exposed. Massive lava, with pillow structure, rests over the Chitradurga granite. The Shimoga schist belt consists of the metavolcanic and metasedimentary rocks and rests over the Archean basement. Paleoproterozoic to Neoproterozoic rocks are exposed within this belt. Towards the close of the Paleoproterozoic period, intense orogenic activity, accompanied by predominantly horizontal forces, took place. The orogeny after the Mesoproterozoic was not intense and developed only broad folds.

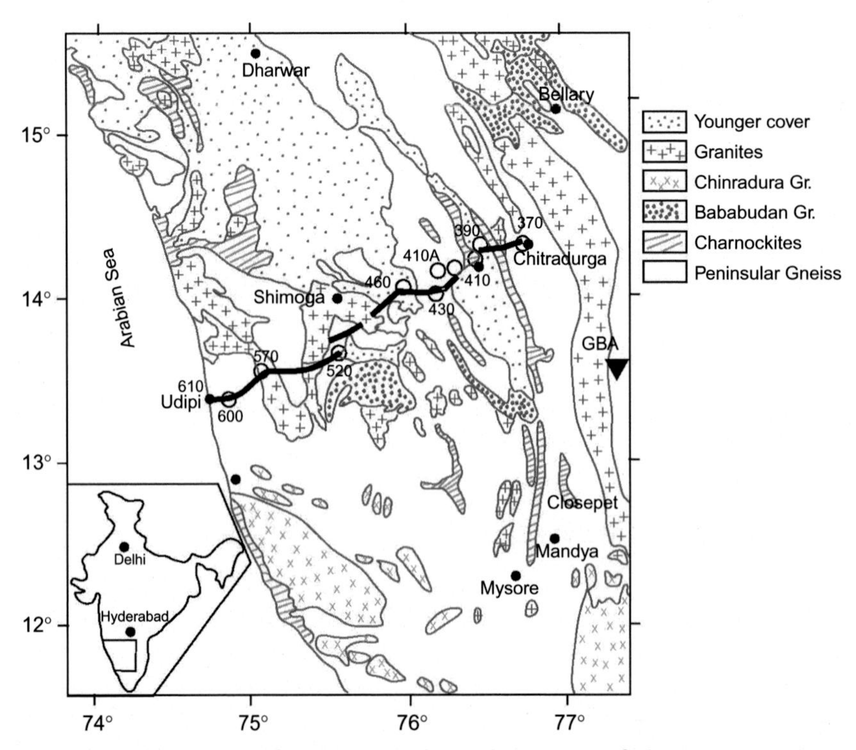

Fig. 2.10 Deep seismic profile plotted on the geological map of Dharwar craton. *(From Kaila, K.L., Roy Choudhury, K., Reddy, P.R., Krishna, V.G., Narain, H., Subbotin, S.I., Sollogub, V.B., Chekunov, A.V., Kharetchko, G.E., Lazarenko, M.A., Ilchenko, T.V., 1979. Crustal structure along Kavali-Udipi profile in the Indian Peninsular Shield from the deep seismic soundings. J. Geol. Soc. India, 20, 307–333).*

2.1.2.2 Seismic Results

Results of only one crustal seismic study, from the Kavali-Udipi profile (Kaila et al., 1979), are available for the Dharwar craton and show that this craton is separated from the Cuddapah basin that lies to its east by an uplifted block (Fig. 2.11). Two low-angle thrust faults, which indicate large deformation, have been delineated at the eastern margin of the Chitradurga schist belt, but no such deformation is seen under the Shimoga schist belt. The basement of the schist belts is at 4–7 km depth and has a velocity of 6.4 km s^{-1}. In the western part, the thickness of the Dharwar formation is up to 10 km at several places. The depth to the Moho discontinuity is 34–41 km in the eastern part of the craton and 38 km in its western part. The velocity in the upper mantle (P_n) is 8.4 km s^{-1}.

Combined interpretation of the seismic and Bouguer gravity values indicated that a near-surface high-density of 2780 kg m^{-3}, corresponding to

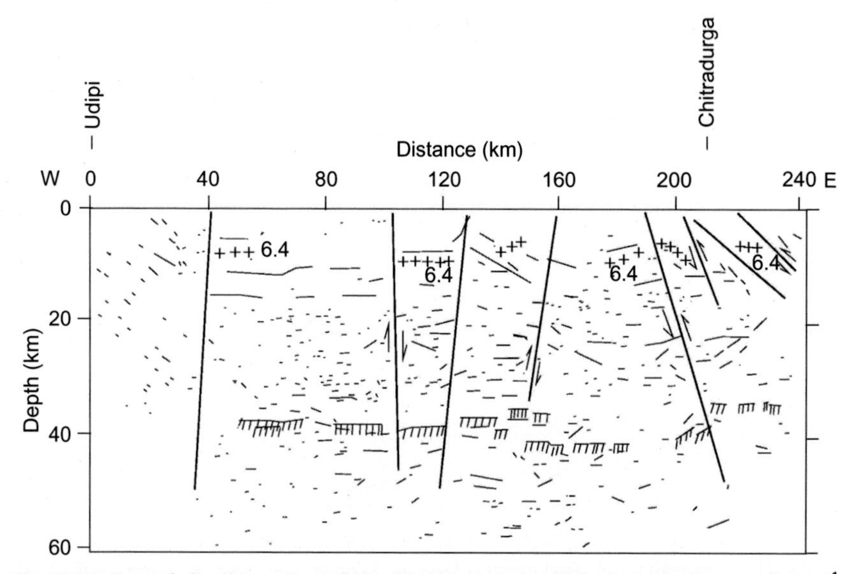

Fig. 2.11 Crustal depth section of the Dharwar craton. Velocity values are in km s^{-1}. *(Modified from Kaila, K.L., Roy Choudhury, K., Reddy, P.R., Krishna, V.G., Narain, H., Subbotin, S.I., Sollogub, V.B., Chekunov, A.V., Kharetchko, G.E., Lazarenko, M.A., Ilchenko, T.V., 1979. Crustal structure along Kavali-Udipi profile in the Indian Peninsular Shield from the deep seismic soundings. J. Geol. Soc. India, 20, 307–333).*

abundantly occurring schistose rocks, is required in the region of the Chitradurga schist belt (Kaila and Bhatia, 1981). Short wavelength anomalies in this region are due to near-surface features of very high density (2980 kg m^{-3}). The Moho boundary is extrapolated to rise to 35 km at the west coast near Udipi.

Modeling of the digitized seismic data for the western Dharwar craton (Sarkar et al., 2001) shows a two-dimensional velocity model (Fig. 2.12) in which the velocity in the upper crust increases from 6.0 km s^{-1} at the surface to 6.18–6.19 km s^{-1} at its boundary. The lower crust is at 22–24 km depth and has a velocity varying between 6.7 and 7.0 km s^{-1}. It overlies the Moho boundary at 37–40 km depth, the upper mantle velocity being 8.40 km s^{-1}. The crust here is not typical of Archean crusts in other parts of the world, in terms of thickness and velocity. It is comparatively thicker and the velocity in the lower crust, though a little higher than normal, does not represent underplating with mafic material. The upper mantle velocity is also higher than normal, probably because the crust of the Dharwar craton is cooler than other Archean crusts (Ray et al., 2003), as indicated by low mantle heat-flow values of 11–16 mW m^{-2}. Its larger thickness is a consequence of its exposure

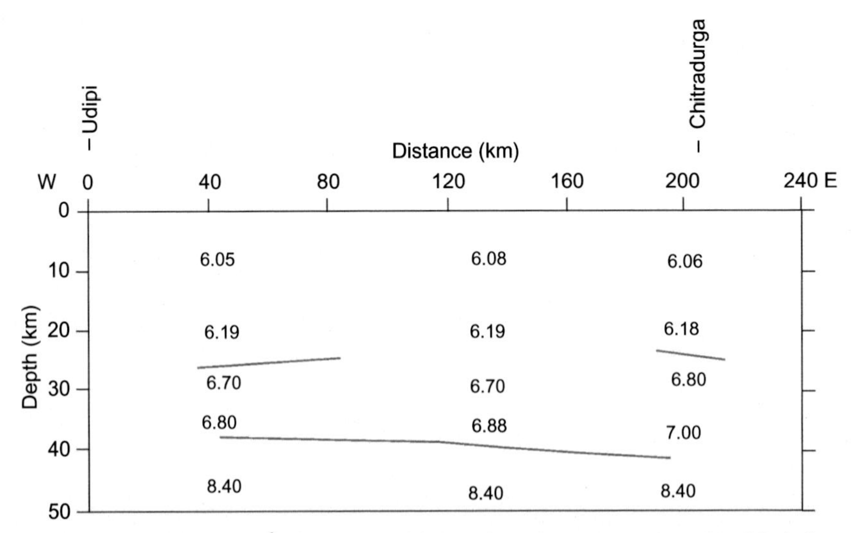

Fig. 2.12 Velocity (km s^{-1})/depth model for the Dharwar craton. *(Modified from Sarkar, D., Chandrakala, K., Padmavathi Devi, P., Sridhar, A.R., Sain, K., Reddy, P.R., 2001. Crustal velocity of western Dharwar craton, South India. J. Geodyn. 31, 227–241, 237).*

to later major tectonic activities responsible for features like Shimoga and Chitradurga schist belts. Larger than normal thickness of the crust in the western Dharwar craton is not true for the entire craton, as receiver function analysis has revealed a 33-km thick crust in the eastern Dharwar craton.

2.1.3 Other Interpretations of East-West Profiles

The velocity-depth model along the Kavali-Udipi profile (as shown in Figs. 2.4 and 2.11) shows many crustal blocks separated by deep faults. Several workers have reinterpreted this model. According to one view the deep faults are younger fractures, faults or dikes that are not related to tectonic evolution of the region (Radhakirhsna and Ramakrishnan, 1993). Another interpretation suggests that the Moho reflections along the profile, though locally offset, are nearly horizontal while the overlying crust has dipping reflections. This geometry of structural discordance with respect to the Moho requires that the Moho, as seen today, formed subsequently. Offsets in the Moho have arisen later, implying that at the time of its formation the Moho constituted a thermomechanical boundary between the rigid crust and a plastic mantle. Widespread horizontality of the Moho is attributed to ductile shearing, evidence of basal crustal weakness and buoyancy-driven sharp density contrast (Roy Choudhury and Hargraves, 1981).

Others suggest that the faults interpreted on the Kavali-Udipi profile are rather speculative and view the depth section after removing the faults (Gibbs, 1986). The redrawn section shows that the discordance in reflection dips is not restricted only to Moho levels, but is also visible in the lower crust (Fig. 2.13). In the eastern part of the profile the east-dipping reflections correspond to prominent thrust faults and the basement of the Cuddapah basin while in the Dharwar craton some of these reflections may correspond with late Archean thrust faults that persist through the crust, and may even offset the Moho reflections. The opposite dipping reflections (indicated by arrows) under the Chitradurga schist belt seem to be close to the boundary between the western and eastern Dharwar craton. These can also be interpreted as a suture that developed during the collision between the eastern and western Dharwar cratons. The dipping reflections may be a product of collision tectonics at these boundaries. The gravity model of this depth section suggests that the easterly dipping reflections to the east of the Cuddapah basin are associated with a sharp gravity gradient (XX′) on the western margin of the eastern ghat fold belt that separates it from the interior of the Indian continent. The easterly dips associated with gravity anomaly YY′ also indicate the separation of the western Dharwar craton from the eastern Dharwar craton (Mishra, 2002).

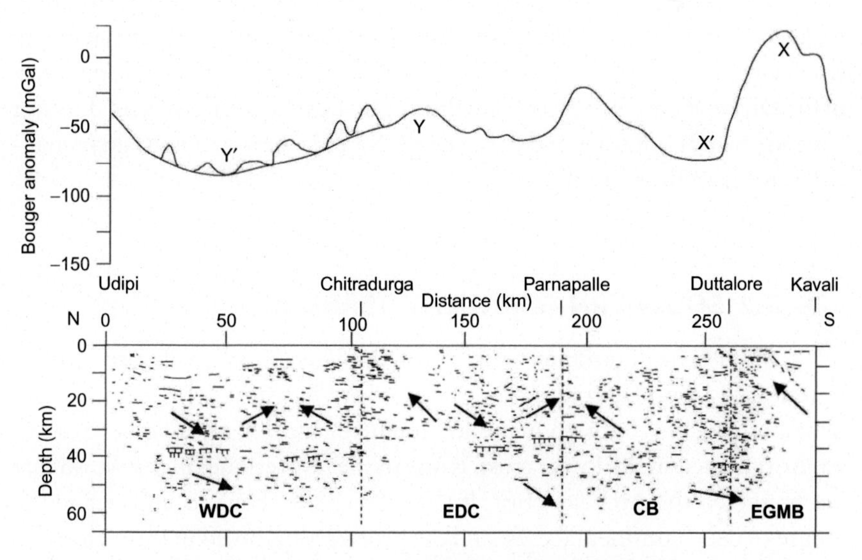

Fig. 2.13 Crustal depth section of the Kavali-Udipi profile without faults and its gravity model. *Arrows* indicate the dips of geological boundaries. *(Modified from Kaila, K.L., Roy Chowdhury, K., Reddy, P.R., Krishna, V.G., Narain, H., Subbotin, S.I., Sollogub, V.B., Chekunov, A.V., Kharetchko, G.E., Lazarenko, M.A., Ilchenko T.V., 1979. Crustal structure along Kavali Udipi profile in the Indian peninsular shield from deep seismic sounding. J. Geol. Soc. India 20, 307–333; Mishra, D.C., 2011. Gravity and Magnetic Methods for Geological Studies. Principles, Integrated Exploration and Plate Tectonics. BS Publications, 938 pp. ISBN: 978-93-81075-27-2.)*

Receiver function modeling of the teleseismic waveforms (Gupta et al., 2003a,b) indicates a thickness of 34–39 km beneath the Cuddapah basin and eastern Dharwar craton and 42–51 km beneath the western Dharwar craton. They also indicate that the Moho is essentially horizontal and the offsets occur in sharp boundaries that coincide with major north-south trending shear zones. The thickest crust occurs beneath the middle Archean greenstone belt, which is the nucleus of the Dharwar craton. This is a major low-strain zone, which has not been subjected to any severe compressive deformation. Since the amphibolite grade metamorphic mineral assemblages in this part equilibrated at the depths of 15–20 km, the crust here would have been exceptionally thick, 57–60 km, at about 3000 Ma. Average Poisson's ratio value for the south Indian crust is 0.24–0.27, which is lower than the global average of 0.27–0.31 for the Archean shields. Lower than normal value of the Poisson's ratio indicates felsic to intermediate composition of the lower crust that requires a delamination process to remove mafic lower crust from the primitive crust or a different process of Archean crustal evolution. Though significant differences are observed between the Archean and Proterozoic crusts at several places (Durheim and Mooney, 1994), no such difference is observed in the south Indian shield. The surface wave dispersion and receiver function studies along a north-south corridor between the Deccan volcanic province and the Dharwar craton (Rai et al., 2003) show that the character of the crust is more or less uniform and there is no evidence for a high-velocity basal layer above the Moho in the central part of the Dharwar craton. This implies that the base of the Dharwar crust does not have any mafic cumulates overlying the Moho and it has remained fairly refractive since its cratonization.

2.2 SOUTHERN GRANULITE TERRANE

The Archean terranes, with exposed lower crustal rocks, are of great significance in understanding the evolutionary history of the Indian continental crust. The southern granulite terrane of India (Fig. 2.1), believed to have originated in the lower crust, is one of the few terranes in the world that has preserved the Archean crust. It provides ample scope to understand the nature of deep continental crust and the processes involved in formation and evolution of the lower crust. The granulite represents rocks from the deep crust, formed at an average pressure and temperature of 7–10 kbar and 700–800°C respectively, corresponding to burial depths of about 20–25 km. These rocks are often formed by thickening in older terranes and are uplifted to the surface by various mechanisms. Various models involving subduction, collision and accretion of microterranes, flower structure and transtensional tectonics have been given to explain the tectonic emplacement of these rocks in the Indian peninsular shield.

2.2.1 Geology and Tectonics

The southern granulite terrane has a complex evolutionary history, from the early Archean to Neoproterozoic (3500–550 Ma), involving repeated multiple deformations, anatexis, intrusions and polyphase metamorphism (Bhaskar Rao et al., 2003). It is a composite of crustal blocks exhumed during different periods and is divided into three major blocks (Fig. 2.1). The northern granulite block occupies the area between the Dharwar craton and the Palghat-Cauvery shear zone and defines transition from a low to high-grade terrane. The region between the Palghat-Cauvery shear zone and the Achankovil shear zone is known as southern granulite block (500–700 Ma) and includes most of the highlands with amphibolites facies gneiss and supracrustal rocks, in addition to the charnockite, granulite and khondalite rocks. The northwest-southeast trending Achankovil shear zone is more prominent toward the southern end of the granulite terrane and indicates Proterozoic tectonic activity.

The network of shear zones (Fig 2.14) in the northern block is termed the Cauvery shear zone system. It consists of the Moyar-Bhavani (MBSZ), Mettur (MEZ) and other shear zones; maximum width of the individual shear zones is around 20 km. All the shear zones exhibit dextral strike-slip movement with a maximum lateral displacement of about 80 km (Chetty et al., 2003). Late Neoproterozoic (750–550 Ma) granites and alkaline plutons intrude most of these shear zones. A few of these zones are associated with layered anorthosite and mafic/ultramafic complexes. The presence of alkaline plutons near most of the faults associated with the shear zones suggests reactivation of earlier faults during the Neoproterozoic period (Kumar et al., 1993; Santosh et al., 2005). The granulite terrane exhibits two periods of high-grade metamorphism (Bhaskar Rao et al., 2003): late Archean in the north and Meso-Neoproterozoic in the south (Madurai block and Kerala Khondalite belt).

The Nilgiri block, consisting mainly of the charnockite, shows 2800 Ma protolith age and 2500 Ma high-grade metamorphic age and indicates a short premetamorphic history (Peucat et al., 1989). In contrast, the protolith and metamorphic age of the Dharwar craton is 3600 and 2500 Ma, respectively.

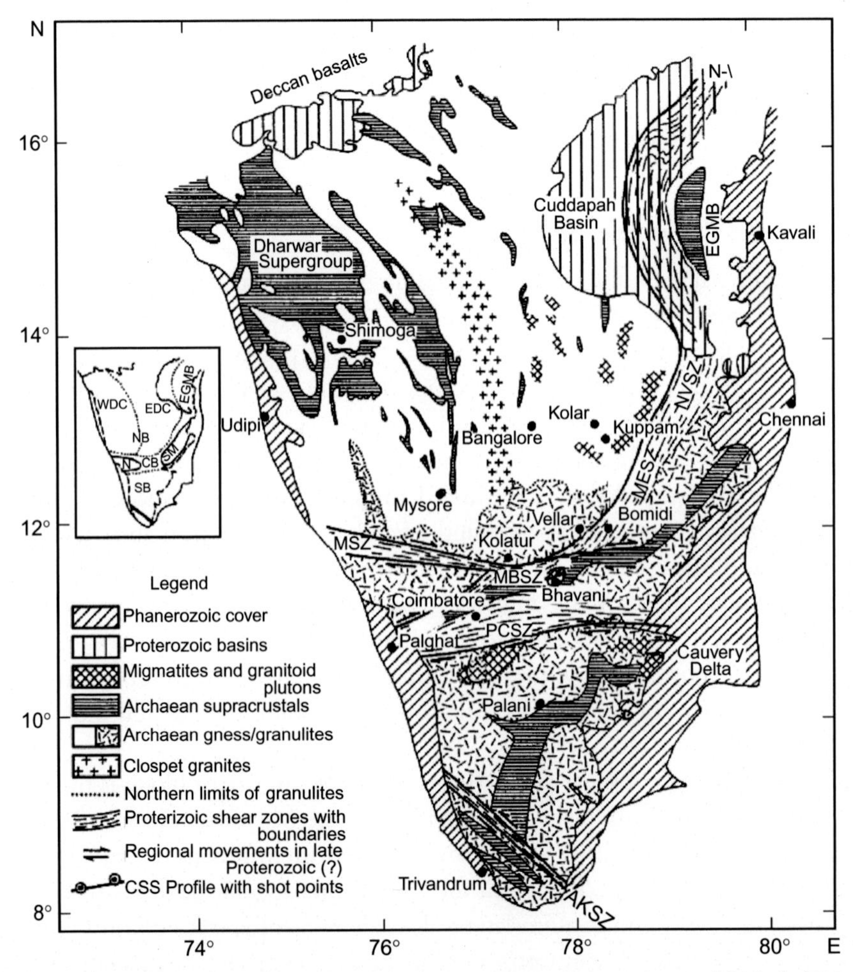

Fig. 2.14 Geology of southern granulite terrane. *EGMB*, eastern ghat mobile belt; *EDC*, eastern Dharwar craton; *WDC*, western Dharwar craton; *MBSZ*, Moyar-Bhavani shear zone; *MSZ*, Mettur shear zone; *PCSZ*, Palghat-Cauvery shear zone; *AKSZ*, Achankovil shear zone. *(From Vijaya Rao, V., et al., 2006. Crustal structure and tectonics of the northern part of the Southern Granulite Terrane, India. Earth Planet. Sci. Lett. 251, 90–103.)*

These two crustal blocks located on either side of the Cauvery shear zone system exhibit distinctly different evolutionary histories. The different geological setting, contrasting lithological and structural features and isotopic data suggest that this shear zone system might represent a suture between the Dharwar craton and the Nilgiri block (Raith et al., 1999). The Nilgiri block is considered to be an allochthonous unit that was thrust onto the Dharwar craton during the late Archean collision. Subduction and collision processes resulted in thickening of the crust and granulite formation in the region (Harris et al.,

1994). Geological formations in the Madurai block show complex folds and have a general northeast-southwest strike in the north, swinging to the northwest-southeast direction in the southern part. Rb-Sr isotopic dates span between 2000 and 500 Ma. A number of anorogenic granites in this block are dated at about 550 Ma and point to a strong impact of Pan-African tectonic, magmatic and metamorphic events (Santosh et al., 2005).

Rifting and associated magmatic history (750–550 Ma) of the region are coincident with the amalgamation and extensional episodes of the Gondwana supercontinent and are attributed to the postcollision extensional processes of the Pan-African orogeny (Collins et al., 2000; Santosh and Yoshida, 2001). Earlier it was believed that the region north of the Palghat-Cauvery shear zone is an Archean-Neoproterozoic terrane boundary and has undergone granulite facies metamorphism during the late Archean, at 2500 Ma, and the southern part during the late Neoproterozoic, at 550 Ma. Geochronological studies, however, suggest that the Oddanchatram shear zone, located further south, demarcates the Archean-Neoproterozoic metamorphic boundary (Ghosh et al., 2004). The anorthosites, generally related to collision and suture zones, are observed at the eastern part of this shear zone. Even though the Madurai block consists of the Archean rocks, the high-grade metamorphism represented by the khondalite rocks is around 550 Ma, which is much younger to the 2500 Ma charnockite rocks to the north. Metamorphism at high pressure (9–12 kb), representing thrusting and collision, is observed in the central region, covering the Cauvery shear zone, whereas ultrahigh temperature metamorphism is noticed to its south (Reddy et al., 2003).

2.2.2 Seismic Results

Receiver function analysis (Rai et al., 2003), along a northwest-southeast corridor starting from the Deccan volcanic province to the granulite terrane, through the western Dharwar craton in the southern part of the Indian shield, shows that the crustal thickness in the two regions is different and the transition between them is relatively sharp. In the volcanic province, the crust is 35- to 38-km thick, similar to the eastern Dharwar craton. Progressive thickening, with an increasing grade of metamorphism, is observed in the western Dharwar craton in the granulite facies block. The present-day Moho is at 60-km depth. Since the exhumed granulite rocks were once buried at a depth of 25–30 km, the crustal thickness during the Precambrian could have been about 90 km.

Deep seismic reflection and refraction/postcritical reflection studies, in the north-south direction, were carried out along a 300-km long Kuppam-Palani profile, over the southeastern part of the eastern Dharwar craton and northern part of the southern granulite terrane and about 220-km long Vattalkundu-Kanyakumari profile over the southern part (Fig. 2.1). Magnetotelluric and gravity studies were also undertaken in an about 100-km wide corridor along with the seismic studies. The northern part of the profile between Kuppam and Bommidi (Fig. 2.15) lies on gneiss of the eastern Dharwar craton, whereas its southern part is over the epidiorite-hornblende gneiss and charnockite of the southern granulite terrane. To the further south the profile is mostly over the fissile hornblende-biotite gneiss. The southernmost part of this profile covers the Moyar-Bhavani (also Mettur) and Palghat-Cauvery shear zones. The Vattalkundu-Kanyakumari profile, lying to the east of the Madurai block, is mostly on the gneiss and charnockite of the southern granulite terrane and crosses the high-grade supracrustal formations of the Achankovil shear zone.

2.2.2.1 Kuppam-Palani Profile

The velocity-depth model of the crust (Fig. 2.15), based on refraction/post-critical reflection data along this profile, shows four velocity layers (Reddy et al., 2003): a surface velocity of 6.1 km s^{-1} that belongs to the exposed Archean gneiss and charnockite, followed by velocities of 6.5–6.55, 6.0–6.05 and 6.9–7.1 km s^{-1} corresponding to the upper, middle and deeper crustal horizons, respectively. Velocity in the top layer increases to 6.3 km s^{-1} at Bommidi, probably due to the presence of high-density rocks in the shallow crust. Velocity inversion in the middle crust, in the form of a 10- to 15-km thick low velocity (6.0–6.05 km s^{-1}) layer, is seen along the entire length of the Kuppam-Palani profile. This layer is attributed to a low-density megashear zone of mantle origin. The high-velocity (6.9–7.1 km s^{-1}) layer at the base of the crust has a thickness of 10–15 km in most parts of the profile, except between Kuppam and Bommidi where it is comparatively thinner. This layer indicates magmatic underplating, which provided a substantial amount of heat into the crust, causing widespread melting of the earlier lower crust. The Moho is deepest (~47 km) at about 50 km south of Kuppam, which indicates a zone of crustal root. It also shows an up warp to the further south. However, to the south of Kolattur all the interfaces, including the Moho, show gradual increase of crustal thickness to the south up to Chennimalai (Moyar-Bhavani shear zone). Here the maximum depth to Moho is about 44 km. Further south, the structural trend reverses and the Moho is at a depth of about 39 km below Palani, indicating a Moho offset of about 5 km (Vijaya Rao et al., 2006). The upper mantle velocity is 8.0–8.05 km s^{-1} throughout the profile.

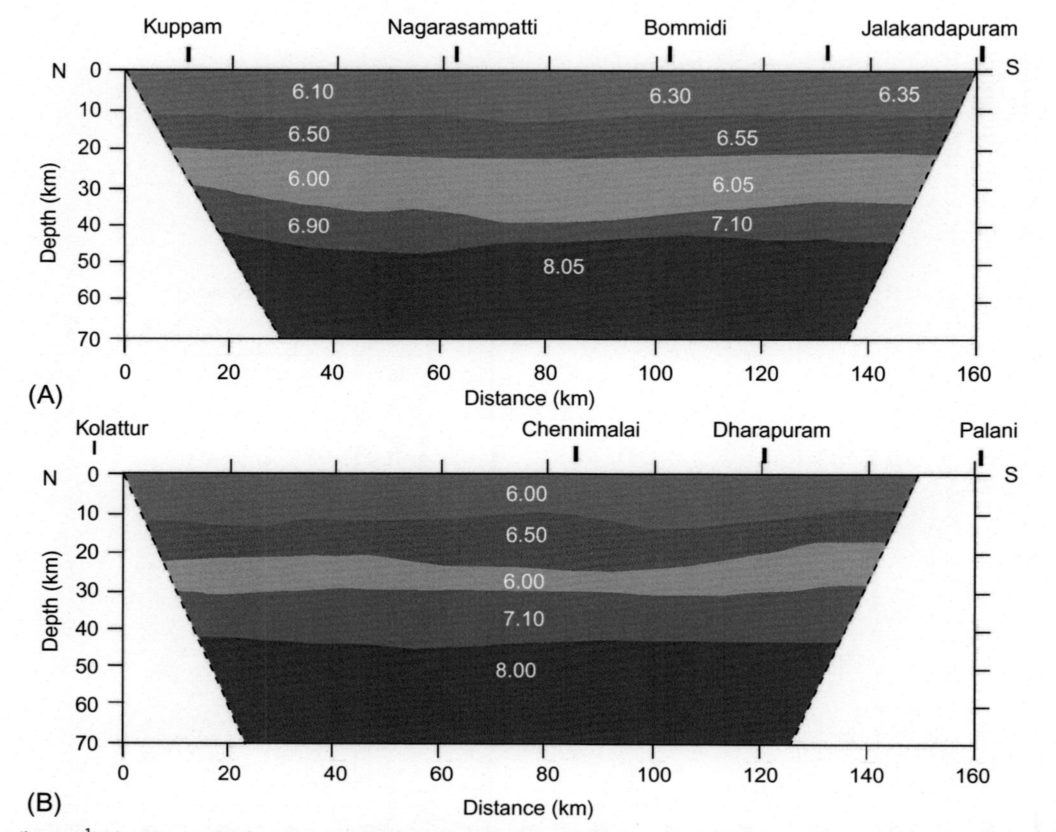

Fig. 2.15 Velocity (km s^{-1})/depth model for the crust along the Kuppam-Palani profile in southern granulite terrane, (A) Kuppam-Kolattur, (B) Kolattur-Palani. *(Modified from Singh, A.P., et al., 2006. Nature of the crust along Kuppam–Palani geotransect (South India) from gravity studies: implications for Precambrian continental collision and delamination. Gondwana Res. 10, 41–47.)*

The Archean terranes generally exhibit poor reflectivity because of low acoustic impedance contrast between the layers, high deformation, long tectonic history and complex structures. Deep reflection studies in the southern granulite terrane show that this is true also for this terrane, where reflectivity is moderate in the northern part and poor in the south. A line drawing of the stack section from deep reflection studies (Reddy et al., 2003) shows the reflectivity pattern in the region. Between Kuppam and Bommidi (Fig. 2.16A) most of the upper crust is poorly reflective, whereas the middle

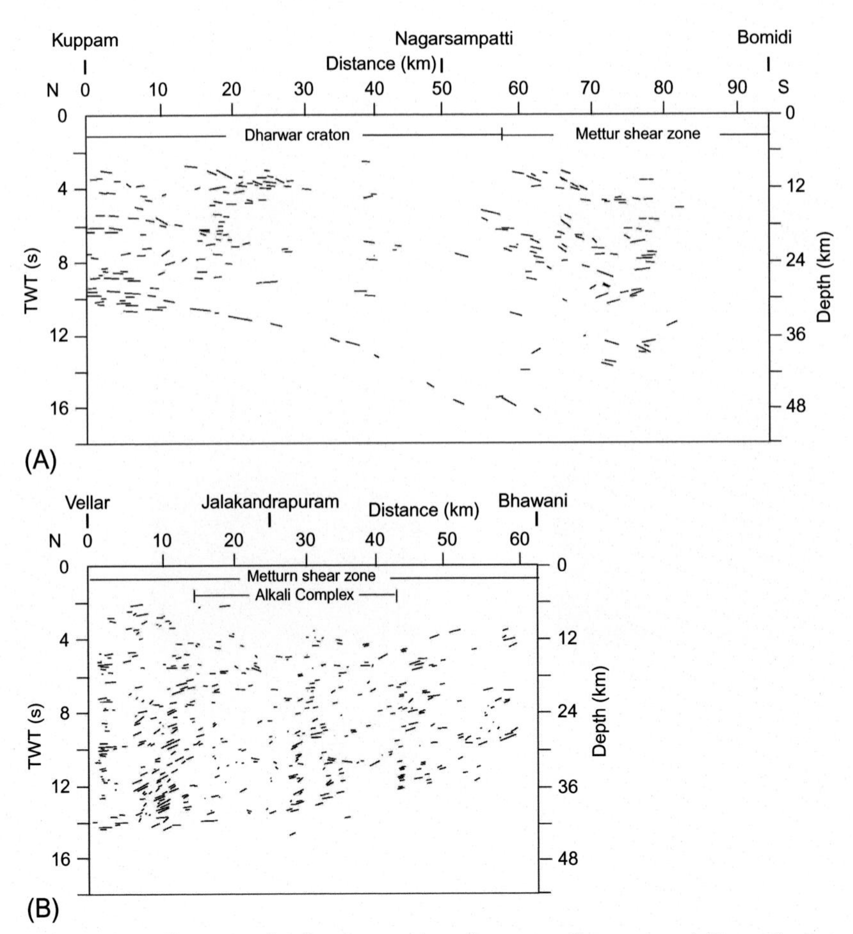

Fig. 2.16 Line drawing of reflection sections between Kuppam and Bhavani along (A) Kuppam-Bommidi, (B) Vellar-Bhavani. *(Modified from Reddy, P.R., Rajendra Prasad, B., Vijaya Rao, V., Sain, K., Prasada Rao, P., Khare, P., Reddy M.S., 2003. Deep seismic reflection/WAR studies along Kuppam-Palani transect in Southern Granulite terrane of India. In: Ramakrishnan, M. (Ed.), Tectonics of Southern Granulite Terrene, Kuppam-Palani Geotransect. Geol Soc. India Mem. 50, 79–106).*

and lower crust show horizontal and dipping reflections at a few locations. A dome-shaped reflection band is observed at around 4 s two-way time (TWT)—about 12 km depth in the northern part of this segment. A strong south-dipping reflection band, which extends up to a distance of 25 km from Kuppam, is also observed around 6 s TWT—about 18 km depth in this part. The most prominent feature of the seismic section is a reflection band, also dipping south, from 10 s TWT (~30 km depth) at the northern end to 15 s TWT (~45 km depth) in the middle of the segment. The base of this reflection band represents the Moho (Reddy et al., 2003). Except for these, no other prominent reflections are seen at any depth up to a distance of about 50 km from Kuppam. A decrease in the middle and lower crustal reflectivity is attributed to a deep-seated fault near the middle of this segment (SP 50). To the north of SP 50 most of the reflections dip to the south but some of the reflections show an updip at the end of this segment. A similar pattern of dips is also seen in the 2-D velocity-depth section (Fig. 2.15). The reflection Moho, usually defined at the bottom of lower crustal reflections, coincides with the Moho delineated by the postcritical reflections.

The top part of the seismic stack section between Vellar and Bhavani (Fig. 2.16B) is also devoid of reflections. Most of the deeper reflections show a northerly dip, opposite to that seen to the north of Bommidi. To the south of Vellar a prominent reflection band is seen at 14 s TWT (~42 km depth). A few minor undulating reflections, probably related to the presence of the granite and alkaline intrusive bodies, can be seen in the middle crust. North-dipping reflections in the southern part (around SP 140a) possibly represent an incomplete limb of a larger dome, the other limb of which cannot be seen due to limitation of the profile length.

To the south of Bhavani, the stack section is of poor quality due to high-amplitude, low-frequency and low-velocity noise, from surface to a depth of 12 s TWT (figure not available). Highly disturbed structural heterogeneity of the shear zones present in that region seem to be responsible for the noise. A few back-scattered signals (secondary arrivals) with reversed move out, between 3.5 and 6.5 s TWT (10- to 20-km depth), indicate the spatial location of the shear zones (Vijaya Rao and Rajendra Prasad, 2006).

Tectonic interpretation of the reflection results, particularly the deep penetrating dipping band of reflections between Kuppam and Bommidi, indicates that within the eastern Dharwar craton the Moho deepens from about 34 km at Kuppam to about 50 km at Bommidi. The seismic picture also suggests a collision of the Dharwar craton and another crustal block (explained in detail in Section 2.2.2.2), probably a part of the present eastern

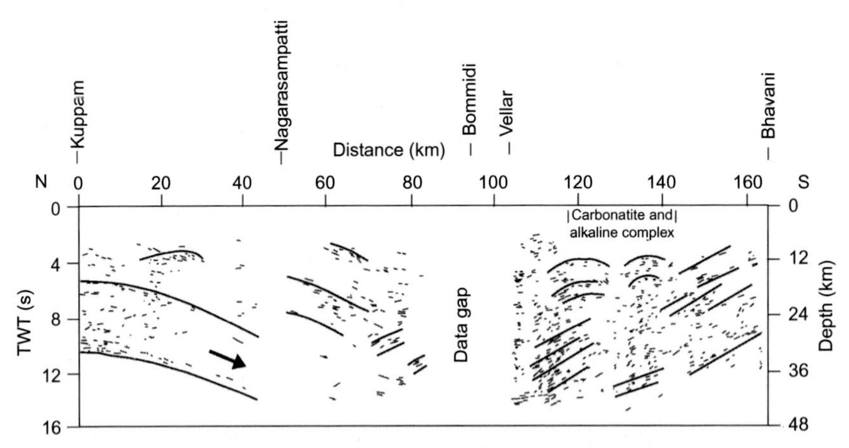

Fig. 2.17 Tectonic interpretation of the seismic reflection section between Kuppam and Bhavani in the southern granulite terrane. *(Modified from Vijaya Rao, V., Sain, K., Reddy, P. R., Mooney, W.D., 2006. Crustal structure and tectonics of the northern part of the Southern Granulite terrane, India. Earth Plan. Sci. Let. 251, 90–103).*

ghat mobile belt, in the south (Fig. 2.17). A volcanic arc exists between these two crustal blocks. This process thickened the crust in the contact zone of the eastern Dharwar craton and Moyar-Bhavani shear zone. The carbonic fluids released from the supracrustal rocks of the subduction zone and volcanic arc environment formed the granulites at the subduction zone. These were later exhumed from the deep crust during the collision process (Vijaya Rao et al., 2006).

In the eastern Dharwar craton the thickness of the lithosphere is 140–160 km and the mantle heat flow value is 16–18 mW m^{-2}. These values in the southern granulite terrane are 100 km and 32–36 mW m^{-2} respectively (Agrawal and Pandey, 2004). This distinct change in geophysical properties is probably responsible for the different thickness of Moho in the two regions, which seems to be a lithological boundary in the former and a rheological boundary in the latter (Vijaya Rao et al., 2007a,b).

The two-dimensional electrical resistivity structure along the Kuppam-Palani profile shows that resistivity values in the crust vary between 2000 and 50,000 Ω m (Fig. 2.18) and indicate distinctly different blocks in the crust that are separated by deep faults (Harinarayana et al., 2003). The blocks to the north of the Moyar-Bhavani shear zone and south of the Karur-Oddanchatram shear zone show high resistivity of the order of 20,000–50,000 Ω m (low conductivity). The intervening region exhibits lower resistivity (higher conductivity) of around 2000 Ω m to a depth of 12 km and 800 Ω m between 12- and 50-km depth.

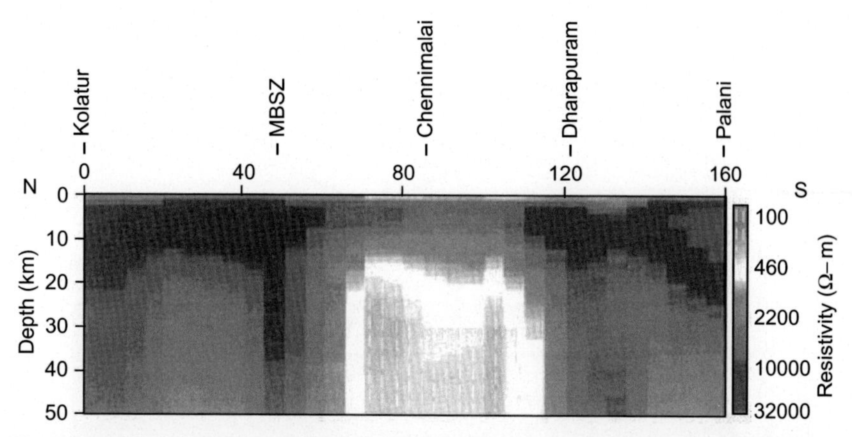

Fig. 2.18 Two-dimensional resistivity model between Kuppam and Palani in the southern granulite terrane. *MBSZ*, Moyar-Bhavani shear zone. *(Modified from Harinarayana, T., Naganjaneyulu, K., Manoj, C., Patro, B.P.K., Begum, K., Murthy, D.N., Rao, M., Kumaraswamy, V.T.C., Virupakshi, G., 2003. Magnetotelluric investigations along the Kuppam-Palani Geotransect, South India—2-D modelling results. In: Ramakrishnan, M. (Ed.), Tectonics of Southern Granulite Terrene, Kuppam-Palani Geotransect. Geol. Soc. India, Mem. 50, 107–124).*

The Bouguer gravity anomaly values between Kuppam and Bommidi change from −45 mGal at Kuppam to −75 mGal at a distance of about 55 km to its south, and increase to −50 mGal at Bommidi. Between Vellar and Bhavani the gravity values are in the range of −60 mGal, except to the north of Bhavani, where they decrease to −70 mGal before increasing again to −60 mGal. Seismic data-based gravity models (Fig. 2.19) along these two segments show a good match between the computed and observed and show two distinctly different densities for crustal blocks of the Dharwar craton and Moyar-Bhavani (Mettur) shear zone (Vijaya Rao et al., 2006). The total magnetic intensity along this section shows a low of about 200 nT over the granitic gneisses of the Dharwar craton and a high value of 800 nT over the Moyar-Bhavani shear zone.

Close to the Moyar-Bhavani shear zone the gravity values show a low of −60 mGal and increase to −25 mGal to the south. Further south, the gravity value decreases to −65 mGal near Palani and indicates a long wavelength gravity high of about 40 mGal over a length of about 110 km. This high indicates the presence of a high-density material at a shallow depth of 6–12 km (Singh et al., 2003). Total magnetic intensity values in the region of the gravity high fluctuate between 1200 and −400 nT and are attributed to mafic intrusions in the region. Interpretation of the gravity and magnetic values along with seismic results has led to modification of the velocity picture (Fig. 2.20) in this region (Vijaya Rao et al., 2006).

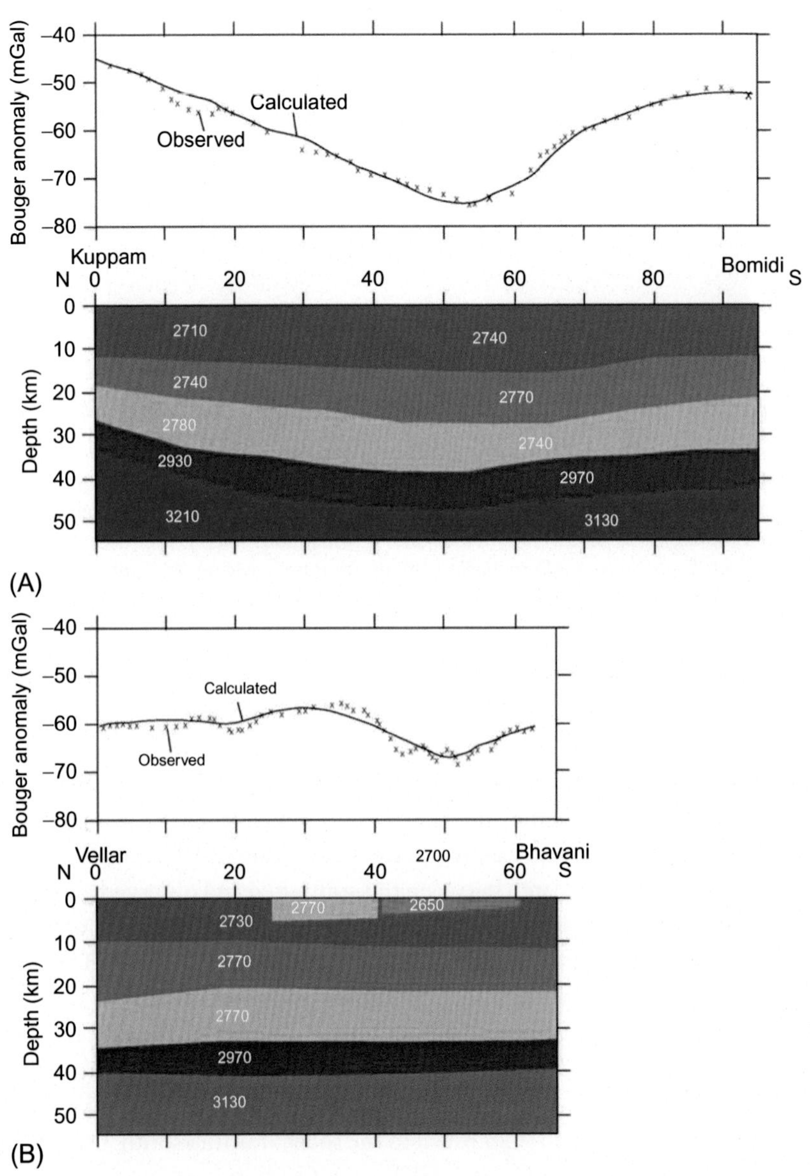

Fig. 2.19 Seismic data-based gravity models between (A) Kuppam and Bommidi, (B) Vellar and Bhavani. *(Modified from Vijaya Rao, V., Sain, K., Reddy, P.R., Mooney, W.D., 2006. Crustal structure and tectonics of the northern part of the Southern Granulite terrane, India. Earth Plan. Sci. Let. 251, 90–103).*

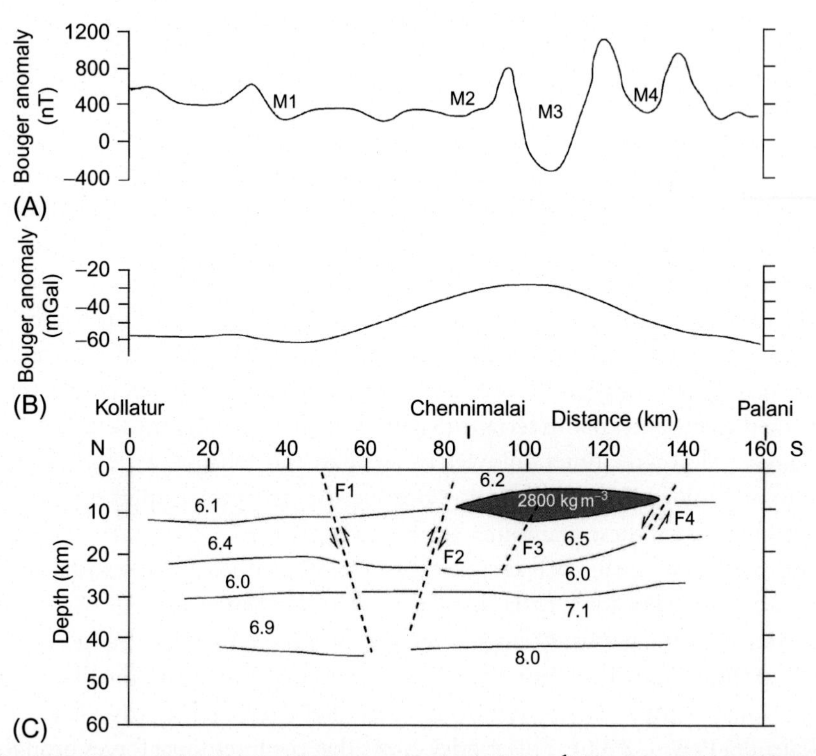

Fig. 2.20 Magnetic (A)/Gravity (B) values and velocity (km s^{-1})/depth model (C) between Kollatur and Palani of the Kuppam-Palani profile. Shaded layer has higher density (2800 kg m^{-3}) as compared to surrounding layers. *(Modified from Vijaya Rao, V., Rajendra Prasad, B., 2006. Structure and evolution of the Cauvery Shear Zone system, Southern Granulite Terrene, India: evidence from deep seismic and other geophysical studies. Gondwana Res. 10, 29–40.)*

2.2.2.2 Evolutionary Model of the Cauvery Shear Zone System

The velocity structure, reflection pattern and Bouguer gravity value between Kollatur and Palani are in agreement on the Moho upwarp in the Moyar-Bhavani shear zone (Fig. 2.20). Combined with the magnetic anomaly (M1) these provide strong evidence for juxtaposition of distinct crustal blocks separated by a deep-seated fault (F1). The faults F2, F3 and F4, inferred from the velocity-depth model, indicate the highly faulted nature of the region. These faults are associated with magnetic anomalies M2, M3 and M4. The spatial locations of these faults coincide with the geologically mapped shear zones. The fault pattern (F2–F4) suggests a graben, which may be related to the Neoproterozoic rifting, in the central part. These faults probably acted as conduits to transport lower crustal material to shallow depth and provided a mechanism for emplacement of the high-density material at a shallow depth of 6–12 km. High-pressure (10–12 kb) granulites, found at the Moyar-Bhavani shear zone, indicate that they have evolved during a compressional regime.

Lateral variation in crustal thickness, gravity pattern and geological attributes on either side of the Moyar-Bhavani shear zone suggest a suture that was formed during subduction and collision during the late Archean. The dome-shaped features, represented by Meso-Neoproterozoic alkaline and granitic plutons indicate intrusions that occurred during later geological cycles. Thrusting and collision seem to be responsible for evolution of the southern granulite terrane as it provided requisite conditions to form granulites and a mechanism to transport them to the surface. The pressure-temperature studies on the granulites of this terrane (Sajeev et al., 2004; Tamashiro et al., 2004) suggest collision tectonics followed by an uplift and collapse of the thickened crust. Petrological studies show evidence for isothermal uplift and rapid cooling rates for these granulites (Ravindra Kumar and Chacko, 1994), indicating that this terrane was tectonically exhumed either due to crustal-scale thrusting or extensional collapse of an over-thickened orogen.

Based on the pressure-temperature conditions, petrological and geophysical results, a schematic model showing different stages for evolution of crust in the Cauvery shear zone system (Fig. 2.21) has been built (Vijaya Rao and Rajendra Prasad, 2006). This model shows that compressional forces brought the Dharwar craton and the southern block closer to each other, leading to a collision between them during the late Archean. Compressional forces operative at that time thickened the crust between these two blocks. When these forces came to a stop, the thickened crust became unstable and led to orogenic collapse under its own weight. Since the crustal roots of a thickened crust undergo gabbro-eclogite phase transition and become denser than the underlying mantle lithosphere, ultimately leading to its delamination, the unstable layer finally peeled off (delaminated) and sank into the mantle. The high velocity of 7.1 km s^{-1} in the lowermost part of this thickened crust, representing magmatic underplating, indicates that subsequent tectonic activities have modified the delaminated lower crust of the region. Higher conductivity between 12- and 50-km depth indicates presence of the volatiles, released during the magmatic underplating, in the lower crust. Alkaline rocks, present between the Moyar-Bhavani and Karur-Oddanchatram shear zones, represent surface manifestations of rifting of these zones.

The faults (F1–F4 of Fig. 2.20) associated with these shear zones acted as conduits for emplacement of the magma, which was generated at a large depth, during the extensional episodes. The mafic dike swarms, present in this

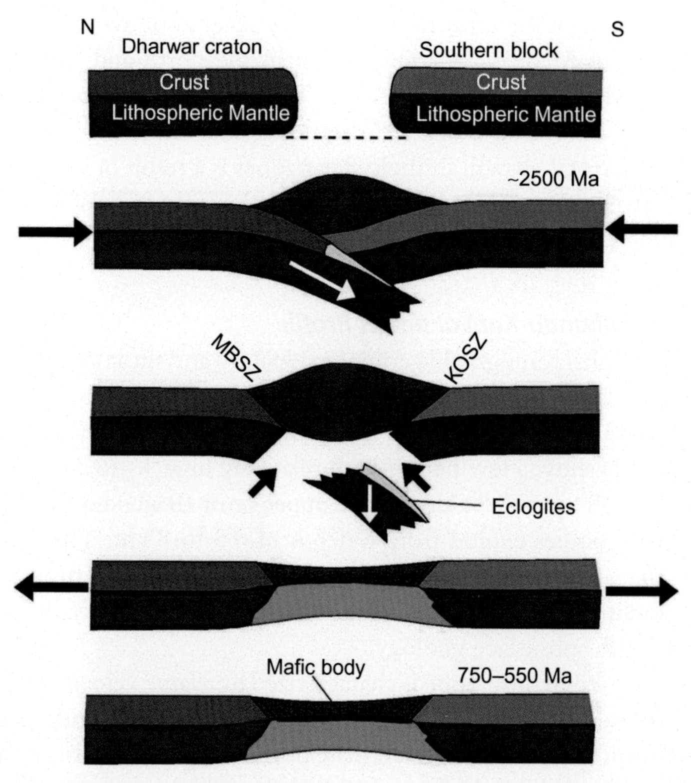

Fig. 2.21 Cartoon showing the evolutionary history of Cauvery shear zone system. *(After Vijaya Rao, V., Rajendra Prasad, B., 2006. Structure and evolution of the Cauvery Shear Zone system, Southern Granulite Terrene, India: evidence from deep seismic and other geophysical studies. Gondwana Res. 10, 29–40).*

region, are in conformity with the process of crustal extension and rifting. These represent the locations where the basaltic magma was transferred from the mantle to the crust in the late Neoproterozoic period. During the period of this rifting earlier fault zones were reactivated leading to elevation of the rift-shoulders, as evidenced by higher elevation of hills in the north and south, respectively. The Karur-Oddanchatram shear zone has played a bigger role in tectonic evolution of the region as compared to other shear zones. Comparatively higher mantle heat flow (Ray et al., 2003) and thinner lithosphere (23–32 mW m^{-2} and 100 km) in the southern granulite terrane, as compared to that in the Dharwar craton (11–16 mW m^{-2} and 163 km) suggests crustal reworking of the Cauvery shear zone system during the late Neoproterozoic rifting (Ravindra Kumar and Chacko, 1994).

Composition of the crust in the Cauvery shear zone system is thus different from the Archean Dharwar craton in the north and the Archean-Proterozoic Madurai block to the south. The shear zones of this system that separate the two crustal blocks to the north and south represent terrane boundaries. Crustal growth in these shear zones is a result of two-way mass transfer between the mantle and the crust, in the form of delamination and underplating (Vijaya Rao and Rajendra Prasad, 2006).

2.2.2.3 Vattalkundu-Kanyakumari Profile

Two-dimensional tomographic compressional (P) and shear (S) wave velocity (V_p and V_s) image and V_p/V_s ratios, based on the refraction/postcritical reflection studies, along a part of this profile across the Achankovil shear zone (Fig. 2.1) show a few bodies with relatively high V_p (6.3–6.5 km s^{-1}) and V_s (3.5–3.8 km s^{-1}) values in the upper crust (Rajendra Prasad et al., 2006). These bodies extend from a depth of 0.5 to 8 km. The bulk rock types in this area are basically high-grade metamorphic charnockites and mafic granulites, which show highly heterogeneous velocity distribution consistent with subsurface geology.

The Achankovil shear zone is characterized by a large velocity gradient at the shallow crustal depths and shows sharp velocity contrast with surrounding units forming fault blocks of alternate horsts and grabens. This shear zone shows anomalously large thickness (~7.0 km) and high average upper crustal V_p (6.4 km s^{-1}) and V_s (3.75 km s^{-1}) values (Fig. 2.22). Large positive velocity perturbations (>0.2 km s^{-1}) below it, and also the other shear zones, indicate that rock types below the shear zones are significantly different from surrounding exposed granulite rocks along the profile and represent a significantly different tectonic regime in the region. These rocks belong to high-grade metamorphic charnockite of middle to lower crustal origin and were formed at very high pressure-temperature conditions. Later, due to rifting during the Neoproterozoic period, these were exhumed to shallow depths forming compositionally different rock assemblages. Large variation of V_s (3.5–3.8 km s^{-1}), V_p/V_s ratio (1.72–1.85) and considerable variation of Poisson's ratio (0.25–0.29) also represent heterogeneity in the crust in this terrane down to 8-km depth. The heat flow value in this region (Ray et al., 2003), between 31 and 55 mW m^{-2} with a mean of 42±8 mW m^{-2} and heat production ranging between 1.11 and 2.63 mW m^{-3} except in the Achankovil shear zone, corroborates the heterogeneity in the upper crust. In the Achankovil shear zone the heat production is anomalously high, with a mean value of 7.0±3.9 mW m^{-3}, and is attributed to large-scale exposed charnockite and basic granulites and their extension to deeper levels.

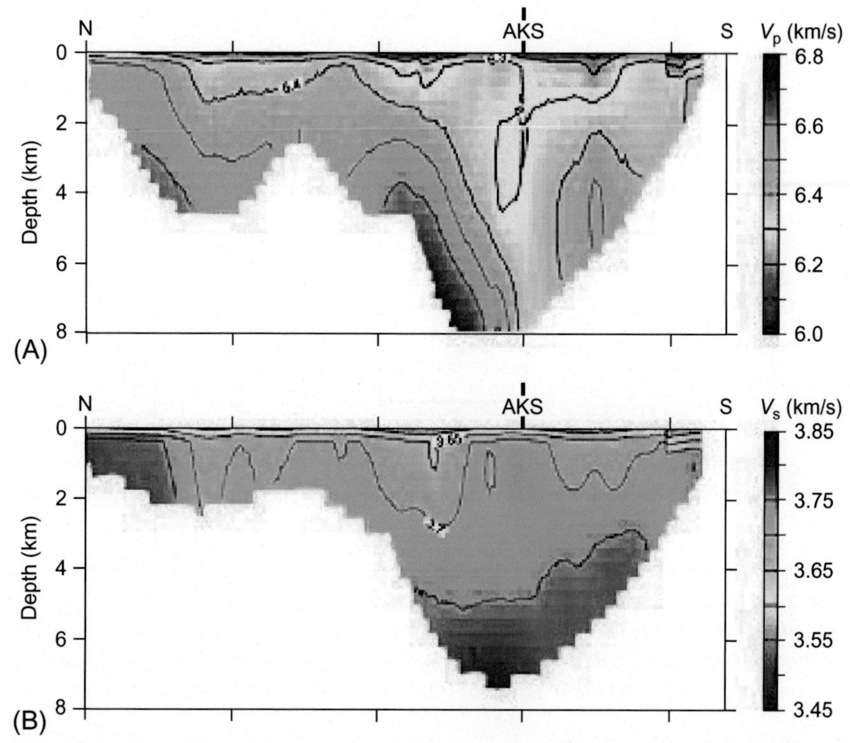

Fig. 2.22 (A) P-wave, and (B) S-wave velocity structure across the Achankovil shear zone. *AKS*, Achankovil shear zone. *(Modified from Rajendra Prasad, B., et al., 2006. A tomographic image of upper crustal structure using P and S wave seismic refraction data in the southern granulite terrain (SGT), India. Geophys. Res. Lett. 33, L14301. https://doi.org/10.1029/2006GL026307.)*

The line drawing of the seismic stack section (Fig. 2.23) over the Madurai block (between Vattalkundu and Kalugumalai) shows conspicuous changes in the reflectivity pattern, which indicate the presence of several structurally distinct crustal blocks (Rajendra Prasad et al., 2007). In general, the upper crust is poorly reflective and there are no recognizable reflections in the shallow region of 0.0–1.0 s TWT. In the extreme northern part of the profile, near Vattalkundu, the reflections dip to the north. In the middle part of profile (A in Fig. 2.23) the south-dipping reflectors are delineated at all depths within the crust. The change in dip direction of the reflectors and geologically reported east-dipping foliations on the surface in the region are indications of reworking of the crust at different times, which are recorded by age data and ultrahigh-temperature metamorphism. The southern granulite terrane has experienced at least seven thermotectonic events including two distinct episodes of metasomatism/charnockitization between 2500 and 2530 Ma and between 550 and 530 Ma (Ghosh et al., 2004).

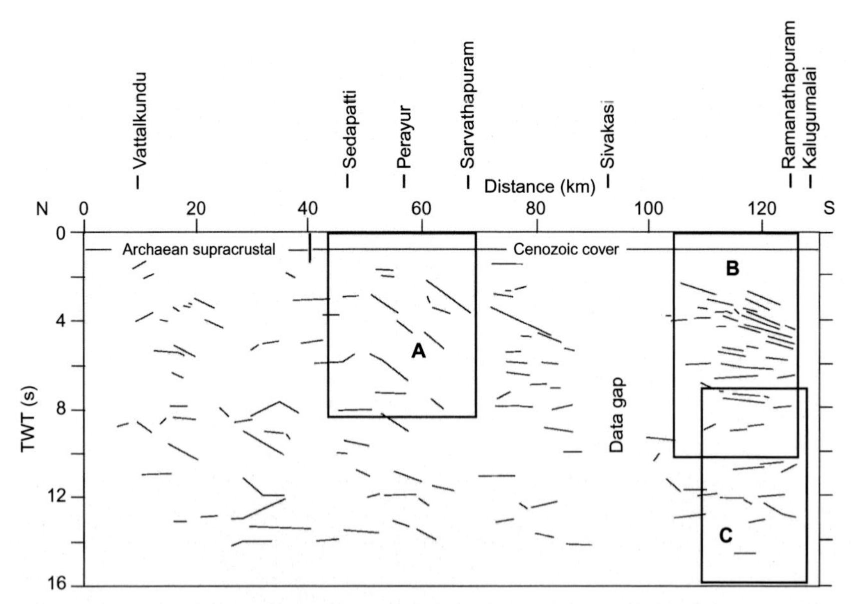

Fig. 2.23 Deep reflection section over the Madurai block. *(Modified from Rajendra Prasad, B., Kesava Rao, G., Mall, D.M., Koteswara Rao, P., Raju, S., Reddy, M.S., Rao, G.S.P., Sridhar, V., Prasad, A.S.S.R.S., 2007. Tectonic implications of seismic reflectivity pattern observed over the Precambrian Southern Granulite Terrene, India. Precambrian Res., 153, 1–10).*

However, it is difficult to speculate on the timing of the reworking and specific dip direction on the basis of stacked sections alone. With the present data set, it can be only speculated that the surface foliations could be due to the youngest thermotectonic episode, i.e., 550–530 Ma. These observations could have significant implications in interpreting the regional tectonic activity (Guru Rajesh and Chetty, 2006). Gentle south-dipping reflections in the uppermost crust and dome-shaped reflections in the lower part represent block uplift accompanied by internal arching of the crust in Madurai block.

The southern part (B in Fig. 2.23) shows a conspicuous increase in reflectivity as compared to the northern part and the reflections are stronger, more continuous and closely spaced. This could be due to major structural/geological change to the south. Thorough recrystallization associated with reworking of the crust along the weak zones and metamorphic surfaces in the area due to ultrahigh-temperature metamorphism seems to be responsible for this change (Rajendra Prasad et al., 2007). This view gets further support from the observation that the Kuppam-Palani transects are less reflective as compared to transects over the Madurai block, which underwent ultrahigh-temperature-pressure metamorphism. In general, the region is characterized by distinct upper- and midcrustal reflectivity, but the reflection Moho identifiable at 13.0–13.5 s TWT (about 40–42 km depth) is diffused (Rajendra Prasad et al., 2007).

The Madurai block is separated from the Kerala Khondalite belt by the crustal scale Achankovil shear zone, defined by northwest-southeast trending foliation fabric with steep dip to the southwest (Guru Rajesh and Chetty, 2006). The monotonous southerly dips in the upper- and midcrustal levels, except for the small northern portion near Vattalkundu and the southwesterly dip of the Achankovil shear zone, suggest that the crustal growth could have been through an accretion process from the south and was affected by northward drift of the Indian plate. This is further supported by northward thrusting or popping up of the deeper crust, which has given rise to the more than 2500-m high Nilgiri massif (Valdiya, 2001).

CHAPTER THREE

Aravalli-Delhi Fold Belt

The Aravalli-Delhi fold belt (Fig. 3.1) is one of the most important geologic terranes in the northwest Indian shield. The trans-Aravalli belt known as the Marwar basin lies to its west. The geological history of this fold belt ranges from Archean to Neoproterozoic. The growth of its crust was through the development of successive Proterozoic orogenic belts on an ensialic basement of the Archean age. The 3300-million-year-old Bhilwara gneissic complex (also known as the Banded gneissic complex, consisting of the Sandmata and Mangalwar complexes and the Hindoli group) represents the oldest cratonic nucleus of the region (MacDougall et al., 1983). Proterozoic Aravalli and Delhi sediments were deposited successively over this Archean basement. The region has witnessed four major regional tectonomagmatic and metamorphic events with ages of ~3000 Ma (Bhilwara gneissic complex), ~1800 Ma (Aravalli orogeny), ~1100 Ma (Delhi orogeny) and 850–750 Ma (post-Delhi magmatic event).

3.1 GEOLOGY AND TECTONICS

The Bhilwara gneissic complex forms the basement of most of the Proterozoic rocks present in the region. This complex has been reworked at about 3000 Ma. The U-Pb isochron age of 3500 ± 200 Ma for a detrital zircon in its schist belt, and Sm/Nd age of 3500 Ma for the biotite gneiss and amphibolites confirms its middle Archean crustal history (MacDougall et al., 1983). Its major constituents are the Sandmata and Mangalwar complexes and Hindoli group of rocks (Sinha-Roy et al., 1995). The Sandmata complex comprises granulite facies rocks, which occur as isolated bodies. These were formed at temperatures of 650–850°C and 8–11 kbar pressures. U-Pb zircon dating of these granulites yielded an age of 1730 Ma (Sharma, 1988). Towards the west, this complex is in thrust contact with the Delhi fold belt. The Mangalwar complex containing metavolcanic and metasedimentary rocks is an older granite-greenstone belt that has amphibolites facies metamorphism; the metavolcanic rocks have originated from the mantle sources (Sugden et al., 1990). The age of this complex is not well constrained.

Structure and Tectonics of the Indian Continental Crust and Its Adjoining Region
https://doi.org/10.1016/B978-0-12-813685-0.00003-0

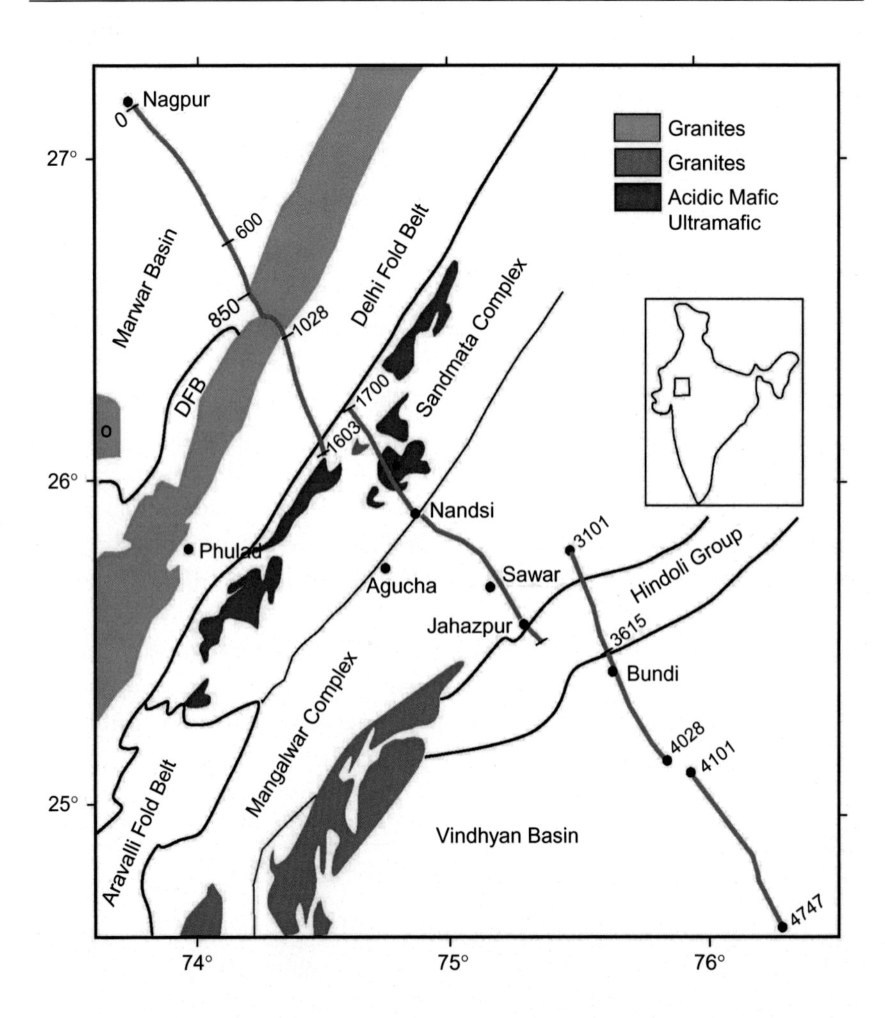

Fig. 3.1 Geological map of the Aravalli-Delhi fold belt with the deep seismic reflection profile plotted on it. *NP*, Neoproterozoic; *MP*, Mesoproterozoic; *PP*, Paleoproterozoic; *MA*, Middle Archean; *LA*, Late Archean. *(Modified from Rajendra Prasad, B., Tewari, H.C., Vijaya Rao, V., Dixit, M.M., Reddy, P.R., 1998. Structure and tectonics of the Proterozoic Aravalli-Delhi Fold Belt in the northwestern India from deep seismic reflection studies. Tectonophysics 288, 31–41.)*

A whole-rock isochron age of 2950 ± 150 Ma is reported for the granite (Untala) that lies in the southwestern part of this complex, whereas the Pb-Zn mineralization associated with it is dated at about 1800 Ma (Deb et al., 1989). While some workers consider this complex as Archean, others consider it as equivalent to Palaeoproterozoic. The Hindoli group constitutes a younger granite greenstone belt lying in the eastern part of the Bhilwara gneissic complex and is sandwiched between the Mangalwar complex and the Vindhyan basin. Its rocks show greenschist to lower amphibolites facies metamorphism. Some workers regard the granite of 2600 Ma (Berach) as its basement while others think that the granite has intruded it and represents plutonism that marks the culmination of the Archean era.

The Paleoproterozoic Aravalli fold belt consists of shelf and deep-sea facies sediments. Ultramafic and associated rocks within this belt represent ophiolitic melange and possibly tectonic slices of obducted oceanic crust emplaced during its evolution. The contact zone is a suture and the main belt has evolved as an aulacogen widening towards the south from a possible triple junction (Sinha-Roy, 1988). Other Aravalli equivalents in this region are the Jahazpur group, the Sawar group and the Agucha formation (MacDougall et al., 1983). Northward widening in the Mangalwar complex and the Hindoli group indicate that they have developed together as an aulacogen and the rocks of these belts are contemporaneous with those of the Aravalli fold belt (Sugden et al., 1990). The absolute ages for the Aravalli rocks are poorly constrained. Rb/Sr isotope analysis yielded an age of 1900 ± 60 Ma for the Aravalli and its equivalent rocks (Choudhary et al., 1984).

The Mesoproterozoic Delhi fold belt comprises a system of half graben and horst, and consists of over 10 km-thick volcano-sedimentary sequences (Sinha-Roy, 1999). It consists of folded and deformed rocks exhibiting polyphase metamorphism, comprising deepwater to platform sediments. Many granitic plutons occur within this belt. Granites exposed within this belt have yielded an Rb-Sr isochron age of 1600 Ma (Choudhary et al., 1984). The deposition of the Delhi group of sediments has taken place in the late Paleoproterozoic period ($\sim$1600 Ma). Basic volcanic and mafic/ ultramafic rocks of this belt are named the Phulad ophiolite (Sinha-Roy, 1999). Similar ophiolite melange is identified in the formations to the further south. The geochemical, geochronological and isotope studies on Phulad ophiolite and related rocks define them as fragments of the Proterozoic island arc complex (Volpe and MacDougall, 1990). The Pb model age of the Delhi group of rocks has been constrained to about 1100 Ma (Deb et al., 1989).

The Vindhyan basin, lying to the southeast of the Aravalli-Delhi fold belt is one of the largest Proterozoic basins in India. The basin margins are demarcated by an arcuate thrust comprising the Aravalli-Delhi fold belt in the northwest and another orogenic belt (Satpura) in the southeast, beyond the boundaries of Fig. 3.1. It was formed as a consequence of collision of the Bundelkhand craton with the Deccan protocontinent in the south and the Mewar craton in the west during the Mesoproterozoic period (Yedekar et al., 1990). The great boundary fault in the northwest and the Narmada-Son lineament in the south bound it. A large number of dikes are reported towards its boundary with the great boundary fault. The basin mainly is composed of shallow marine deposits, with a maximum thickness of 5 km accumulated during 1700–700 Ma. A kimberlite pipe (Majhgawan) intrusion within the Vindhyan basin has been dated at 1100 Ma (Kumar et al., 1993).

The Marwar basin to the west of the Delhi fold belt consists of flat, undeformed clay evaporite sequences. Its basement consists of the granites (850 Ma) and an igneous suite of 740 Ma (MacDougall et al., 1983).

Some of the several tectonic and evolutionary models that have been proposed for development of this region, based on its geology, are described here.

(i) One model assumes that the Bhilwara gneissic complex and the Aravalli fold belt was a combined protocontinent in the east and a Precambrian oceanic plate consisting of a midoceanic ridge exited towards the west. The intervening trough was the site of the Delhi group of sediments. Eastward subduction of the oceanic plate and the ridge squeezed the Delhi sediments against the protocontinent consisting of the Aravalli fold belt and Bhilwara gneissic complex in the east and culminated in the Delhi orogeny (Sychanthavong and Desai, 1977).

(ii) Another model shows extensive remobilization in the Northwestern Indian shield resulting in its evolution through interaction of different crustal blocks, which led to development of the fold belts in the region. The fold belts represent a sequential development due to ensialic rifting, generation of the oceanic crust and its consumption similar to the Phanerozoic plate tectonic regime. The westward movement of an Archean craton and its collision with another craton to the west, during the Proterozoic, resulted in eastward underthrusting of the intermediate oceanic crust. Two separate rifting events resulted in development of the Aravalli and Delhi sedimentary basins. In the first stage of rifting the Aravalli and equivalent basins were formed. Then, eastward underthrusting of the oceanic crust developed a fold belt during the Palaeoproterozoic period (~1800 Ma). In the second stage of rifting, the ocean opened and Delhi sedimentation took place.

Around 1300 Ma, this oceanic crust subducted westwards, resulting in the development of a trench, accretionary carbonate turbidite and a magmatic arc with an ophiolite mélange. The compressional phase closed the rift, deforming and metamorphosing the sediments, and formed the Delhi fold belt (Sinha-Roy, 1988).

(iii) The third model disagrees with the plate tectonic model. According to it there is not much evidence for accretion of the oceanic crust during the entire Precambrian evolutionary history of this region, except for a few minor components in the southern part of the Delhi fold belt. Even the total volume of the mantle magmatism, with a few exceptions, is not very significant. On the other hand, the lithological character of the Precambrian rocks indicates repeated recycling of the crustal material through sedimentation, metamorphism, partial melting, and formation of the tectonic slices. There is a distinct westward polarity in the evolution of different tectono-stratigraphic units in relation to the Archean basement rocks outcropping in the eastern plains. The history of crustal evolution subsequent to cratonization of the crust, at around 850 Ma, is marked by outpouring of dominantly felsic lava/tuff and intrusion of the granitoids in a typically anorogenic setting, followed by development of the large cratonic basins (Roy, 2000). This pattern of tectonic evolution also suggests that inception of the Vindhyan basin to the east of the Aravalli-Delhi fold belt predated the western volcanic-sedimentary anorogenic succession (felsic volcanics/granitoids and the Marwar basin).

(iv) The fourth model also disagrees with the plate tectonic model and suggests that occurrence of the low-grade mono-metamorphosed Aravalli fold belt, between the high-grade Delhi fold belt on the northwest and upper amphibolites facies of the Bhilwara gneissic complex on the southeast, does not favor the Aravalli orogeny as an intermediate orogenic event between the pre-Aravalli and Delhi orogeny. It suggests an alternate model in which development of the large oceanic basins (characterized by Aravalli-Delhi sediments) is the result of ductile extension of hot sialic material rather than brittle rupture and ductile spreading followed by contraction of the crust (Sharma, 1988).

The Aravalli-Delhi fold belt has undergone a phase of neotectonic activity as well. Some of the opinions about this activity are: (1) the block uplift of the fold belt is due to a warp-like tilt towards the west before the beginning of the Tertiary period (Heron, 1953); (2) the Neocene and Quaternary tectonics rejuvenated the old basement faults and segmented blocks of the crust were uplifted along these faults to form the north Aravalli Ranges (Dassarma, 1988); (3) the post-Neocene tectonics in the central Aravalli mountain range rejuvenated the regional Precambrian faults in response to flexural bending of the Indian plate by a movement generated at the back due to locking of the plate by the early Pleistocene time (Sen and Sen, 1983).

3.2 SEISMIC RESULTS

Deep seismic reflection studies to delineate the crustal structure and understand the tectonic evolution were carried out along the 400-km long Nagaur-Kunjer profile (Fig. 3.1) across the Aravalli-Delhi fold belt. This northwest-southeast profile starts at Nagaur, in the Neoproterozoic Marwar basin, and traverses through the Mesoproterozoic Delhi fold belt, middle Archean Sandmata and Mangalwar complexes, and late Archean Hindoli group and terminates near Kunjer in the Meso-Neoproterozoic Vindhyan basin.

Refraction and postcritical reflection studies, carried out in a small part of the profile, to determine the velocity configuration of the crust under the Delhi fold belt and Sandmata complex (Tewari et al., 1997a) show a crust with velocities of 5.75, 6.20, 5.80 and 6.60/7.30 k ms^{-1}, followed by an upper mantle velocity of 8.40 km s^{-1} (Fig. 3.2). The lower velocity of 5.80 km s^{-1},

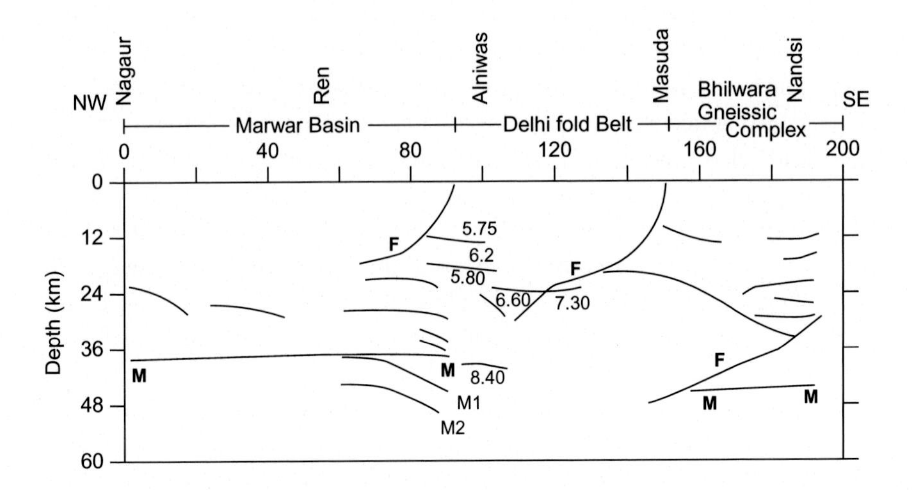

Fig. 3.2 Velocity-depth model of the crust below the Marwar basin and the Delhi fold belt. *(After Tewari, H.C., Dixit, M.M., Madhava Rao, N., Venkateswarlu, N., Vijaya Rao, V., 1997. Crustal thickening under the Paleo-Mesoproterozoic Delhi Fold Belt in north-western India evidence from deep reflection profiling. Geophys. J. Int. 129, 657–668.)*

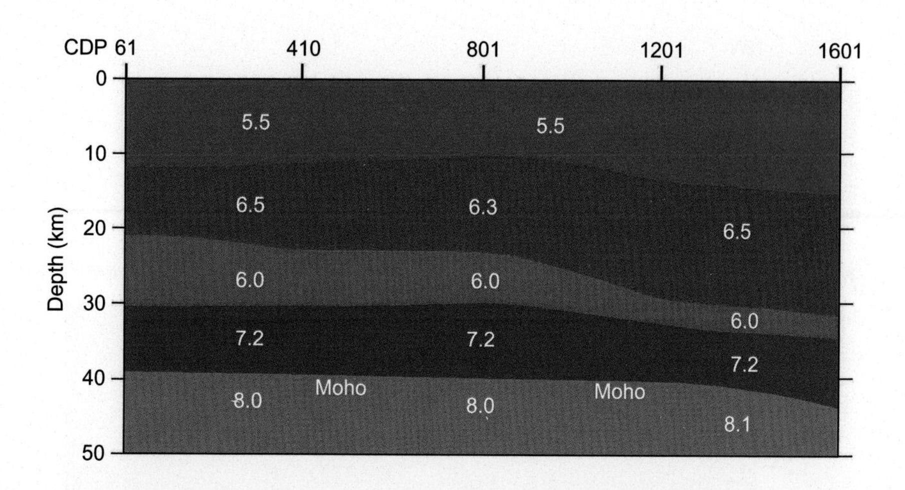

Fig. 3.3 Velocity (km s^{-1})-depth model for the crust below the Marwar basin along the profile of Fig. 3.1. *(After Satyavani, N., Dixit, M.M., Reddy, P.R., 2001. Crustal velocity along the Nagaur-Rian sector of the Aravalli fold belt, India, using reflection data. J. Geodyn., 31, 429–443.)*

below the velocity of 6.20 km s^{-1}, represents a velocity reversal at the base of the upper crust. Below the Delhi fold belt the lower crustal velocity is 6.60 km s^{-1}. The velocity of 7.30 km s^{-1} replaces this velocity at its contact with the Sandmata complex. Another crustal velocity model (Fig. 3.3), based on the reflection stack velocity, for the Marwar basin is more or less similar to the previous model except that a low velocity of 6.00 km s^{-1} is inferred in the lower crust of the Marwar basin (Satyavani et al., 2001). Velocities in the upper and lower Vindhyan sediments are 4.6–4.8 and 5.1–5.3 km s^{-1} respectively (Rajendra Prasad and Vijaya Rao, 2006).

The sedimentary basin configuration of the Marwar basin (between Nagaur and Ren of Fig. 3.2) shows a three-layered velocity model (Vijaya Rao et al., 2007a) above the basement of velocity 5.9 km s^{-1} (Fig. 3.4). The maximum thicknesses are: 300 m for the top layer of Quaternary to Tertiary sediments of velocity 2.2 km s^{-1}, 250 m for the second layer of Marwar sediments of velocity 4.2 km s^{-1} and 500 m for the third layer, possibly an igneous suite, of velocity 4.8 km s^{-1}. The gravity highs along the seismic profile approximately correspond to the basement ridges and the gravity lows are due to the Marwar sediments. This analysis also suggests that the Marwar basin might have formed as a consequence of a plume activity.

The deep reflection studies along the Nagaur-Kunjer profile generated a migrated stack section of the data (Fig. 3.5) that showed variation in the crustal reflectivity pattern down to 18 s TWT, about 55-km depth. The

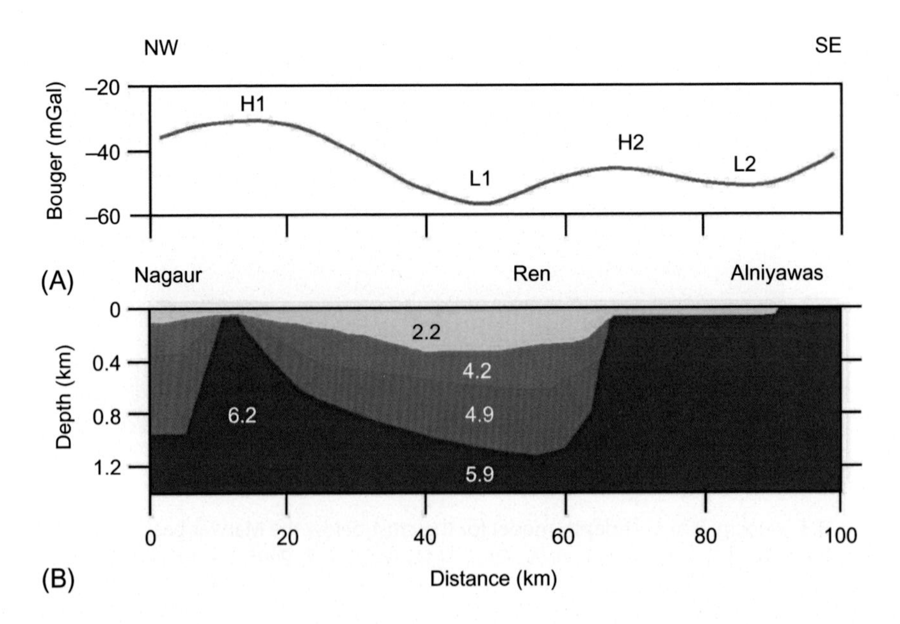

Fig. 3.4 Sedimentary basin velocity (km s^{-1})-depth model for the Marwar basin part of the profile of Fig. 3.1. (A) Bouger anomaly along the profile in Marwar basin, H's and L's indicate highs and lows. (B) Depth model of the sedimentary basin. *(Redrawn after Vijaya Rao, V., Sain, K., Krishna, V.G., 2007. Modelling and inversion of single sided refraction data from the shot gathers of multifold deep seismic reflection profiling—an approach for deriving the shallow velocity structure. Geophys. J. Int. 169, 507–514; Vijaya Rao, V., Sain, K., Rajendra Prasad, B., 2007. Dipping Moho in the southern part of the Eastern Dharwar craton, India, as revealed by the coincident seismic reflection and refraction study. Curr. Sci. 93, 330–336.)*

upper crust, down to about 8 s TWT, is generally poorly reflective except in small parts of the profile, whereas reflectivity in the lower crust varies considerably, indicating its complex pattern along the profile. In this complex pattern, the dipping reflections indicate an older crust that has undergone deformation while the horizontal reflections indicate a younger, undeformed crust. A line diagram of the migrated seismic depth section shows prominent reflecting horizons along the profile (Fig. 3.6). Changes in the dip direction and steeply dipping reflections, cutting across the nearly horizontal reflections at various depths in the crust, demarcate the boundaries between different crustal units.

Changes in the reflectivity pattern also delineate the crustal structure. Differences in this pattern identify the boundaries of various crustal blocks. Based on these differences the profile is divided into five blocks: (a) moderate to highly reflective Marwar basin, (b) moderate to poorly reflective Delhi

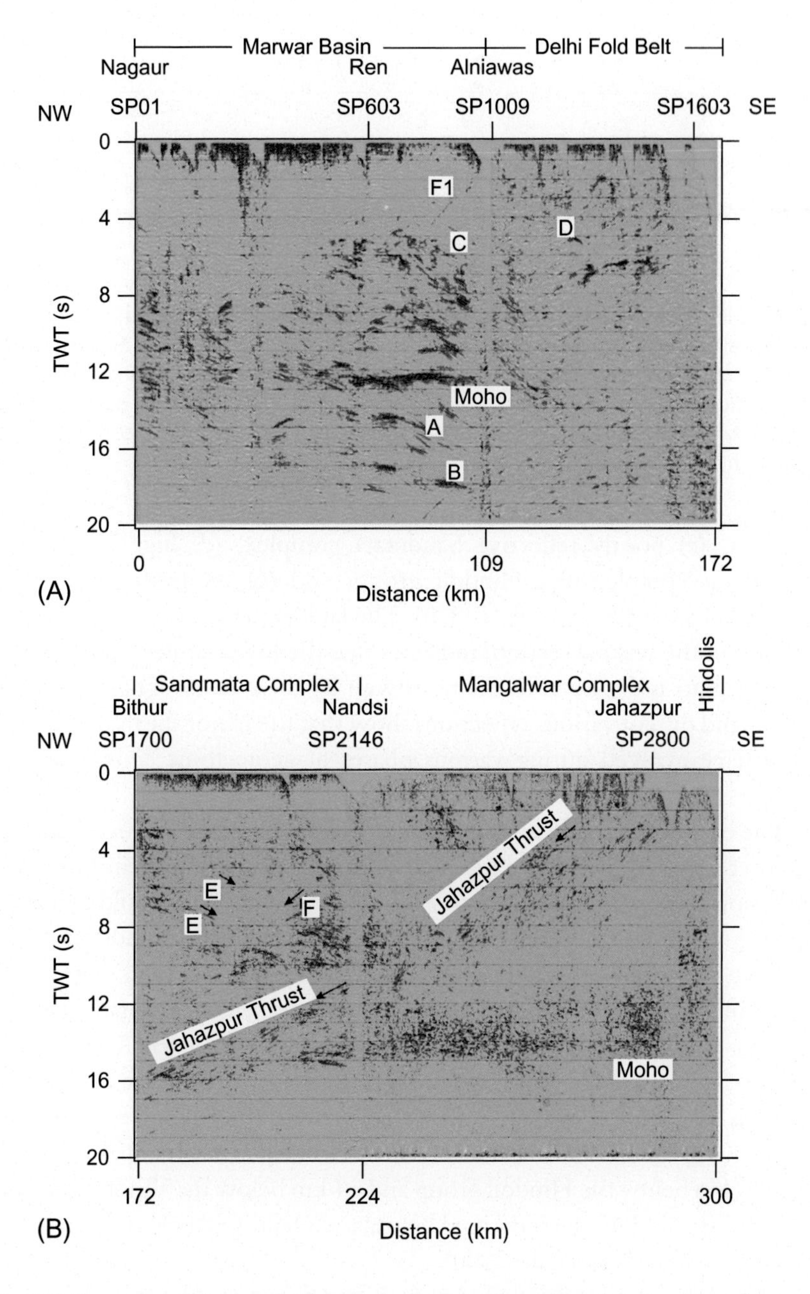

Fig. 3.5 Examples of the true scale of migrated depth section showing the reflectivity pattern in (A) Marwar basin and Delhi fold belt and (B) Jahazpur thrust fault. The distances are from Nagaur, SP 0. *(Modified from Mandal, B., Sen, M.K., Vaidya, V.V., Mann, J., 2014. Deep seismic image enhancement with the common reflection surface (CRS) stack method: evidence from the Aravalli-Delhi fold belt of northwestern India. Geophys. J. Int. 196, 902–917.)*

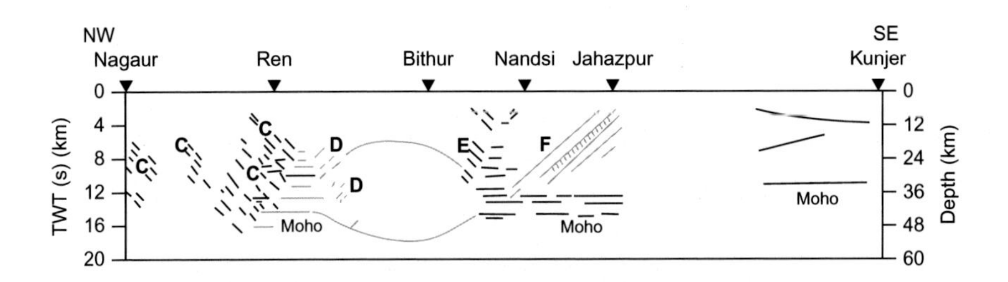

Fig. 3.6 Line drawing of the deep reflection section between Nagaur and Kunjer across the Aravalli-Delhi fold belt. *MB*, Marwar basin; *DFB*, Delhi fold belt; *SC*, Sandmata complex; *MC*, Mangalwar complex; *HG*, Hindoli group; *VB*, Vindhyan basin. *(After Vijaya Rao, V., Rajendra Prasad, B., Reddy, P.R., Tewari, H.C., 2000. Evolution of Proterozoic Aravalli-Delhi Fold Belt in the northwestern Indian Shield from Seismic studies. Tectonophysics 327, 109–130.)*

fold belt, (c) poorly reflective Sandmata complex, (d) highly reflective Mangalwar complex and Hindoli group, and (e) moderately reflective Vindhyan basin (Tewari et al., 1997b). The fault pattern, based upon the line drawing of the seismic section, indicates that the lower crust of most of the tectonic units is displaced to the northwest by these faults. The reflectivity pattern and dips of various reflections show that the major shear/dislocation zones have evolved during various phases of geotectonic activity in the region.

The base of the nearly horizontal, laterally continuous reflection bands in the lower crust is generally identified as the Moho boundary in the high-resolution seismic reflection studies. The Moho reflections could be either dipping or horizontal. While the dipping Moho reflections indicate collision, the horizontal reflections are regarded as a younger Moho that is generated in the postcollision extension process, which incorporates ductile flow in the lower crust due to crust-mantle interaction. Along the Nagaur-Kunjer profile both types of reflections are seen at the Moho between 12.5 and 15 s TWT. These reflections correspond to depths of about 38 km below the Marwar basin, 46 km below the Mangalwar complex, 40 km below the Hindoli group and 34 km below the Vindhyan basin. The Moho could not be identified below the Delhi fold belt and Sandmata complex (Vijaya Rao et al., 2000).

The Moho, which could not be identified below the Sandmata complex in the earlier studies, was imaged by Mandal et al. (2014) at about 50-km depth. The reflectivity pattern (Rajendra Prasad et al., 1998) reveals that the lower crust of the Marwar basin and Delhi fold belt show similar reflectivity patterns, indicating that the former is an extension of the latter. The crust of the Delhi fold belt is bound in the northwest by a listric fault down to the upper crust and to the southeast by a down to the lower crust fault, which separates the Bhilwara gneissic complex from it. The lower crust in most of the Delhi fold belt and Bhilwara gneissic complex has only one prominent dome-shaped reflection, its deepest point being at about 10 s TWT (~30-km depth). The top of this reflection is at about 6 s TWT (~18-km depth) in the middle crust. This reflection acts as a shield due to which the lower crust of the gneissic complex is devoid of any other reflections.

The eastern part of the Bhilwara gneissic complex, i.e., the Hindoli group, has a different reflectivity pattern as compared to the rest of the complex. A highly reflective crustal scale thrust zone (Jahazpur thrust) dipping at about 30 degrees towards the northwest is present here. This thrust zone is an approximately 25-km thick pile of parallel dipping reflections, with 80-km lateral extension, and originates at the Moho at about 45-km depth. Since the seismic section does not provide any information on the top part of this section, the reflections within this zone appear to terminate at a depth of about 5 km near Jahazpur. Lower crustal reflectivity to the east of this thrust zone is very high, while to its west the same is poor. There are no common features in the reflectivity pattern to the east and west of this thrust. The western boundary fault of the thrust zone is a major crustal boundary that separates the Sandmata complex from the Hindoli group. Despite the fact that the reflectivity pattern does not show a clear-cut boundary separating the moderately reflective Vindhyan basin and the highly reflective Hindoli group, the lower crust of the two regions is different. Reflectivity in the sedimentary part and the upper crust of the Vindhyan basin is very high and indicates gradual thickening of the sediments from the northwest towards the southeast.

In the Marwar basin a prominent southeast dipping discontinuous band of reflections extends from a depth of 8–10 km down to the Moho and beyond into the mantle (*C* of Figs. 3.5A and 3.6). These reflections terminate at the western boundary of the highly folded and deformed Delhi fold belt against opposite dipping reflections at the boundary of the Sandmata complex (*D* of Figs. 3.5A and 3.6). Opposite dipping reflections are also seen at the boundaries of the Mangalwar complex (*E* and *F* of Figs. 3.6 and 3.7). Correlation of the surface features with the subsurface reflection geometry suggests that this divergence in the reflection pattern is a manifestation of the collision process, developed in a compressional environment. A similar

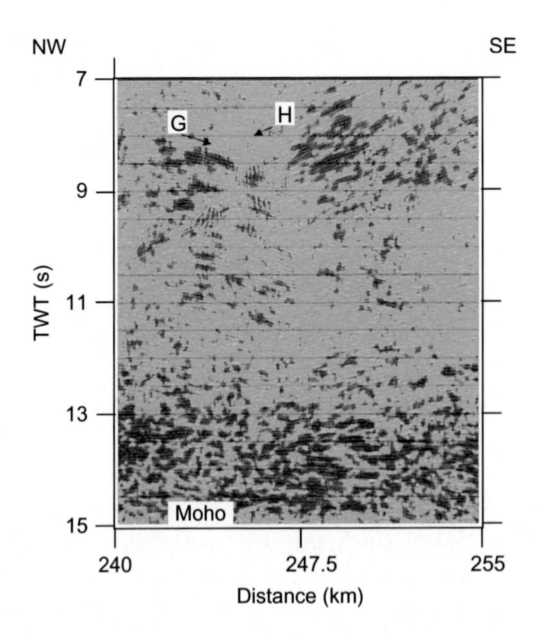

Fig. 3.7 Example of prominent reflection bands (E, F) showing opposite dips. *(Modified from Mandal, B., Sen, M.K., Vaidya, V.V., Mann, J., 2014. Deep seismic image enhancement with the common reflection surface (CRS) stack method: evidence from the Aravalli-Delhi fold belt of northwestern India. Geophys. J. Int. 196, 902–917.)*

reflectivity pattern is observed in many Proterozoic fold belts as well as in the Phanerozoic regions in several parts of the world, hinting at similarity in the plate tectonic process from the Proterozoic to Phanerozoic (Rajendra Prasad et al., 1998; Mandal et al., 2014).

The migrated near-offset reflection images on the western margin of the Delhi fold belt reveal several steeply dipping isolated reflections and represent the Delhi thrust fault (Krishna and Vijaya Rao, 2011). Joint interpretation of seismic near-offset and coincident refraction/wide-angle reflection sections have effectively delineated both velocity stratification and geometry of complex deep crustal structures. The complexity of the dipping and the subhorizontal middle to lower crustal reflectivity, revealing a lamellar structure of short isolated reflectors, suggests small heterogeneities. The unusual reflections represent a set of local velocity variations imprinted on the background of deep crustal velocity structure. The crustal scale Delhi thrust fault on the western margin of the Delhi fold belt, delineated as a stack of reflectors dipping 25–30 degrees SE, probably developed during the Mesoproterozoic collision episode related to the Delhi orogeny and is contemporaneous with the formation of Rodinia. These structural features have survived through geological periods.

The lower crust is generally ductile in nature and therefore the existence of the faults at that depth is rare. However, faults do survive in the lower crust of subduction/suture zones due to different environment and strain hardening processes that act at these depths. The Jahazpur thrust that extends into the lower crust, in a region where the Moho is at a depth of 40–46 km, therefore has a great significance in the tectonic evolution of this region. The nature and dimension of this thrust shows operation of thick-skinned tectonics in the region. Its ramp and flat geometry, as against listric geometry for the extensional faults, indicates that it has developed in a compressional environment.

In a nutshell, the seismic structure shows that the lower and upper crusts of the region are dissected by a number of dislocation zones that bound different tectonic domains (Sinha-Roy, 2000). The dome-shaped reflection beneath the Delhi fold belt and Sandmata complex is a composite structure of more than one generation in which the northwest dipping reflection probably represents the structural pattern of the Delhi fold belt and the southeast dipping reflection that of the Aravalli fold belt.

The Bouguer gravity anomaly in the region is characterized by a broad gravity high of about 70 mGal between shot points 850 and 2800 (over the Delhi fold belt and Sandmata complex) and has prominent lows on either side (Fig. 3.8). This anomaly is observed all along the strike (northeast-southwest) of the 700 km-long Aravalli-Delhi mountain belt and was earlier attributed to a horst structure. The faults bounding the horst were considered to extend into the mantle (Reddi and Ramakrishna, 1988). Mandal et al. (2014) consider that they correspond to the palaeo-subduction zone formed during the Mesoproterozoic period. The crustal structure of the palaeo-subduction zones of the Jahazpur thrust and Delhi thrust belt remained frozen and no major thermal and tectonic effects overprinted them, probably because the temperature remained cold near the subduction zones.

The crustal density model (Fig. 3.9) constrained by the seismic reflection section shows a high-density (3100 kg m^{-3}) intrusive body, corresponding to the high velocity of 7.30 km s^{-1}, at the base of the crust under the Delhi

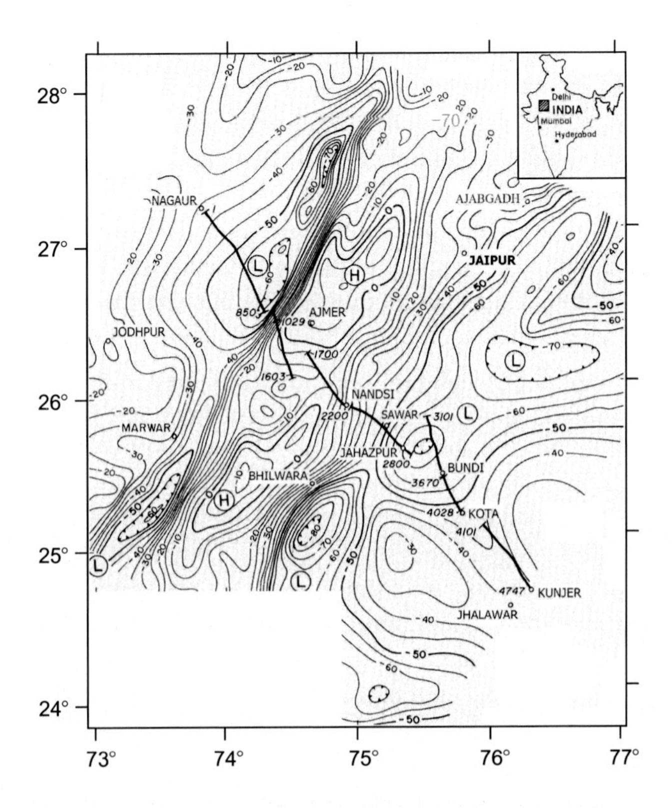

Fig. 3.8 Bouguer gravity anomaly map of the region. *(After Reddi, A.G.B., Ramakrishna, T.S., 1988. Subsurface structure of the shield area of Rajasthan-Gujarat as inferred from gravity. In: Roy, A.B. (Ed.), Precambrian of the Aravalli Mountain, Rajasthan. India. Geol. Soc. India, Mem. 1, 279–284.)*

fold belt and Sandmata complex (Rajendra Prasad et al., 1998). This body follows the pattern of seismic reflections and becomes shallower in the middle. The Moho depth in this region, where the reflection data could not provide it, was determined as about 48 km based on the dip of the reflections from both side and the gravity modeling.

The total magnetic intensity values along the seismic profile (Fig. 3.10) show high values and prominent anomalies (M1–M4) in the Alniawas-Nandsi region (Sandmata complex). High magnetic intensity in this region is due to the presence of the granulites. These anomalies coincide with major fault zones derived from the seismic studies. At both ends of the profile the magnetic anomaly is more or less constant, indicating that the Marwar and Vindhyan basins are relatively nonmagnetic (Tewari et al., 1998; Tewari, 1998).

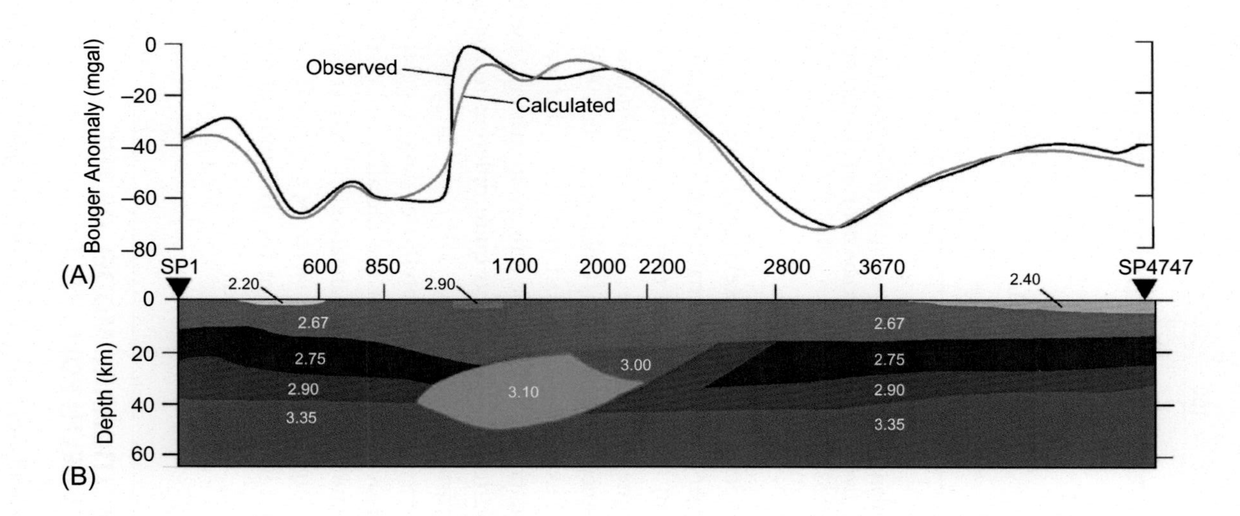

Fig. 3.9 Crustal density model along the Nagaur-Kunjer profile. Density is in grams per cubic centimeter (example 3.35 g cm^{-3} to 3350 kg m^{-3}). (A) Observed and computed Bouger anomaly along the profile. (B) Depth model based on combined analysis of seismic and gravity data. *(After Rajendra Prasad, B., Tewari, H.C., Vijaya Rao, V., Dixit, M.M., Reddy, P.R., 1998. Structure and tectonics of the Proterozoic Aravalli-Delhi Fold Belt in the northwestern India from deep seismic reflection studies. Tectonophysics 288, 31–41.)*

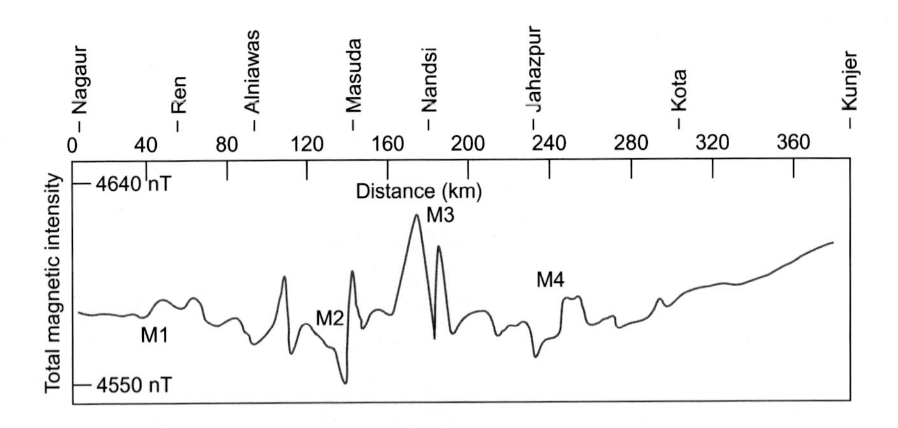

Fig. 3.10 Magnetic data along the Nagaur-Kunjer profile. *(After Tewari, H.C., 1998. The effect of thin high velocity layers on seismic refraction data: an example from Mahanadi basin, India. PAGEOPH 151, 63–79.)*

The geoelectrical section between Kekri and Kota (Jahazpur thrust region of Nagaur-Kunjer profile) shows that a northwest-dipping conductor (resistivity 50 Ω m) in the middle crust is present in this region (Fig. 3.11). It is disrupted near Jahazpur, where it extends to shallower depths and displays a steep upward trend (Gokaran et al., 1995). This pattern is the geoelectrical expression of the Jahazpur thrust that was imaged in the seismic reflection data. Similarly to the reflectivity pattern, the resistivity characteristics of various layers are also quite different to the east and west of the Jahazpur thrust.

The seismic, gravity and magnetotelluric data suggest that the Jahazpur thrust has great significance for the metallogeny of the region. Presence of the base metal deposits at Jahazpur and Sawar, close to the exposure of the fault, suggests that it has acted as a conduit for transportation of ore fluids. High conductivity of the Jahazpur thrust is due to sulfide deposits, while its high reflectivity is due to mylonitization (Rajendra Prasad et al., 1999).

3.3 THE EVOLUTIONARY MODEL OF THE ARAVALLI-DELHI FOLD BELT

Based on the results of seismic, gravity, magnetic, geoelectrical, geological, geochemical and geochronological studies, several evolutionary models have been suggested for the region of the Aravalli-Delhi fold belts

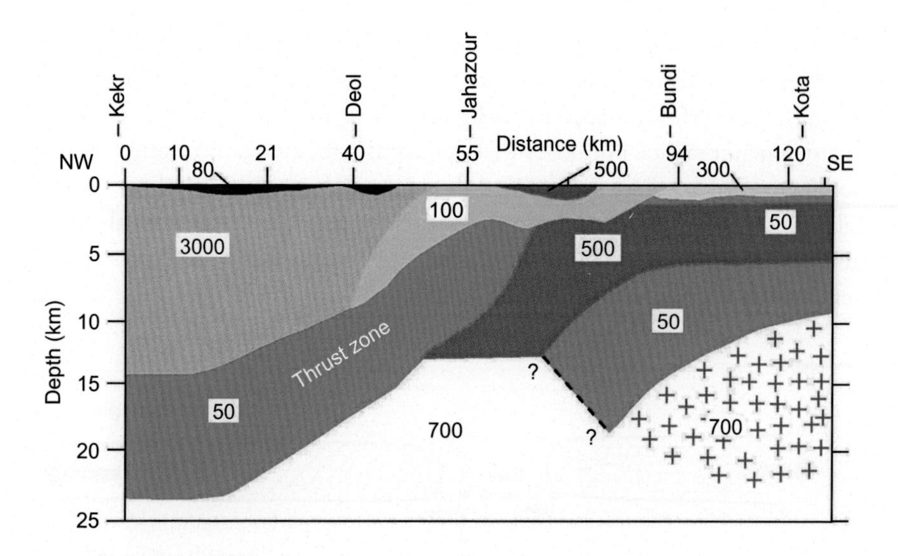

Fig. 3.11 Resistivity section in the region of Jahazpur thrust along the Nagaur-Kunjer profile. *(Modified after Rajendra Prasad, B., Vijaya Rao, V., Reddy, P.R., 1999. Seismic and magnetotelluric studies over a crustal scale fault zone for imaging a metallogenic province of Aravalli-Delhi Fold Belt region. Curr. Sci. 76, 1027–1031.)*

(Roy, 2000; Sinha-Roy, 2000; Tewari, 1998; Tewari et al., 2000; Tewari and Vijaya Rao, 2003). Except for the model of Roy (2000), the others use the reflectivity pattern as a tool to decipher the structure and understand the tectonic significance of the crustal blocks. Specific reflectivity patterns are attributed to the particular tectonic regimes. These models believe in a plate tectonic framework that involves repeated rifting, generation of oceanic crust, sedimentation, its consumption at the subduction zones and consequent formation of fold belts during the Proterozoic period, similar to those suggested for the Phanerozoic period. These plate tectonic models are more or less similar to each other.

According to the plate tectonic models, a thermal plume developed under this region during the Palaeoproterozoic, leading to rifting of the continental crust, opening of an ocean and generation of a triple junction. During this process, the Bhilwara gneissic complex was separated into the northern and southern blocks. Geochemical signatures of the amphibolites, which separate the Bundelkhand craton lying to the east from the Mewar craton to the west (Fig. 3.12), resemble low-K tholeiites to calc-alkaline basalts of ocean floor. Sedimentation took place around 2000 Ma wherein the Aravalli and Hindoli sediments were deposited on the marginal basins of the Mewar and Bundelkhand cratons, respectively. Emplacement of granites during 2000–1900 Ma appears to be related to this episode. The Jahazpur thrust, a major crustal boundary that demarcates the two contrasting tectonic domains of greenschist facies Hindoli group and amphibolites facies Mangalwar complex, plays a major role in this model. The geochemical signatures suggest the presence of an ocean in the region of the Aravalli fold belt, which separated the Bundelkhand craton (crustal block to the east of

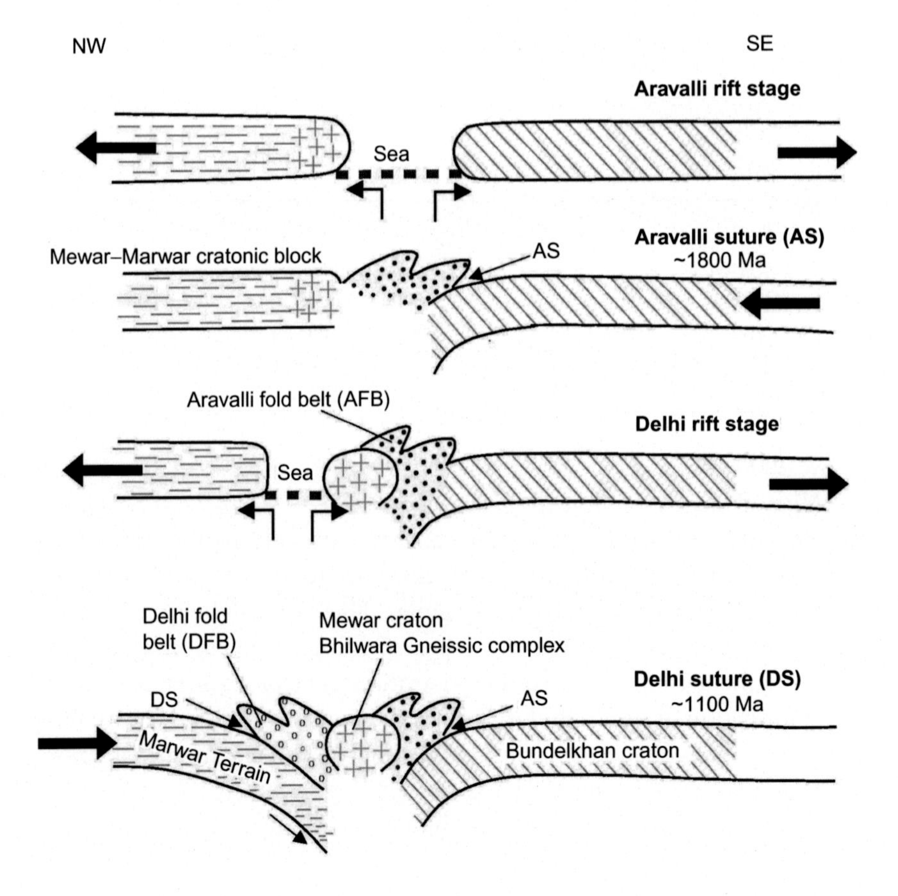

Fig. 3.12 Cartoon showing the evolutionary model of the Aravalli-Delhi fold belt. *(From Vijaya Rao, V., Rajendra Prasad, B., Reddy, P.R., Tewari, H.C., 2000. Evolution of Proterozoic Aravalli-Delhi Fold Belt in the northwestern Indian Shield from Seismic studies. Tectonophysics 327, 109–130.)*

the Bhilwara gneissic complex) from the Mangalwar complex to the west. The compressional forces, operative during the Proterozoic, moved the eastern craton to the west. Due to this westward movement, intervening oceanic crust, along with the sediments, was subducted under the western craton. Further convergence created an imbricated fault zone (Jahazpur thrust). Most of the subduction zones and orogenic boundaries, like this thrust, are highly reflective and conductive. The subhorizontal reflections of the lower crust appear to obliterate the lower part of the Jahazpur thrust and overprinted it. Therefore, these reflections may be a younger feature developed during the postorogenic processes. However, the Jahazpur thrust, where the palaeo signatures of the earlier compressional episode are still present, is an older feature formed during the Palaeoproterozoic collision (Mandal et al., 2014). Base metal mineralization, observed at a surface location of this thrust zone, is possibly due to the subduction-related activity. It is a well-known fact that large-scale metallogenic provinces, world over, are associated with the subduction zones.

In the next stage of convergence, collision between the Bundelkhand craton to the east and the Marwar craton to the west resulted in the Aravalli orogeny. The opposite dipping reflections (E and F of Fig. 3.7) indicate collision signature and characterize different crustal blocks on either side of this orogeny. The collision resulted in shortening and thickening of the crust. As the collision continued, the initial thrust became steeper and the colliding blocks were effectively locked. This led to development of the Aravalli suture between the two cratons. This type of suturing at the subduction zones is common to many ancient collision boundaries of the stable, colder Precambrian regions and electrically conductive zones (BABEL Working group, 1990).

A thickened crust in a subduction/collision zone provides an ideal pressure and temperature regime for generation of the granulites. The carbonic fluids released from the sediments of a subducting slab also help in their formation. The Sandmata granulites, generated under such an environment, were thrusted up due to collision and are emplaced as tectonic wedges within the basement gneisses. These occur as imbricate zones between various rocks and are represented by a reflection band in the seismic section (E of Fig. 3.7). The pressure-temperature regime of the granulites of this region suggests that they are the products of a volcanic arc environment (Sarkar et al., 1989). Continental collision is the main mechanism for generation of many granulite grade metamorphic belts, as the volcanic arc environment has characteristic properties appropriate for generation of the granulites (Bohlen, 1987).

Bimodal volcanism in the Delhi fold belt (Deb et al., 1995) suggests that subsequent to the Paleoproterozoic Aravalli orogeny, the region underwent another episode of rifting during the Mesoproterozoic period. This rifting separated the Bhilwara gneissic complex as a rift fragment from the Marwar craton and also opened an ocean; the joint Bundelkhand-Aravalli-Bhilwara gneissic complex Craton was to the east and the Marwar craton to the west of this ocean (Fig. 3.12). In this basin, the Delhi sediments were deposited. The compressional phase that followed this rifting led to eastward subduction of the Marwar craton along with the oceanic crust of the Delhi fold belt. The southeast dipping reflection band (C of Fig. 3.6) indicates this eastward subduction. A thick high-velocity (7.3 km s^{-1}) crust, high density of 3100 kg m^{-3} for the lower crust and large depth to the Moho boundary (~50 km) that is characteristic of the island arc regions are seen here. Continuous subduction of the western block (Marwar craton) resulted in development of this island arc. Absence of reflections in the Delhi fold belt is possibly due to diapiric rise of the high-density magmatic material at the island arc, which has disturbed the normal crustal structure. Rise of the magmatic material at the island arc is also manifested as the gravity high. Further convergence of the Marwar and Bundelkhand-Aravalli-Bhilwara gneissic complex Cratons, with the island arc between them, resulted in collision and development of the Delhi fold belt around 1100 Ma (Fig. 3.12). The opposite dipping reflection bands at this junction correspond to the Delhi orogeny. The intervening oceanic crust was transformed into the continental setting by a subduction process, forming the ophiolite. Subsequent to the collision, the crustal material was partially melted at depth under the high pressure-temperature regime emplacing the Erinpura granites along the western boundary of the Delhi fold belt. The discontinuous character of the reflections in the present-day Marwar basin is due to the post-Delhi magmatic events responsible for development of the granite of 850 Ma and igneous suite of 750 Ma. Presence of an 1800 Ma old (Aravalli) collision pattern in the form of dipping reflections down to 18 s TWT and island arc signatures in the Aravalli-Delhi fold belt suggest that originally a thick crust was formed, which has not been significantly altered even after the Proterozoic collision. This indicates the thermal and tectonic stability of the region. Evolutionary periods of the Aravalli (~1800 Ma) and Delhi (~1100 Ma) fold belts correlate well with the global orogenic activity and are likely to correspond to the supercontinental episodes. The geological age in these belts becomes younger from east to west and indicates that the period of tectonic activity in the region has extended for a long duration.

The Vindhyan basin in the southeast and Marwar basin in the northwest are situated on the fringes of the Bundelkhand and Mewar cratons, respectively. These basins are neither deformed nor metamorphosed, which suggests that they are formed in the intracratonic part subsequent to collision and suturing of the cratons and received shallow water deposits derived from erosion of the uplifted orogenic belts.

The orogenic process involving compression, collision and the thrusting of crustal blocks generated ductile shear/fault zones in the region, which subsequently were mylonitized. Most of these shear zones are thrust faults and have good reflectivity, as seen close to the Aravalli and Delhi terrene boundaries (Fig. 3.6). Mylonitic zones, in different parts of the world, are found to be highly reflective. Structural and petrologic characteristics of the shear zones also generate high-amplitude reflections, which originate from top and bottom of a fault zone as well as within it. A bipolar gravity anomaly that shows a large negative anomaly over the trench and a high positive anomaly over the island, similar to the one seen in this region, is normally associated with the subduction zones. This type of anomaly indicates that the collision and suturing took place between a continent and an island arc. In the gravity model (Fig. 3.9) the negative anomaly corresponds to the underthrusted older craton, while the positive anomaly characterizes the overthrusted younger mobile/fold belt. The positions of the Mewar and Marwar blocks of the crust can be explained through movement and underthrusting of various cratonic blocks of the region (Fig. 3.13). The boundary separating the two seems to be an intracratonic zone (Sinha-Roy, 2000).

The evolutionary model of Roy (2000), however, believes that the occurrence of a komatitic basaltic body, in association with the deformed

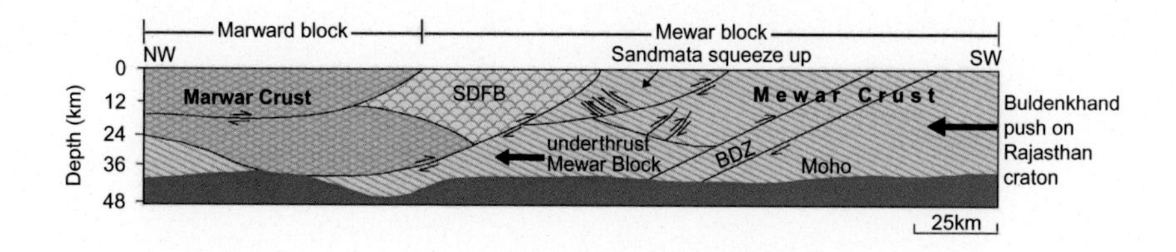

Fig. 3.13 Cartoon showing the intracontinental underthrusting of cratonic blocks responsible for evolution of the Aravalli-Delhi fold belt. *(After Sinha-Roy, S., 2000. Plate tectonic evolution of crustal structure in Rajasthan. In: Verma, O.P., Mahadevan, T.M. (Ed.), Research Highlights in Earth System Sciences. Ind. Geol. Cong., Roorkee, India, 17–25.)*

gneiss-granulite rocks of the Sandmata complex, is an expression of underplating of the mafic-ultramafic magma and implies that the middle-lower crustal dome-shaped feature, observed in the seismic studies, is the signature of Mesozoic-Cainozoic age and not Precambrian age. According to this model significant geomorphologic changes in the region during the late Phanerozoic, high heat flow of 64 mW m^{-2} (Panda, 1985), crosscutting gravity patterns and a 70 Ma thermally induced Pb-loss event in the granites of the Aravalli Mountain (Sivraman and Raval, 1995) are definite proofs of rejuvenation of the Aravalli crust.

CHAPTER FOUR

East Coast Sedimentary Basins of India

The east coast of India consists of several deltaic sedimentary basins that were created due to drainage of the major rivers into the Bay of Bengal. These rivers have large catchment areas and hence their own deltas. Except for the Ganga delta (Bengal basin), which is highly tide dominated, the other deltas are more or less similar in their mode of formation and have arcuate shapes.

The Gondwana rift basins in central and eastern India (Godavari, Mahanadi and Damodar grabens of Fig. 4.1) were formed due to the breakup of east Gondwanaland during the upper Carboniferous-lower Cretaceous period. Similar basins are also observed in Australia and Antarctica, the other constituents of east Gondwanaland. Thus, these basins constitute a transcontinental rift system with conjugate structures on either side.

Evolution of the sedimentary basins on the east coast of India has a long, complex magmatic and tectonic history. It is controlled by many tectonic pulses that include the breakup of east Gondwanaland, separation of India from Antarctica and Australia, evolution of the northeast Indian ocean and collision of the Indian plate with the Eurasian plate (Curray et al., 1982). The paleontological, stratigraphic and structural observations favor a common intra-Gondwana rift system that serves as the basis for many recent Gondwana reconstructions. Intrusion of the volcanic rocks and neotectonic activity in the coastal basins and the magmatic activities of the eastern ghat have also been considered in these reconstructions (Burke and Dewey, 1973). It is believed that the plate containing India, Antarctica and Australia separated from the rest of Gondwanaland at about 200 Ma. The Indian plate separated from the Antarctica and Australia plates at about 147 Ma in the late Jurassic. The east coast basins were intracontinental pull-apart basins until the Jurassic, and thereafter became pericratonic due to the breakup of Gondwanaland (Rao, 1993). The eastern ghat belt was largely responsible for the evolution of these basins during the Jurassic period. Basin-penetrating faults limited these basins on the west. During the late Jurassic, horsts and grabens were formed in these basins

Structure and Tectonics of the Indian Continental Crust and Its Adjoining Region
https://doi.org/10.1016/B978-0-12-813685-0.00004-2

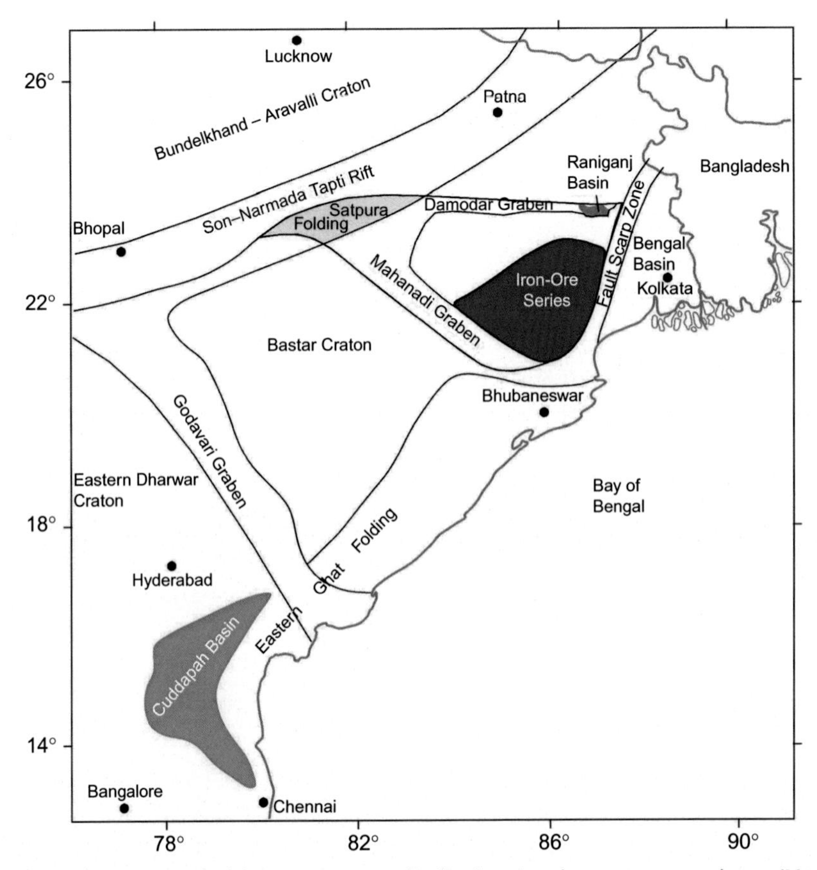

Fig. 4.1 Tectonic map of the eastern part of India showing the east coast grabens. *(Modified from Rao, V.K., 2002. Crustal structure and evolution of Godavari graben (Chinthalpudi sub-basin) and Krishna Godavari basin—an integral approach. Ph.D. Thesis, Osmania University, Hyderabad, India.)*

due to tensional forces, leading to subsidence accompanied by the local uplifts (Sastri et al., 1974). The geological setting of the various basins and cratons (Fig. 4.1) shows that the east coast basins are separated from the southern and central Indian basins by various cratons.

4.1 THE EAST COAST BASINS

The eastern ghat has several petrophysical and geochemical similarities to areas in Antarctica that were once adjoining it. These indicate juxtaposition of the east coast basins of India with the grabens in Antarctica before the rifting and breakup of Gondwanaland. An upwelling mantle plume (Kerguelen) was the thermal source involved in the process of this breakup. The concept of mantle plume-generated triple junctions, along which the continents break and separate, is supposed to be responsible for development of sedimentary basins on the east coast of India (Burke and Dewey, 1973). In a triple junction, two of the three arms open up to the ocean and finally join to form a plate boundary, while the third arm fails to open and becomes a rift. Major rivers flow down to the newly opened basins along these rifts and form deltas. Four such junctions, located close to the mouths of rivers flowing into the Bay of Bengal, have been identified on the east coast of India. The sedimentary basins here have developed on the failed arms of these triple junctions.

During the northerly movement of the Indian plate, its east coast remained a passive continental margin favorable for stable shelf sedimentation. Late Carboniferous to Triassic sedimentation in the rift basins on the east coast is fluvial and continental in nature. Early Cretaceous marine transgression facilitated a large amount of sedimentation along these basins. These basins also experienced widespread volcanic activity (Rajmahal Traps, dated at 117 Ma) between the late Jurassic and early Cretaceous (Baksi, 1995) and also between the early Cretaceous and Eocene (Chatterjee, 1986). In the deep wells of the Godavari basin the volcanic rocks are also encountered at the top of the upper Cretaceous and were identified as the Deccan Trap volcanics (~65 Ma).

Formation history of the east coast marine basins shows four important stages: (1) prerift stage, (2) rift stage (rifting of the continents during the upper Cretaceous associated with the volcanic eruption and deposition of the clastic sediments both on land and offshore), (3) postrift carbonate platform stage (development of the Paleogene sediments representing thick carbonate buildup under stable shelf conditions and establishment of the continental shelf-slope divide), and (4) middle-late Tertiary marine clastic/deltaic sedimentation stage (deposition of thick Neocene sediments under deltaic to marine conditions following a major hiatus during the Pliocene-early Pleistocene).

Evolution of most of the sedimentary basins is related to the growth of their crust. In recent years magmatic underplating has been considered to be a major mechanism of crustal growth. According to this mechanism, new material is added to the base of the existing crust during periods of crustal extension when the heat flow is high (White and McKenzie, 1989). Earlier it was thought that this mechanism is dominant for crustal growth and its differentiation in the Precambrian only, but now it is considered active from the Proterozoic through the Phanerozoic. Verification of this mechanism and estimating its significance to crustal growth is difficult in the Tertiary and Quaternary basins because the material that represents the lower crust and upper mantle is not exposed in these basins (Fountain et al., 1990). Seismic imaging of the crust is the only tool able to provide some clues to the genesis and emplacement of such material, which is seen as a high-velocity layer at the base of the crust. Therefore to understand the evolution of the east coast basins of India, knowledge of the velocity-depth configuration and structure of the crust is essential.

Extensive seismic studies were carried out in the West Bengal, Mahanadi and Godavari basins for exploration of the hydrocarbons. To augment these efforts, seismic studies to understand the crustal configuration were carried out along several profiles in these basins. Seismic refraction and postcritical reflection data in the Mahanadi and Godavari basins were collected in analog form, while in the West Bengal basin these data are in digital form. Since in most parts of these basins the hydrocarbon exploration data could not clearly delineate the basement of the sedimentary formation, the same was also configured in addition to the velocity-depth configuration of the crust. Results of these studies along each of these previous three basins, described in the following paragraphs, throw some light on their evolution.

4.2 THE WEST BENGAL BASIN

The Bengal delta is one of the largest deltaic regions in the world. This delta, situated in the northeast of the Indian peninsula, forms the Bengal basin and comprises the Indian and Bangladesh provinces. It is one of the largest pericratonic foreland basins in the world and is formed due to the breakup of east Gondwanaland and subsequent separation of Antarctica and Australia from India, evolution of the northeast Indian Ocean and finally collision of the Indian plate with the Eurasian plate (Curray et al., 1982). These processes are also related to the Kerguelen hot spot activity that extruded volcanics, in the form of the Rajmahal Traps, and concealed the earlier Gondwana sediments. The basin is extensively covered by alluvium (Quaternary). The western part of this delta is known as the West Bengal basin (Fig. 4.2) and is bounded on the west by the Archean complex of the Indian Peninsular shield and east-west trending intracratonic upper Carboniferous to lower Triassic Gondwana exposures. In the east this basin extends into the Bangladesh plains and to the south it is open to the Bay

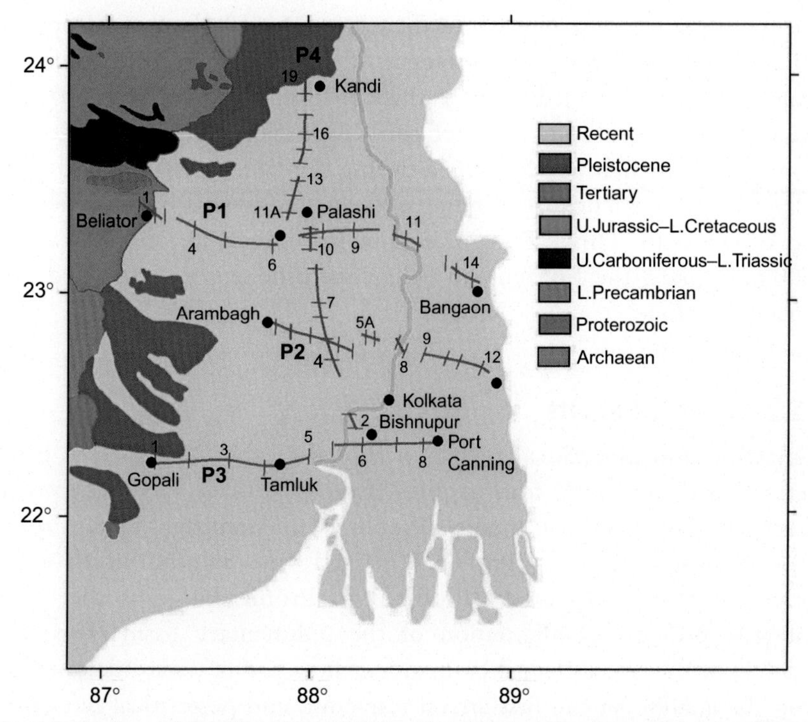

Fig. 4.2 Geology of the West Bengal delta, also showing the deep seismic profiles P1, P2, P3 and P4. *(Modified from Krishna, V.G., Vijaya Rao, V., 2005. Processing and modelling of short-offset seismic refraction-coincident deep seismic reflection data sets in sedimentary basins: an approach forexploring the underlying deep crustal structures. Geophys. J. Int. 163, 1112–1122.)*

of Bengal. The Rajmahal Trap (Cretaceous), overlying the Archean basement, is present to the northwest.

4.2.1 Geology and Tectonics

The West Bengal basin is divided into three structural parts: western scarp zone, middle stable shelf, and eastern deep basin (Sen Gupta, 1996). Thickness of the sediments in the deeper part of the basin is about 14 km. Intense tectonic activity in a continental environment, resulting in formation of the Gondwana basins on the east coast of India during the late Carboniferous period, has resulted in prominent southeast regional dips in the Bengal basin (Tiwari, 1983). The basin is affected by tensional forces, which resulted in outpouring of the basaltic volcanic rocks Rajmahal Traps) during the late Jurassic and early Cretaceous periods. Within the basin these volcanic rocks are concealed below the sediments. Lithology of a deep well at Palashi (Fig. 4.2) shows the presence of about 1000-m thick Rajmahal Traps at a depth of about 2600 m. A thick column of the Tertiary and Quaternary sediments overlies the Cretaceous formations. The only large tectonic feature, which has developed during the late Cretaceous and early Paleocene time, is a flexure trending in the northeast-southwest direction, also known as the Hinge Zone or Eocene shelf break, at a depth of about 5200 m. During the Eocene and Paleocene time this flexure acted as a continental slope.

4.2.2 Seismic Results

Refraction and postcritical reflection studies in the West Bengal basin were carried out along four profiles (Fig. 4.2): three in the east-west direction; Gopali-Port Canning (Profile P1), Arambagh-Taki (Profile P2), Beliatore-Bangaon (Profile P3); and one in the north-south direction—Bishnupur-Kandi (Profile P4). Profile P4 joins the three east-west profiles. Configuration of the sedimentary basin (Figs. 4.3 and 4.4) of the West Bengal basin along these profiles were determined from the analysis of the first arrival refraction and postcritical refraction data, respectively (Kaila et al., 1992, 1996). Along the east-west profiles five prominent refractors (velocities 1.7–1.9, 2.4–3.0, 3.7–3.9, 4.6–5.3 and 3.8–4.0 km s^{-1}) are identified overlying the basement, which has a velocity of 5.8–6.2 km s^{-1}.

Well velocity studies by the hydrocarbon industry have identified the velocity of 4.6 km s^{-1} with the Rajmahal Traps; therefore the velocity of 4.6 km s^{-1} encountered at about 4 km depth and increasing to 5.3 km s^{-1} at the bottom of the layer in the seismic sections is identified with these traps. However, in deeper parts of the basin to the east of the Hinge Zone, in a deep well drilled to 5600 m depth at the Eocene shelf edge, only the Paleocene formation is reached and the trap formation is not even touched. Here, the Rajmahal Traps are expected to be at a much larger depth and the velocity of 4.6 km s^{-1} cannot be identified with these traps. The velocity models show a velocity inversion from 4.6–5.3 km s^{-1} (Rajmahal Traps) to 3.8–4.0 km s^{-1} close to the Hinge Zone. The deep well lithology shows the presence of Gondwana sediments below the Rajmahal Traps. The layer of velocity 3.8–4.0 km s^{-1} is, therefore, identified with the Gondwana sediments.

The depth to the basement in this basin increases from west to east, with a maximum depth of about 14 km at the eastern end of the profiles, close to

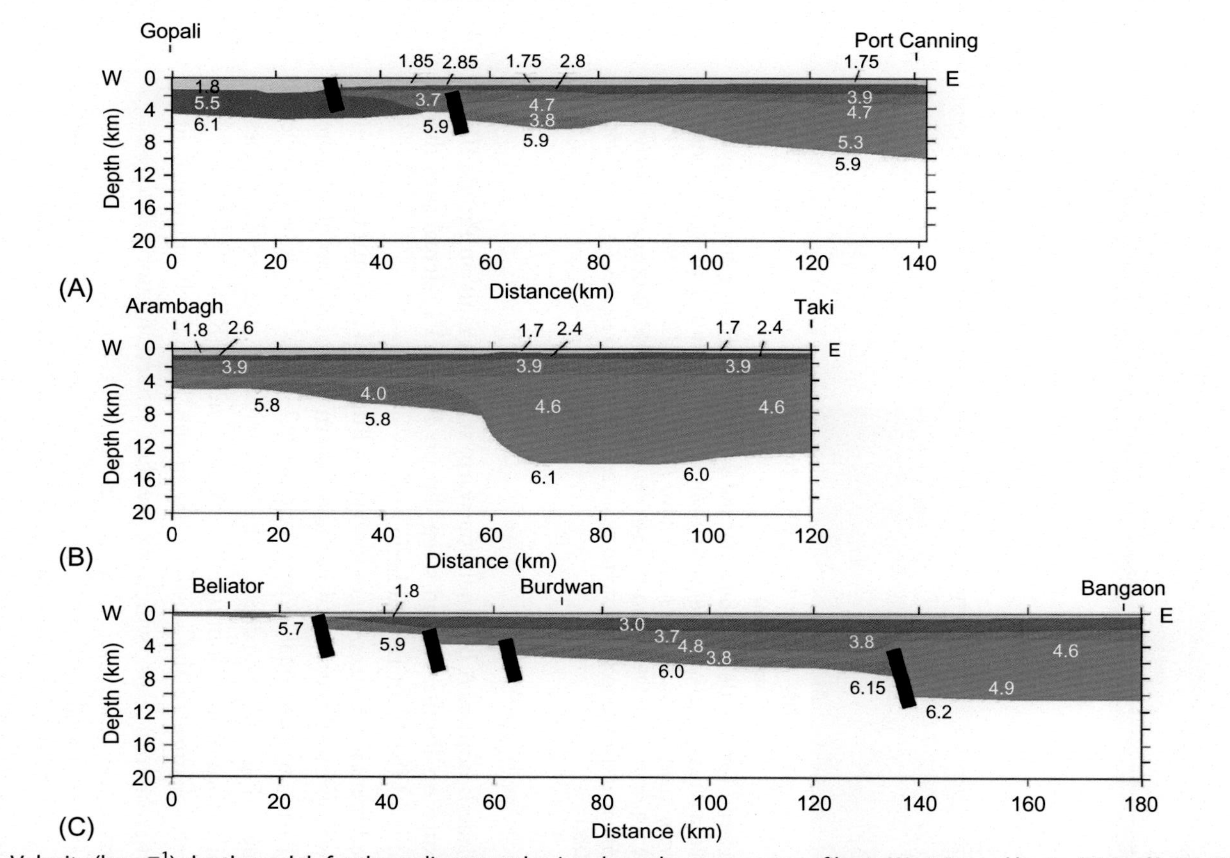

Fig. 4.3 Velocity (km s^{-1})-depth models for the sedimentary basins along the east-west profiles in West Bengal basin: (A) Profile P1, (B) Profile P2, and (C) Profile P3. *(Modified from Kaila, K.L., Reddy, P.R., Mall, D.M., Venkateswarlu, N., Krishna, V.G., Prasad, A.S.S.S.R.S., 1992. Crustal structure of the West Bengal basin, India, from deep seismic sounding investigations. Geophys. J. Int. 111, 45–66; Rao, I.B.P., Murty, P.R.K., Rao, P.K., Murty, A.S.N., Rao, N.M., Kaila, K.L., 1999. Structure of the lower crust revealed by one- and two-dimensional modelling of wide-angle reflections, West Bengal basin, India. PAGEOPH 156, 701–718.)*

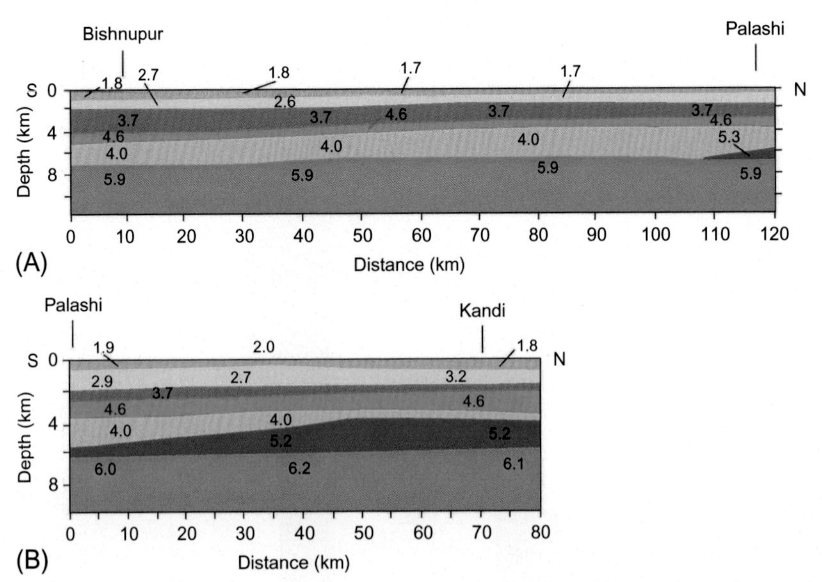

Fig. 4.4 Velocity (km s^{-1})-depth models for the sedimentary basin along the north-south profile in West Bengal basin, between (A) Bishnupur-Palashi and (B) Palashi-Kandi. *(Modified from Kaila, K.L, Murty, P.R.K., Madhava Rao, N., Rao, I.B.P., Koteswar Rao, P., Sridhar, A.R., Murthy, A.S.N., Vijaya Rao, V., Rajendra Prasad, B., 1996. Structure of the crystalline basement in West Bengal basin, India, as determined by DSS studies. Geophys. J. Int. 124, 175–188.)*

the Bangladesh border. A basement up warp, seen as a fault on Profile III, sharp flexure on Profile P2 and steep dip on Profile P1, demarcates the Hinge Zone in the West Bengal basin.

The sedimentary basin configuration along Profile P4 is similar to that on the east-west profiles except that in the northern part of the profile, to the north of Palasi, an additional layer of velocity 5.2–5.3 km s^{-1} is seen above the basement (Fig. 4.4). The depth to the basement along this profile reduces consistently from south (Bishnupur, ~7 km) to north (Kandi, ~5 km).

A typical feature of the seismic data in this basin is the presence of high-energy later arrivals, which are neither refractions nor reflections. These arrivals are nearly parallel to the first arrival refraction phases but at a constant time interval behind them (Fig. 4.5), and are identified as reflected refractions that belong to the family of free-surface multiples. These multiples are generated due to a very high-velocity gradient in the shallow sedimentary layers and completely mask most of the later reflection phases, including those from interfaces in the crust. Analysis of these arrivals

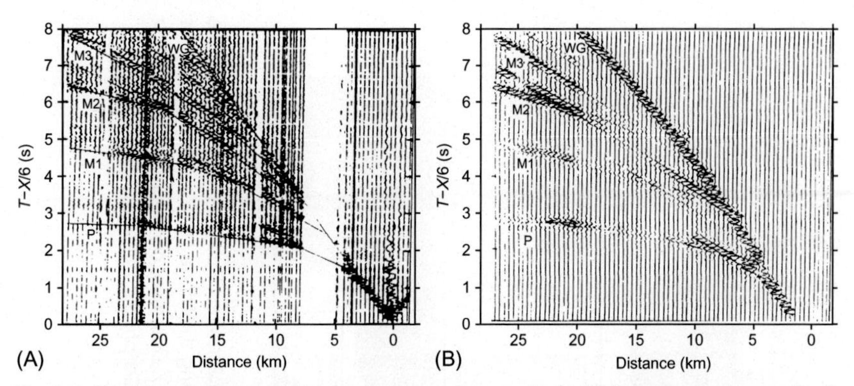

Fig. 4.5 (A) Seismogram showing high-energy later arrivals. (B) Represents the synthetic seismograms for the model (Fig. 4.6). *((A) After Sarkar, D., Reddy, P.R., Kaila, K.L., Prasad, A.S.S.S.R.S., 1995. Multiple diving waves and high-velocity gradients in the Bengal sedimentary basin. Geophys. J. Int. 121, 969–974.)*

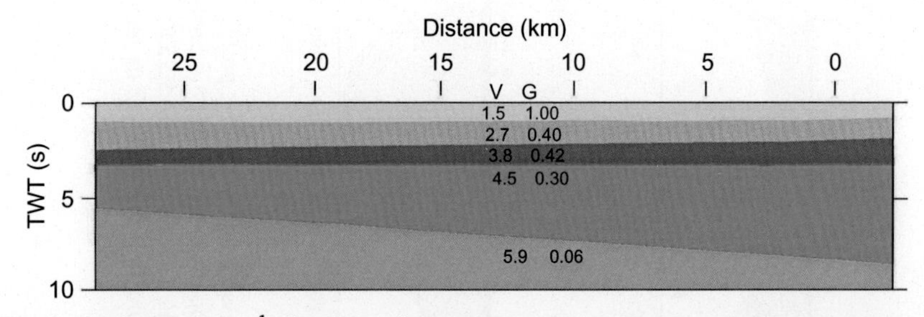

Fig. 4.6 Velocity (km s^{-1})-depth configuration up to the basement in the western part of Profile P1 showing high-velocity gradients (G) that cause near-surface multiples shown in Fig. 4.5. *(After Sarkar, D., Reddy, P.R., Kaila, K.L., Prasad, A.S.S.S.R.S., 1995. Multiple diving waves and high-velocity gradients in the Bengal sedimentary basin. Geophys. J. Int. 121, 969–974.)*

(Sarkar et al., 1995) has resulted in modification of the earlier sedimentary basin configuration of the West Bengal basin (Fig. 4.6).

The tomographic models of the sedimentary basin (Damodara et al., 2017) show the variation in the seismic velocity between 1.5 and 5.8 km s^{-1} (Fig. 4.7). Similarly to the earlier results, the models along the three east-west profiles show that the thickness of the sediments increases to the east. The deeper part of the basin appears to be more affected by the tectonic activity as compared to its upper part. The basement (velocity 5.9 km s^{-1}) depth at the eastern part of the Profiles P1, P2 and P3 is 12, 16 and 14.5 km, respectively. The largest thickness of the

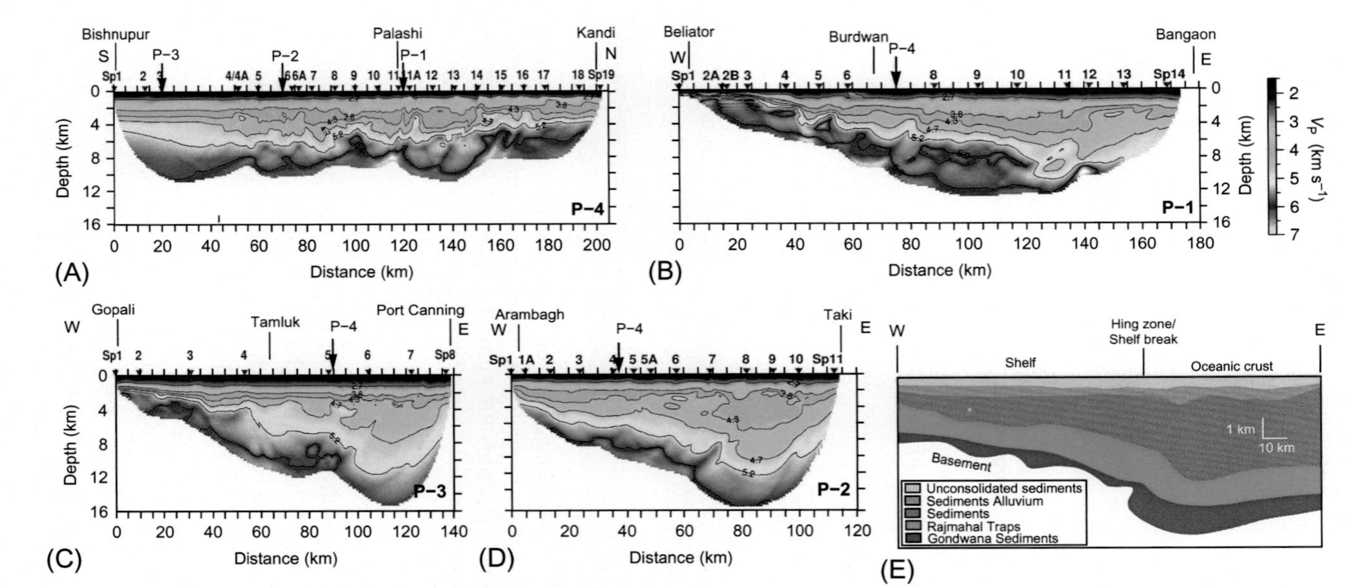

Fig. 4.7 The tomographic model of sedimentary thickness along the seismic profiles as shown in Fig. 4.2 in West Bengal. Figures (A)–(D) are along the profiles as written P1–P4, respectively. The subplot (E) represents the cartoon diagram showing the hinge zone and deepening of the basement. *(Modified after Damodara, N., Vijaya Rao, V., Sain, K., Prasad, A.S.S.S.R.S., Murty, A.S.N., 2017. Basement configuration of the West Bengal sedimentary basin, India as revealed by seismic refraction tomography: its tectonic implications. Geophys. J. Int. 208, 1490–1507.)*

sediments is on profile II, located in the middle parts of the basin. There is a sudden increase in thickness of the sediments along all the three east-west profiles. This regional feature is identified as the Hinge Zone and marks the end of the continental shield. Comparison of the velocity structure with the lithostratigraphy of the deep well data (Das et al., 1993) suggests that the 1.8 km s^{-1} velocity represents the unconsolidated sediments, 2.1 km s^{-1} alluvium, 2.6–3.2 km s^{-1} the Pliocene, Neocene and Oligocene sediments. The 3.6–4.3 km s^{-1} velocity is attributed to the Eocene-Paleocene sediments (Sylhet limestone) and the 4.7–5.5 km s^{-1} velocity to the basaltic traps (Rajmahal). The velocity of 4.5–5.5 km s^{-1}, confined to the deeper parts of the basin, may represent deeper traps, while the velocity of 5.8–5.9 km s^{-1} may represent the crystalline basement, similar to other regions in India.

Damodara et al. (2017) also present a fence diagram showing a pseudo three-dimensional basement picture of the west Bengal basin (Fig. 4.7). This diagram shows the extension of the Hinge Zone in the northeast direction at least up to a distance of 150 km. The hinge zone also appears to coincide with thinning of the crust and magmatic underplating at its base.

The velocity-depth models for the crust along all the east-west profiles show a low-velocity layer of about 3 km thickness (velocity 5.6–5.8 km s^{-1}) in the upper crust (Fig. 4.8). This layer, with its top at a depth varying between 9 and 15 km, dips to the east in conformity with the basement dip, and lies between the basement (velocity 5.8–6.2 km s^{-1}) and another 7 to 14 km thick layer of higher velocity (6.4–6.6 km s^{-1}). Another layer of velocity 6.8–7.0 km s^{-1} (lower crust) and thickness varying between 5 and 10 km lies below the latter. The Moho, with a velocity of 8.05–8.1 km s^{-1}, is at a depth varying between 36 and 26 km on Profile P3 and about 33 km on Profile P1. A prominent dome-shaped feature, of about 40 km width, is observed in the eastern part of Profile P3 (Fig. 4.8).

Along the other profiles the depth to the Moho is of similar order but the dome-shaped feature is not seen, as these profiles do not extend far enough to the east to cover it. The Moho configuration indicates that the crust is akin to continental in nature, at least in the western and middle parts of the West Bengal basin. The large thickness of the sediments and broad unwrap in the Moho, seen together with outpouring of basaltic lava during the Cretaceous, indicate rifting of the crust in this region even though a high-velocity rift pillow in the lower crust that is normally associated with the rift basins could not be seen by Rao et al. (1999). Configuration of the crust along the north-south Profile P4 is similar to that obtained on the east-west profiles. The Moho is at a depth of

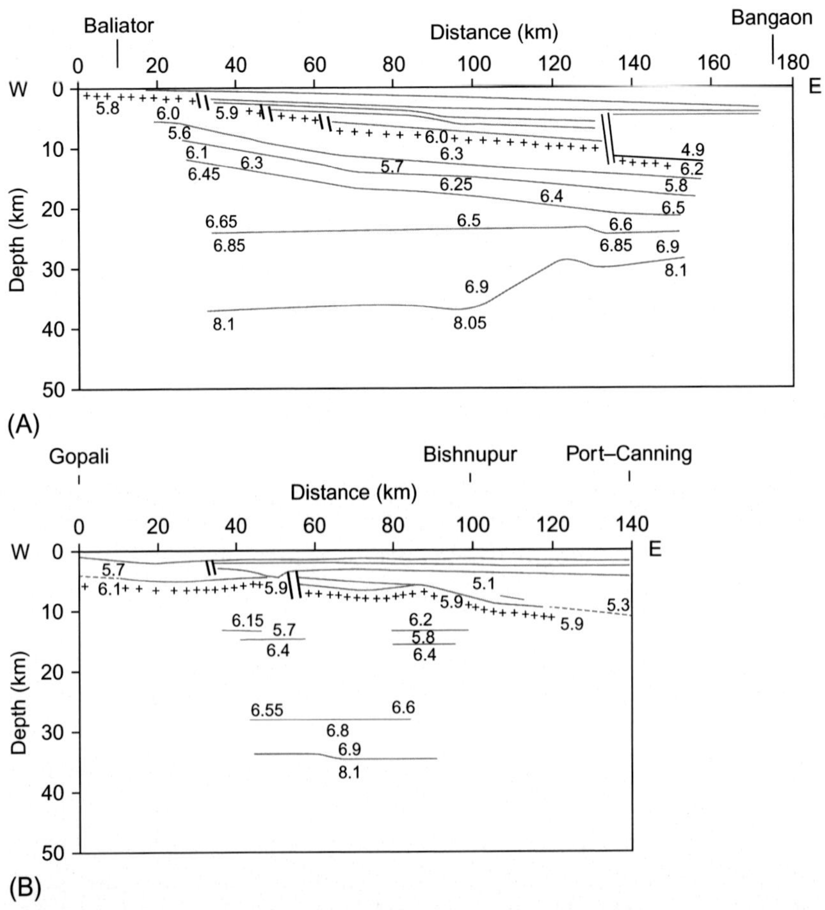

Fig. 4.8 Velocity (km s^{-1})-depth models for the crust along the east-west profiles (A) Profile P1 and (B) Profile P3. *(Modified from Kaila, K.L., Reddy, P.R., Mall, D.M., Venkateswarlu, N., Krishna, V.G., Prasad, A.S.S.S.R.S., 1992. Crustal structure of the West Bengal basin, India, from deep seismic sounding investigations. Geophys. J. Int. 111, 45–66.)*

about 35 km in the south and 38 km in the north (Fig. 4.9). The Moho and the basement configuration of the West Bengal basin indicate a north-south elongation and a Moho up warp toward the eastern part of the Profile P3, parallel to 87°E, the average axis of the Kerguelen plume path in the Bengal basin (Kaila et al., 1996; Rao et al., 1999).

A reinterpretation of the postcritical reflection data of Profile P3 (Mall et al., 1999) shows an additional reflection at the base of crust, which was identified on most of the seismograms as the Moho (M_2). In this model

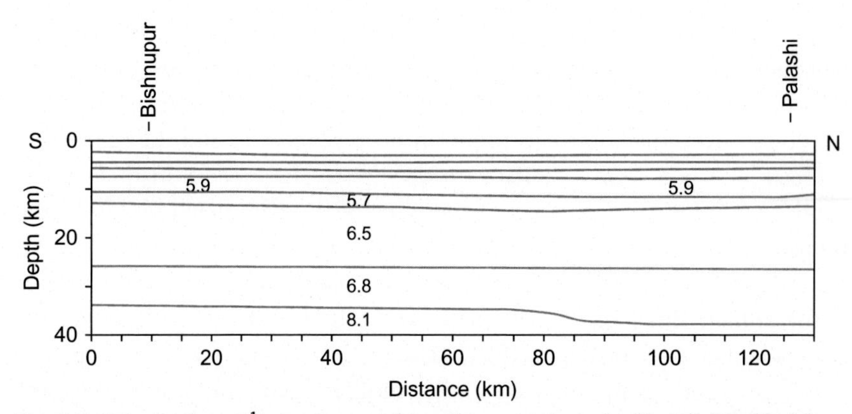

Fig. 4.9 Velocity (km s^{-1})-depth model for the crust along Profile P4. *(Modified from Rao, I.B.P., Murty, P.R.K., Rao, P.K., Murty, A.S.N., Rao, N.M., Kaila, K.L., 1999. Structure of the lower crust revealed by one- and two-dimensional modelling of wide-angle reflections, West Bengal basin, India. PAGEOPH 156, 701–718.)*

the M_1 reflections, which were earlier considered as reflection from the Moho, were identified as originating from the top of a high-velocity (7.5 km s^{-1}) underplated mantle layer in the lower crust that was associated with mantle plume-related rifting (rift pillow) during the Cretaceous (Fig. 4.10). According to this model, the tectono-thermal episodes due to passage of the Indian continent over the Kerguelen plume, separation of Gondwanaland and the heavy load of the sedimentary basin have contributed to deformation of the deep tectonic configuration in the West Bengal

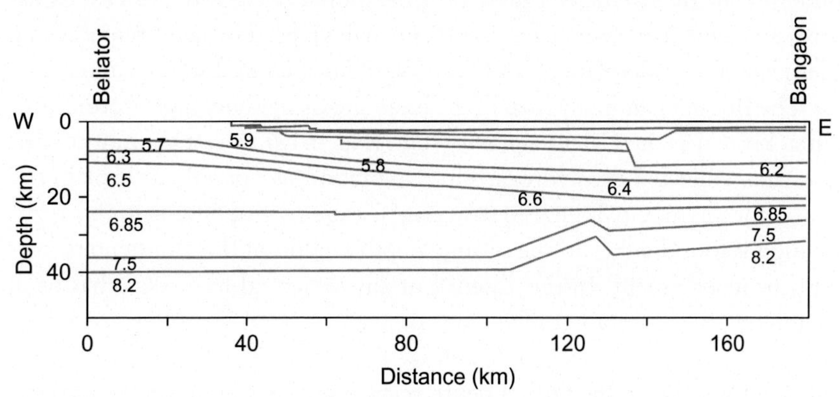

Fig. 4.10 Crustal velocity (km s^{-1})-depth model along Profile P3 showing the underplated layer (velocity of 7.5 km s^{-1}) above Moho. *(Modified from Mall, D.M., Rao, V.K., Reddy, P.R., 1999. Deep sub-crustal features in the Bengal basin, Seismic signatures for plume activity. Geophys. Res. Lett. 26, 2545–2548.)*

basin. The plume also provided mantle material for extrusion of the Rajmahal Traps through the lithosphere (either oceanic or continental) and contributed to deformation of the crust and lithosphere through secondary convection. Basaltic rocks found in deep bore wells and the subsurface velocity of 4.6–5.3 km s^{-1} that corresponds to these rocks in the seismic depth sections is an indication of the trace of the Kerguelen plume in the Bengal basin. The plume path could either take a swing towards 85°E or towards 90°E ridges because of the change in plate motion due to the impact of the plume. The basaltic rocks were also found in the deep bore wells in Bangladesh to the east (Khan, 1991). The model of Mall et al. (1999) appears to be more convincing than the earlier models mainly because it is able to show a high-velocity rift pillow, normally associated with rift basins, at the base of the crust.

In another one-dimensional velocity model for the crust of the West Bengal basin, the free-surface multiples were first used to generate a velocity model for the sedimentary basin and then attenuated to develop the subcritical reflections through the use of suitable velocity filters (Krishna and Vijaya Rao, 2005). This model (Fig. 4.11), based on combined interpretation of the seismic refraction/postcritical reflections and near vertical reflection data, shows two low-velocity layers in the sedimentary section (down to about 8 km depth) and two low-velocity layers in the crust, one in the upper crust between 10- and 12-km depth and the other in the lower crust between 30- and 32-km depth. The high-velocity (7.1 km s^{-1}) lower crustal layer delineated at depth varying between 32 and 34 km represents magmatic underplating due to the Kerguelen mantle plume activity, the surface exposures of which are observed as the Rajmahal Traps. The layer representing the magmatic underplating could not be delineated by earlier workers.

The Bouguer anomaly map of the basin shows a gravity high around Calcutta, known as the Calcutta gravity high (Fig. 4.12). A vast amount of seismic data, collected for hydrocarbon exploration, is not able to explain the reason for this gravity high. No structural feature corresponding to this gravity high is found either in the seismic depth section of the sedimentary column or its basement. In the absence of any structural features, it has been attributed to a combined effect of compact shale of the Oligo-Miocene age and a thick section of high-density limestone of Eocene age (Tiwari, 1983). However, in the crustal depth section of this region, the Moho dips from the east to the west (Fig. 4.8), which would normally create a gravity low. For generation of a gravity high the Moho dips should have been in the reverse direction. But a sharp increase in the thickness of the low-density

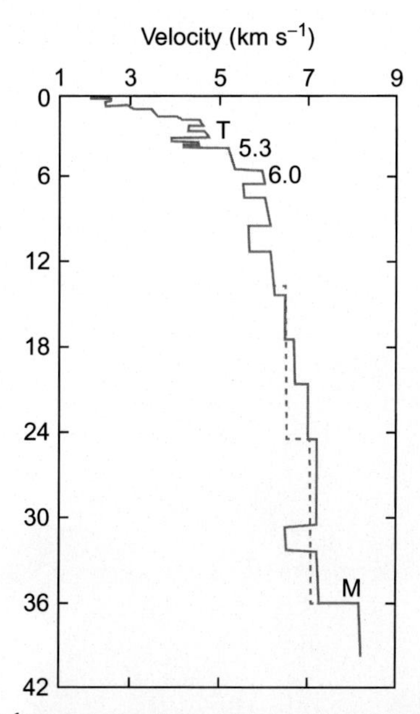

Fig. 4.11 Velocity (km s^{-1})-depth model for the crust in the Bengal basin showing the low-velocity layers in the sedimentary basin and crust. *(From Krishna, V.G., Vijaya Rao, V., 2011. Velocity modelling of a complex deep crustal structure across the mesoproterozoic south Delhi Fold Belt, NW India, from joint interpretation of coincident seismic wide-angle and near-offset reflection data: an approach using unusual reflections in wide-angle records. J. Geophys. Res. 116, B01307, DOI: 10.1029/2009JB006660.)*

sediments to the east of the Hinge Zone in the top part of the basin causes the gravity values to decrease to the east, which offsets the gravity low created by the Moho dip. The combined effect of the Moho dips to the west and thickening of the sediments to the east appears to have resulted in creation of the Calcutta gravity high (Rao et al., 1999). This gravity feature was also explained as due to thickening of the sedimentary column to the east and lateral structural variation in the middle/lower crustal layers, coupled with the structural trends in the basement (Prasad et al., 2002).

4.3 THE MAHANADI BASIN

The Mahanadi delta, located around the confluence of the river Mahanadi with the Bay of Bengal, is a classical arcuate delta. Lower Gondwana sediments (lower Triassic to upper Carboniferous), laterites (Pliocene-Pleistocene) and the eastern ghat group of rocks, consisting of the granite/

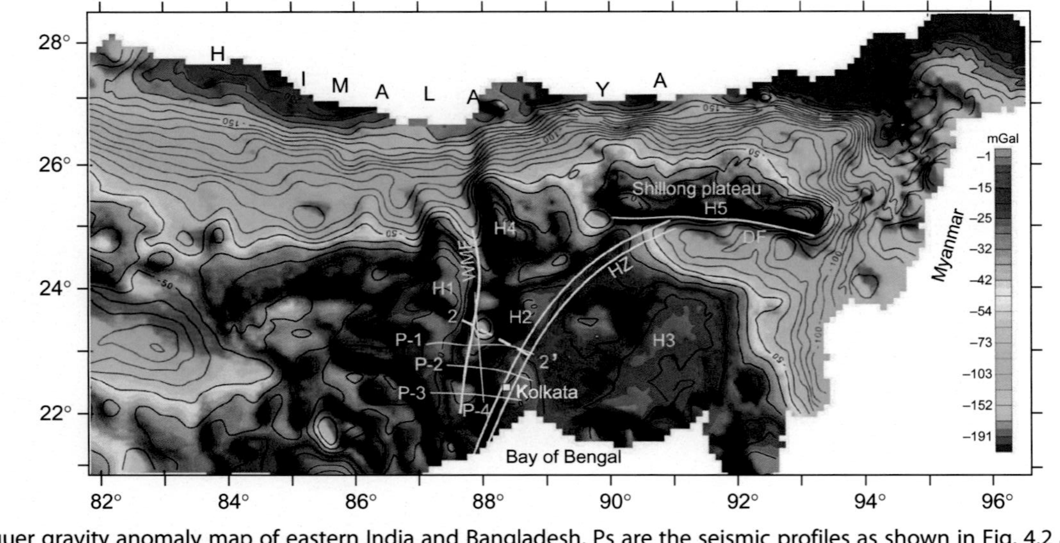

Fig. 4.12 Bouguer gravity anomaly map of eastern India and Bangladesh. Ps are the seismic profiles as shown in Fig. 4.2 and H1, H2... are the gravity highs. *(Source: http://icgem.gfz-potsdam.de/ICGEM/.)*

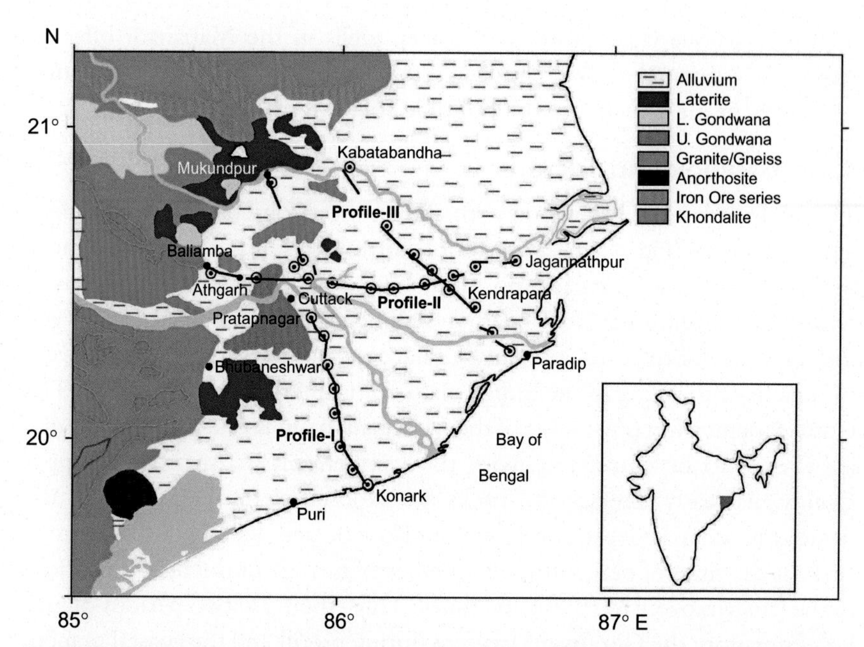

Fig. 4.13 Geological map of Mahanadi region with seismic profiles recorded for the crustal studies plotted on it. *(After Kaila, K.L., Tewari, H.C., Mall, D.M., 1987b. Crustal structure and delineation of Gondwana rift in the Mahanadi delta area, India from deep seismic soundings. J. Geol. Soc. India 29, 293–308.)*

gneiss (Archean), khondalite (Precambrian metamorphic) and charnockite/anorthosite (Precambrian metamorphic), are exposed to the west of the delta (Fig. 4.13). The Mahanadi basin, which is a part of this delta, extends both onshore and offshore and has undergone several stages of rifting and subsidence, followed by sediment deposition and uplift due to tensional forces during the late Jurassic (Sastri et al., 1974). The eastern ghat group forms the basement of the Mahanadi basin (Fuloria, 1993) and is in fault contact with the Iron Ore Group of rocks (Archean) to the north. Similar to the other basins on the east coast of India, widespread volcanic activity along the rift zones of this basin is also associated with the breakup of India from Antarctica and Australia during the early Cretaceous. Sedimentation, both in the onshore as well as the offshore basins, accompanies the volcanism. Until the Jurassic this basin was an intracontinental pull-apart basin and became pericratonic after the breakup of Gondwanaland (Rao, 1993). Geologically the onshore part is significantly different from the offshore part. Both basins came into existence during the Jurassic but the earliest sedimentation started during the Cretaceous. Thickness of the Tertiary sediments, underlain by early Cretaceous volcanic rocks, in deep wells in the Mahanadi offshore region is estimated to be between 2000 and 4000 m. The lowermost sequences in these wells are the volcanic rocks (Mahalik, 2000).

4.3.1 Geology and Tectonics

The Mahanadi basin is covered with alluvium that obscures its deeper geological features (Fig. 4.12). A series of northwest-southeast regional faults that occur at its western and northern boundary gave birth to grabens in which the Gondwana coal-bearing sediments were deposited between the upper Carboniferous and early Cretaceous period. These grabens were formed prior to rifting of the Indian plate from Antarctica and Australia. The Cuttack depression (Fig. 4.14) of the Mahanadi basin is one such graben that is believed to be contiguous with the Lambert rift in Antarctica. Upper Gondwana (early Cretaceous) rocks are exposed at the beginning of the delta. The coastal depressions of Puri-Konark and Paradip were formed later, at the time of pericratonic movements during continental separation, in the late Mesozoic to Paleogene times. Thus, there are two periods of graben formation: the Gondwana grabens during prerift and the coastal grabens after rifting of the continents (Jagannathan et al., 1983).

The first significant subsidence, onset of major sedimentation and marine transgression began in the late Cretaceous. A hinge line separating the continental shelf from the continental slope was established during the Paleocene and marine carbonates were deposited over the gently sloping shelf of upper Cretaceous period. During the late Oligocene-early Miocene (28–20 Ma), the basin experienced a major regression and uplift followed by hiatus and erosion. Transgression and subsidence took place in the middle Miocene and thick Mio-Pliocene clastic sediments were deposited over the Paleocene shelf sediments. Owing to the basin subsidence and a high rate of sedimentation the Mio-Pliocene sediments were affected by numerous growth faults and rollover structures. Subsequently Pleistocene glaciations brought extensive regression along the east coast. This was accompanied by subaerial weathering and laterites formation.

4.3.2 Seismic Results

Due to the hydrocarbon potential of the Mahanadi basin, all types of geophysical studies were carried out here. A tectonic map (Fig. 4.14), based on the results of extensive aeromagnetic, gravity and seismic studies related to hydrocarbon exploration, shows the faults, depressions and ridges within the basin and divides it into several depressions separated by shallow basement

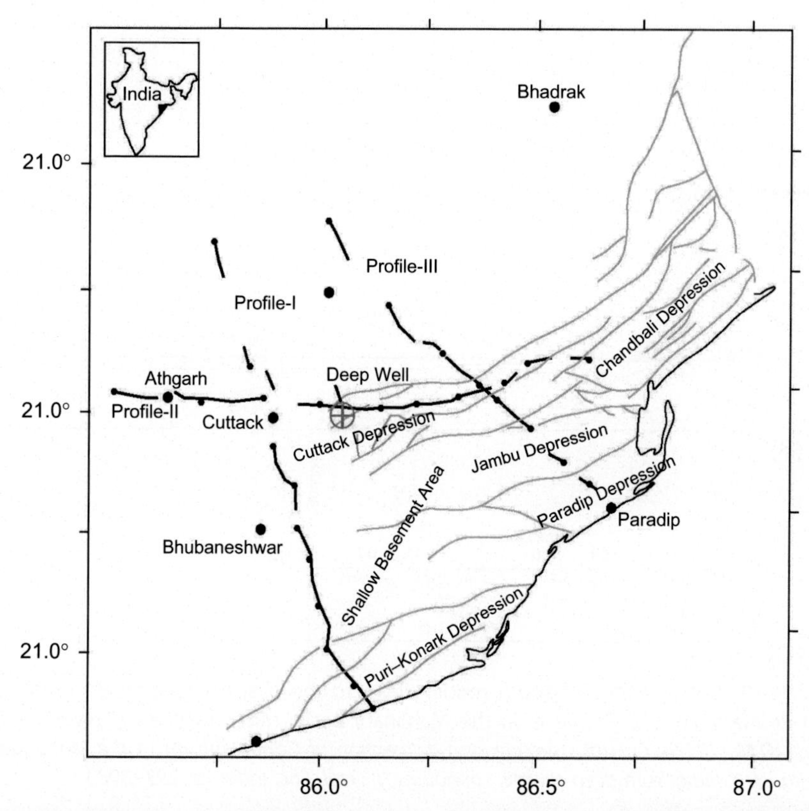

Fig. 4.14 Tectonic map of the Mahanadi basin based on the hydrocarbon exploration data, along with seismic profiles recorded for the basement and crustal studies. *(After Baishya, N.C., Singh, S.N., 1986. DSS profiling in Mahanadi on shore areas: a case study. In: Kaila, K.L., Tewari, H.C. (Eds.). Deep Seismic Soundings and Crustal Tectonics. A E.G., Hyderabad, pp. 1–9.)*

ridges (Baishya and Singh, 1986). This map shows that the coastal depressions of Puri-Konark and Paradip are separated from the Cuttack depression by a shallow basement ridge. However, the basement configuration, particularly in deeper parts of the basin, is not brought out clearly in hydrocarbon-related studies. To understand the evolution of the basin and decipher a broad basement configuration, long-range seismic refraction and postcritical reflection data in analog form were collected along three profiles (Fig. 4.13). Profiles I and III were recorded across the Cuttack depression while Profile II is along it. Profile I also cuts the Konark depression and Profile III the Paradip depression. Due to the short length of the profiles, coverage of the crust-mantle boundary (Moho) is very limited.

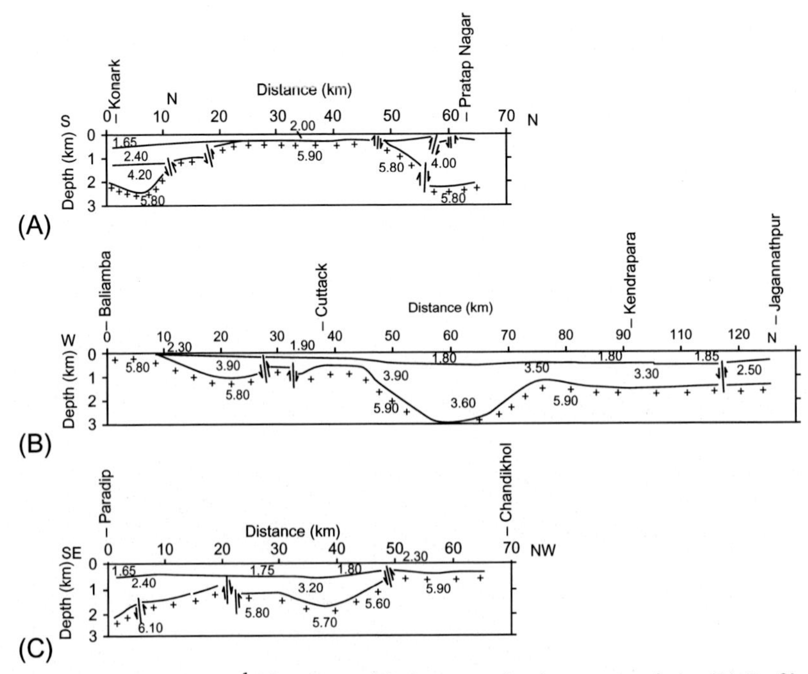

Fig. 4.15 Velocity (km s^{-1})-depth model down to the basement along (A) Profile I, (B) Profile II, and (C) Profile III in the Mahanadi basin. *(After Kaila, K.L., Tewari, H.C., Mall, D.M., 1987b. Crustal structure and delineation of Gondwana rift in the Mahanadi delta area, India from deep seismic soundings. J. Geol. Soc. India 29, 293–308.)*

Results along these profiles (Fig. 4.15) provide configuration of the Precambrian basement (Kaila et al., 1987). A four-layered model with velocities of 1.65, 2.40 and 4.20 km s^{-1} overlying a basement of velocity 5.80 km s^{-1} represents the sedimentary basin configuration in the Konark depression. In the Paradip depression the layer with velocity of 4.2 km s^{-1} is absent and velocity in the basement is 6.10 km s^{-1}. Maximum depth to the basement is about 2400 m in the Konark depression and 2100 m in the Paradip depression. Inversion of the first arrival refraction data (Behera et al., 2002) along Profile I shows more or less similar velocities, as determined earlier, for the sediments (1.75, 2.40 and 4.00 km s^{-1}) over a basement of velocity of 5.95 km s^{-1}, for the Konark depression. The Cuttack depression shows a three-layered velocity model with velocities of 1.80–1.90 and 3.30–3.90 km s^{-1} overlying a basement of velocity 5.9 km s^{-1}. Maximum depth to the basement in this depression is estimated as 2800 m.

However, subsequent to these studies a deep well was drilled, to a depth of 2993 m, in the Cuttack depression. This well did not reach the

Precambrian basement. It also showed several thin volcanic layers, two of which are significantly thick (about 100 and 200 m, respectively) and separated by Cretaceous sediments; the first is at a depth of about 500 m and the second at about 1800 m. Based on the drilling results the earlier model of this depression was modified (Fig. 4.16) to show the presence of two thin high-velocity volcanic layers, the top one of velocity 5.20 km s^{-1} and the deeper of velocity 4.80 km s^{-1}, within the low-velocity sediments of velocity 3.50 km s^{-1} (Tewari, 1998). This model shows a maximum sedimentary thickness of more than 3000 m and indicates that the presence of high-velocity volcanic layers within the low-velocity sediments did not allow a correct estimation of depth in the earlier analysis. While the deeper volcanic layer definitely belongs to the Rajmahal Traps (~117 Ma), doubts remain about the age of the shallower volcanic layer, which could even belong to the Deccan Trap event (~65 Ma). However, the model of Behera et al. (2004) shows that the shallow structure down to the basement is highly complex with numerous faults forming alternate ridges and depressions and predicts a 4–5 km thick sedimentary cover in the deepest part of the Cuttack depression. The thin lid of the high-velocity volcanic rock at the shallow depth underlain by a thick (1.75 km) pile of Gondwana sediments probably represents a sill-type structure.

The initial model for the crust of the Mahanadi basin (Kaila et al., 1987) showed four velocities below the basement of velocity 5.9 km s^{-1}. This model had a layer of high velocity of 6.45 km s^{-1} at an average depth of 6 km and velocities of 6.0, 7.1 and 8.1 km s^{-1} at 17.5, 20.5 and 34.5 km, respectively. The last is identified as the Moho boundary. No explanation

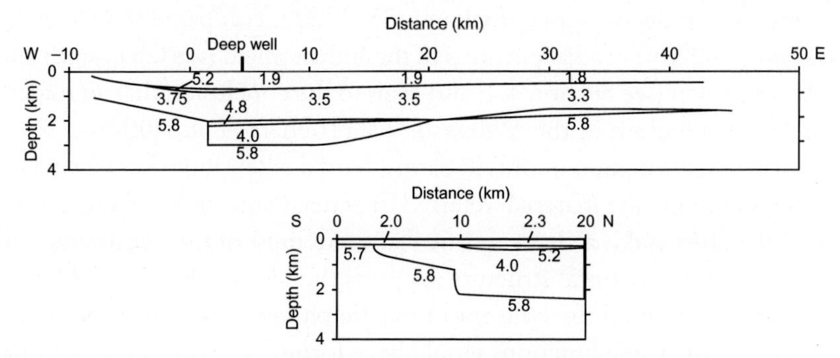

Fig. 4.16 Velocity (km s^{-1})-depth model for the Cuttack depression showing thin high-velocity layers (L1, L2) between the sedimentary layers. *(After Tewari, H.C., 1998. The effect of thin high velocity layers on seismic refraction data: an example from Mahanadi basin, India. PAGEOPH 151, 63–79.)*

is offered for the low-velocity layer (6.0 km s^{-1}) in the upper crust, but the layer with the velocity of 7.1 km s^{-1} is identified as an underplated layer at the base of the crust.

Reinterpretation of these data after digitization (Behera et al., 2002, 2004) modifies the earlier models of the Mahanadi basin and shows that the crust consists of four velocity layers below the basement. This model indicates that intense volcanism has affected the entire crustal column in the form of undulating intra crustal boundaries including the Moho. The lateral variation of velocity (5.9–6.1 km s^{-1}) and density (2.7–2.75 g cm^{-3}) of the basement rocks is due to a change in its composition. It inferred that the rocks of the Bhubaneswar ridge are crystalline (granite/gneiss) in nature, while those of the Chandikhol ridge mainly comprise khondalite rocks of eastern ghat orogeny (high-grade metamorphic).

The basement (velocity 5.9–6.1 km s^{-1}) is followed by velocities of 6.5, 6.0, 7.1 and 7.5 km s^{-1} at average depths of 5.5, 16, 20 and 27 km, respectively (Fig. 4.17). The crust is sitting over an upper mantle of velocity of 8.1 km s^{-1}, the Moho being at an average depth of 37 km. The high velocity (6.5 km s^{-1}) at a shallow crustal average depth of 5.5 km represents a magmatic body in the upper crust. The fluids released during the process of magmatism normally get trapped at some particular depth within the crust, and this may have formed the low-velocity (6.0 km s^{-1}) layer at the midcrustal level. This interpretation also presents elaborate crustal models for each of the three profiles (Fig. 4.16). In these models, it is assumed that the velocity of 7.1 km s^{-1} is a part of the lower crust while the velocity of 7.5 km s^{-1} represents the underplated layer at the base of the crust. The reduced crustal thickness (35–37 km) compared to its average thickness (~42 km) in adjacent areas of the Indian shield (see Chapter 2) and Godavari graben (see Section 4.4) also leads to the conclusion that an underplated crust is present in the Mahanadi delta (Behera et al., 2004).

During the formation of the rift basins, crustal extension takes place with the addition of mantle material, referred to as magmatic underplating, at the base of the crust and is a characteristic feature of most of the continental rift zones. This type of crustal structure supports the idea of Burke and Dewey (1973), who proposed the concept of mantle plume-generated triple junction. One of the triple junctions would have formed between greater India, Antarctica and Australia along the eastern margin of India. An ocean opened along two arms separating Antarctica and Australia from India and the third arm failed to open and formed the Mahanadi rift basin. The tectonic

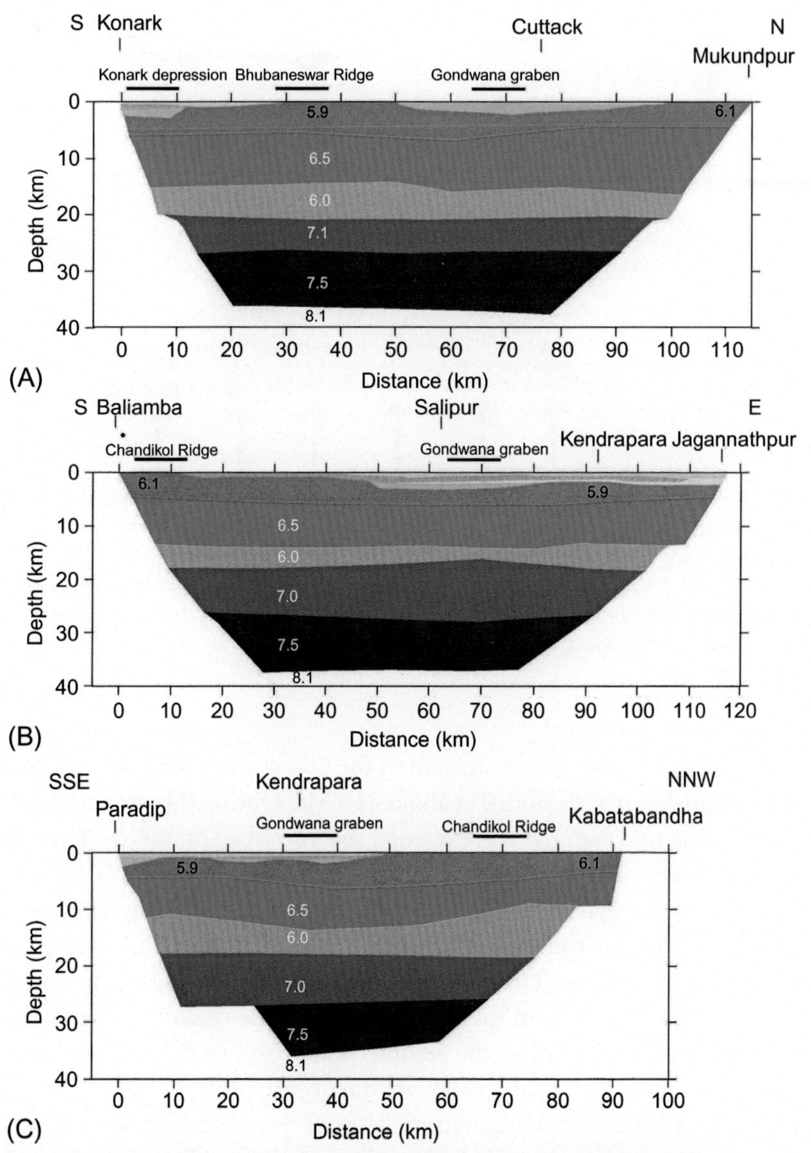

Fig. 4.17 Crustal velocity (km s^{-1})-depth models along (A) Profile I, (B) Profile II, and (C) Profile III of Fig. 4.13 in the Mahanadi basin. *(Modified from Behera, L., Sain, K., Reddy, P.R., 2004. Evidence of under plating from seismic and gravity studies in the Mahanadi delta and its tectonic significance. J. Geophys. Res. 109, B12311, 1–25.)*

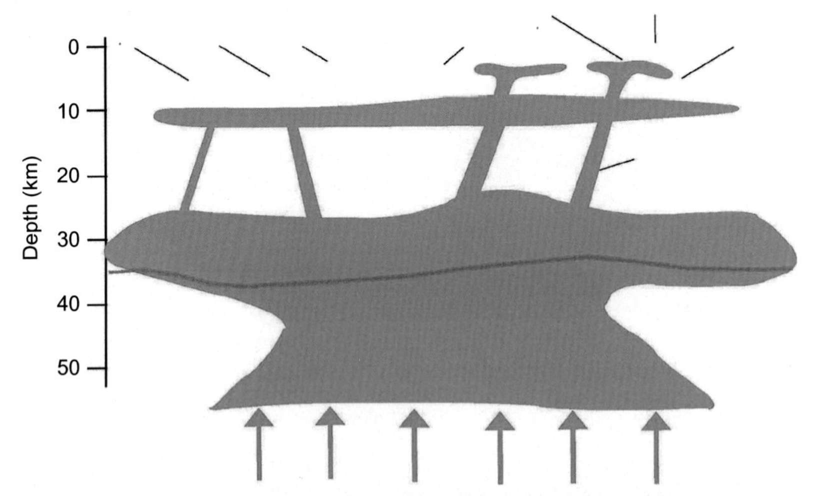

Fig. 4.18 Schematic evolutionary model of the Mahanadi basin showing the underplated crust, the magmatic body in the upper crust and volcanic extrusion within the sedimentary basin. The Moho is defined by the *thick black line. (From Behera, L., Sain, K., Reddy, P.R., 2004. Evidence of underplating from seismic and gravity studies in the Mahanadi delta of eastern India and its tectonic significance. J. Geophys. Res. Solid Earth 109, B12. https://doi.org/10.1029/2003JB002764.)*

activities for such an event are related to the breakup of Gondwanaland due to the Kerguelen mantle plume at about 117 Ma. During this process, a large quantity of mantle material was extruded and emplaced on the surface as the Rajmahal Traps. This evolutionary model is presented in Fig. 4.18.

The Bouguer gravity anomaly map of the Mahanadi basin (Fig. 4.19) shows large lows at the coastal depressions of Puri-Konark/Paradip and the Cuttack depression. These lows are separated by a high in the middle. A gravity high also exists in the northern part of the region. Interpretation of the gravity data, based on the seismic results, shows that, except in the northern part, the gravity anomalies are due to variation in the sedimentary thickness and basement configuration. Most of the northern anomaly, in an east-west direction, is due to the presence of about 3.5-km thickness of khondalite (density 2900 kg m^{-3}) below the surface (Tewari et al., 1988). Later interpretations (Fig. 4.20) have taken into account the density values in the entire crust. Due to its higher velocity, the density of the underplated layer in the lower crust is taken as 3050 kg m^{-3}. Combined interpretation of the seismic and gravity data (Behera et al., 2004) shows that high-density (2800 kg m^{-3}) intrusive mafic material that was transported to shallower

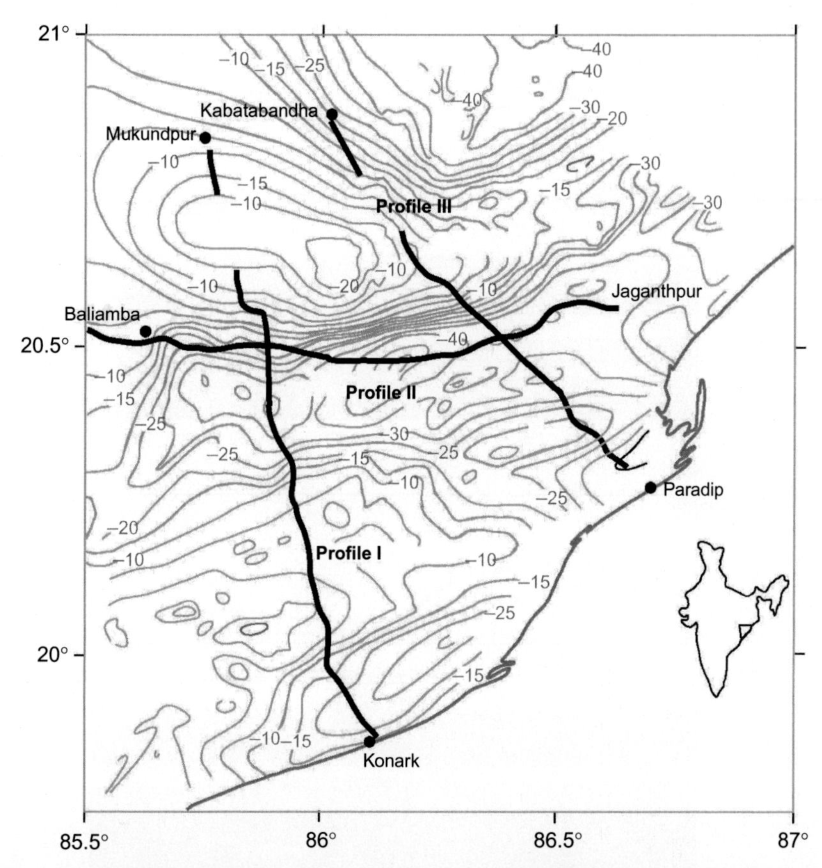

Fig. 4.19 Bouguer gravity anomaly map of the Mahanadi basin. *(Modified after Tewari, H.C., 1998. The effect of thin high velocity layers on seismic refraction data: an example from Mahanadi basin, India. PAGEOPH 151, 63–79.)*

depths because of extensional force of the upwelling Kerguelen plume resulted in an anomalously high velocity of 6.5 km s^{-1} in the upper crust.

The heat flux values in the Mahanadi basin (77 mW m^{-2}) are higher as compared to those in the stable shield and other cratonic regions (30 mW m^{-2}) of India. These indicate that presence of the underplated layer, with a density of 3050 kg m^{-3} (Fig. 4.20), is due to emplacement of the mafic material in this rift basin.

Petrophysical and geochemical similarities between the lower crustal rocks exposed along the eastern ghat hills and those over part of east Antarctica indicate the juxtaposition of the Mahanadi basin of India and the Lambert graben in Antarctica before rifting and breakup. The thermal

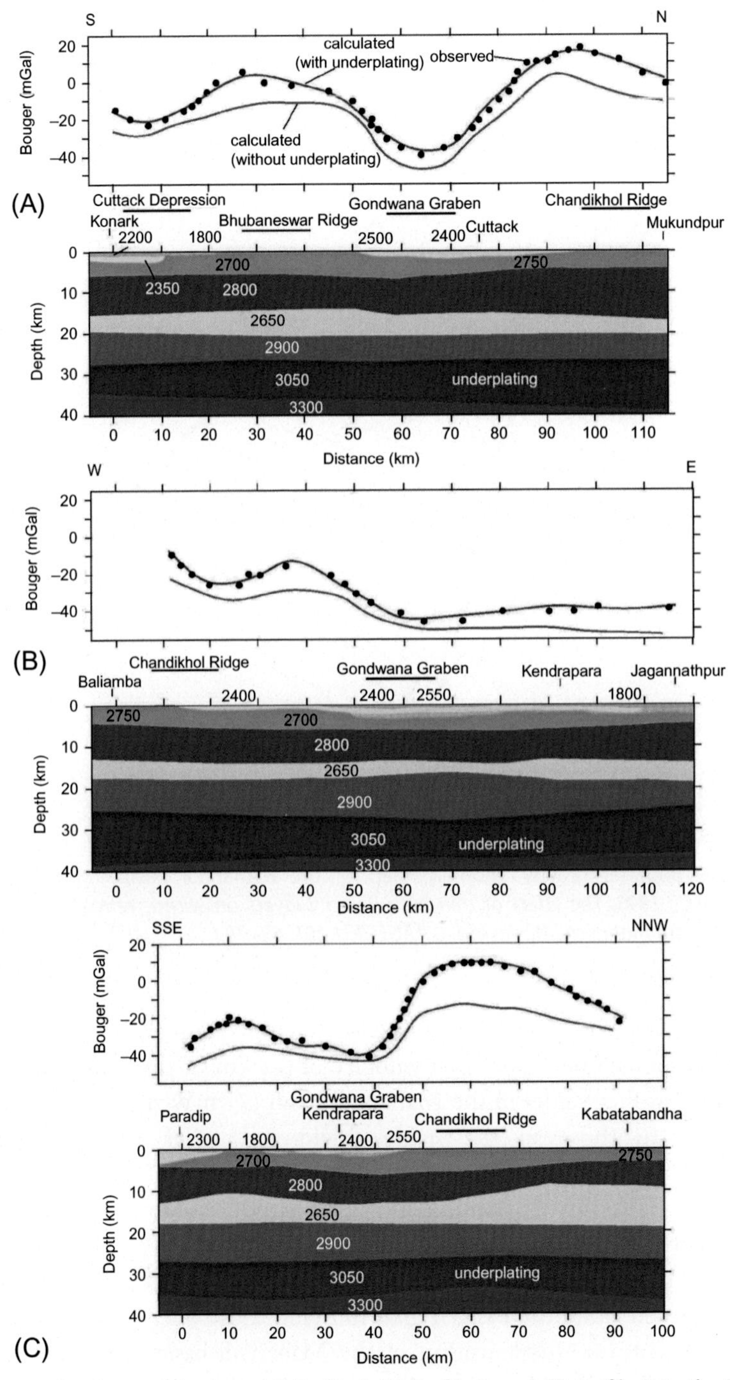

Fig. 4.20 Gravity models along (A) Profile I, (B) Profile II, and (C) Profile III in the Mahanadi basin. Densities are in kilograms per cubic meter. *(Modified from Behera, L., Sain, K., Reddy, P.R., 2004. Evidence of under plating from seismic and gravity studies in the Mahanadi delta and its tectonic significance. J. Geophys. Res. 109, B12311, 1–25.)*

source required for the mafic underplated material may be related to the Kerguelen plume, which was located for a long time under the India-Antarctica region and erupted during the early Cretaceous. This resulted in the breakup of greater India from Antarctica followed by a large-scale magmatism closely correlated with the outburst of Rajmahal volcanism in the eastern part of India and the magmatic activities along the continental shelf of east Antarctica. Because of the extensional force, caused by upwelling of the plume head, part of the high-velocity and high-density mafic material might have been transported to the shallow depth through weak zones/conduits and formed as intrusive (Behera et al., 2004).

4.4 THE GODAVARI RIFT BASIN

The Godavari rift basin (Fig. 4.21) is an intracontinental rift, which is preserved within the older geological sequence of the Indian peninsular

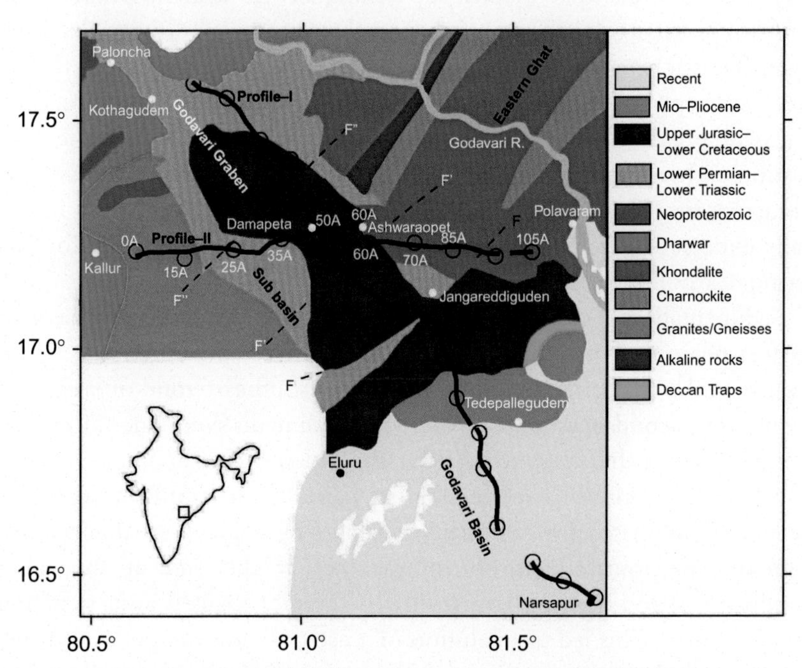

Fig. 4.21 Geological map of the Godavari basin with seismic profiles plotted on it. Profile I is along the Chinthalpudi (SP 7–17) and Godavari Coastal (SP 0–6) subbasins and Profile II is across the Chinthalpudi subbasin. *(After Rao, G.S.P., Tewari, H.C., Rao, V.K., 2000. Velocity structure in parts of the Gondwana Godavari Graben. J. Geol. Soc. India 56, 373–384.)*

shield. This northwest-southeast trending linear feature has a width varying between 50 and 100 km and extends as a graben from close to the east coast of India to about 400 km inside the Indian shield. Most of the sediments in this basin consist of the Proterozoic and Gondwana (Permian-Jurassic) rocks. It has developed along a zone of weakness inherited from the Archean orogeny and is bounded by normal faults. The Archean complex consisting of the Dharwar (~2800 Ma) and eastern ghat (1625 ± 75 Ma) suite of rocks forms its basement, which is in fault contact with the younger sediments along the basin margins.

4.4.1 Geology and Tectonics

The Godavari rift is considered to be a paleo-rift but does not have the characteristic properties of either an old or a new rift. Opinions about the age of this rift vary. Some consider it Paleoproterozoic (Datta et al., 1983) while others think that diapiric upwelling of the asthenosphere is the main cause for its development during the Gondwana (lower Permian-lower Triassic) period (Mishra et al., 1987). It neither fits with the rift valley concept nor is compatible with the idea that it is a faulted remnant of a widespread rift valley (Chatterjee and Ghosh, 1970). Tectonic reactivation and vertical movements during the Paleozoic are associated with this active rift (Jagannathan et al., 1983). Volcanic activity, which is normally expected during rifting, is not seen here either during or before the Permo-Carboniferous period.

Sedimentation in this rift graben took place in four successive phases. In the first phase the Paleoproterozoic sediments were deposited as a consequence of the rift activity along a northwest-southeast zone of weakness. During the second phase, Neoproterozoic sediments were unconformably deposited along the same trend over the Paleoproterozoic and Archean rocks. Lower Gondwana sedimentation started in the third phase. Before the end of this phase a few subbasins, separated by ridges, were differentially uplifted. The fourth phase of rift activity at the end of the lower Gondwana period (lower Permian-lower Triassic) resulted in its structural reorganization. This led to evolution of a narrow, linear upper Gondwana basin, confined to the central part of the graben. Consequent to the upper Gondwana deposition, upper Cretaceous sedimentation took place in the northwestern part of the graben. This was followed by peneplanation and extrusion of the Deccan Trap volcanic flows at the end of the Cretaceous (Sastri, 1995).

The Godavari rift basin is divided into four subbasins. The first three are part of the main Godavari graben where the Gondwana group of rocks is deposited over the Archean gneiss basement. Earlier it was believed that only lower Gondwana rocks are exposed in this basin but its stratigraphy was later revised to show the existence of upper Gondwana rocks in large parts of the Chinthalpudi subbasin (Laxminarayana, 1995). This subbasin has the true character of a graben, as both its limbs are defined by faults. In one of the deep wells, drilled to 2955-m depth within this subbasin, the Archean basement of the eastern ghat was met at 2935 m depth (Agrawal, 1995).

The southwest-northeast trending coastal subbasin is covered by alluvium and is oriented transverse to the main Godavari graben. It is further divided into several depressions by the basement ridges of Bapatala, Kaza and Tanuku (Fig. 4.22). While the Tanuku and Kaza ridges separate

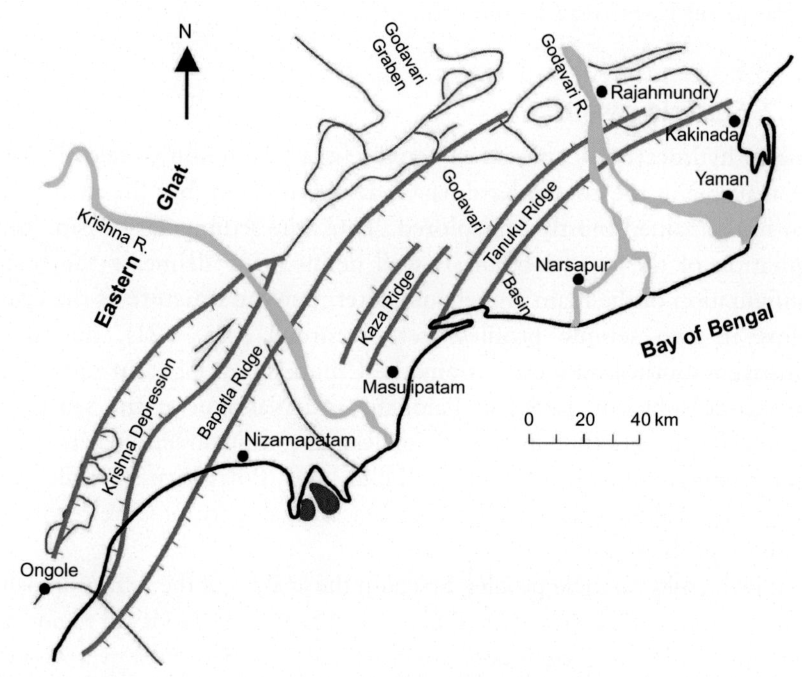

Fig. 4.22 Tectonic map of the coastal Krishna-Godavari subbasin showing the ridges and depressions. *(After Rao, G.S.P., Tewari, H.C., Rao, V.K., 2000. Velocity structure in parts of the Gondwana Godavari Graben. J. Geol. Soc. India 56, 373–384.)*

different parts of the coastal subbasin, the Bapatala ridge separates the coastal subbasin from the main Godavari graben. The Mio-Pliocene sandstone, with thickness varying between 200 and 400 m, is exposed to the south of this ridge. A narrow belt of the Deccan Trap is also exposed close to the contact of the coastal subbasin with the main Godavari graben. Deep wells drilled for hydrocarbon exploration show that the trap thickness varies between 10 and 375 m (Mohinuddin et al., 1993). Rocks of the eastern ghat form the basement of this subbasin. Rb-Sr dating of these basement rocks yielded an isochron age of 854 ± 20 Ma and a whole rock sample of biotite from a deep well yielded an isochron age of 395 ± 6 Ma. These ages reflect the intrusion time of granitic magma and greenschist metamorphism into the country rock and are related to the tectono-thermal and folding events, and indicate that the eastern ghat was deformed during the 395 ± 6 Ma event (Rao, 2002). This event is likely to have provided the thermo-tectonic perturbation leading to creation of a depression and deposition of subsequent Gondwana sediments in the Godavari rift basin (Rao Rammohan et al., 1999).

4.4.2 Seismic results

Due to hydrocarbon exploration activity, a large amount of seismic data are available in the coastal Krishna-Godavari subbasin, but the main graben has remained mostly unexplored. Thus, the sedimentary basin configuration of the coastal basins is well defined. To delineate the basin configuration of the main graben and determine the structure of the crust below it, two seismic profiles were recorded (Fig. 4.21), one in a northwest-southeast direction along the Chinthalpudi subbasin and across the coastal subbasin, between Paloncha and Narsapur at the sea coast (Profile I), and another in a more or less east-west direction across the Chinthalpudi subbasin, between Kallur and Polavaram (Profile II). Refraction and postcritical reflection data, in analog form, were recorded on these profiles.

Initial results on these profiles, based on the analysis of the refraction data, identified the outcrop velocities (Kaila et al., 1990) in the Godavari region as: alluvium, 1.8 km s^{-1}; sandstone, 3.8–4.0 km s^{-1}; upper and lower Gondwana, 2.3–2.6 km s^{-1}; another layer in lower Gondwana, 3.5 km s^{-1}; granitic gneiss, 4.3–5.5 km s^{-1} and khondalite, 4.8–5.6 km s^{-1}. Refraction data obtained by the hydrocarbon industry gives the velocity of a formation within the lower Gondwana as 4.0 km s^{-1} and the Proterozoic formations as 4.6 km s^{-1}. Earlier results of the refraction studies were modified after reprocessing of the first arrival refraction data of Profile I in the region of the coastal subbasin (Tewari et al., 1996), between the upper Gondwana exposures at the Bapatala ridge and the sea coast at Narsapur, and show a three-layered configuration, with velocities of 2.8 km s^{-1} (upper and lower Gondwana), 4.2 km s^{-1} (lower Gondwana) and 5.4 km s^{-1}, respectively, at the upper Gondwana exposures. The velocity-depth section to the south of the Gondwana exposures shows about 1000 m thick alluvium layer of velocity 1.8 km s^{-1} followed by a 200–400 m thick layer of velocity 4.0 km s^{-1} (sandstone/Deccan Traps), 1000–1200 m thick layer of velocity 2.8 km s^{-1} (Gondwana sediments), and 200–2400 m layer of velocity of 4.2 km s^{-1} (Gondwana/older sediments) over a basement with velocity of 5.4 km s^{-1}.

Since the velocity in the basement (5.4 km s^{-1}) is somewhat lower than that normally found in the Archean rocks, it is possible that the basement of the Gondwana group may either be the Proterozoic or eastern ghat suite of rocks. This section shows that the sedimentary thickness at the Bapatala and Tanuku ridges is less than at other places. Presence of the Gondwana sediments in this region indicates that the Godavari graben continues under the coastal subbasin at least up to the seacoast. A correlation of seismic section with the deep well data suggests that the Deccan Trap volcanic rocks in this region extruded through a volcanic vent or a zone of basement faults along the present east coast (Reddy et al., 2002). Volcanic rocks equivalent to the Rajmahal Traps were also encountered in a well close to the Tanuku ridge within the coastal subbasin (Biswas, 1996).

The velocity-depth model based on refraction data along Profile I in the Chinthalpudi subbasin (Fig. 4.23) shows a five-layered model with velocities of 2.3–2.4, 3.2, 4.0, 5.5–5.7 and 5.9–6.1 km s^{-1}, respectively, for this subbasin (Rao et al., 2000). The first three of these velocities belong to the upper and lower Gondwana sediments. The 3.2 km s^{-1} velocity layer is present only in a limited part of the Profile I, to the immediate north of the Bapatala ridge, where the upper Gondwana sediments are exposed. The velocity of 5.5–5.7 km s^{-1} represents the basement of the Gondwana sediments. Largest thickness of these sediments along this profile is about 2600 m. Profile II extends beyond the limits of the Godavari graben. The velocity-depth configuration within the graben on this profile is almost similar to that along Profile I. The maximum sedimentary thickness is

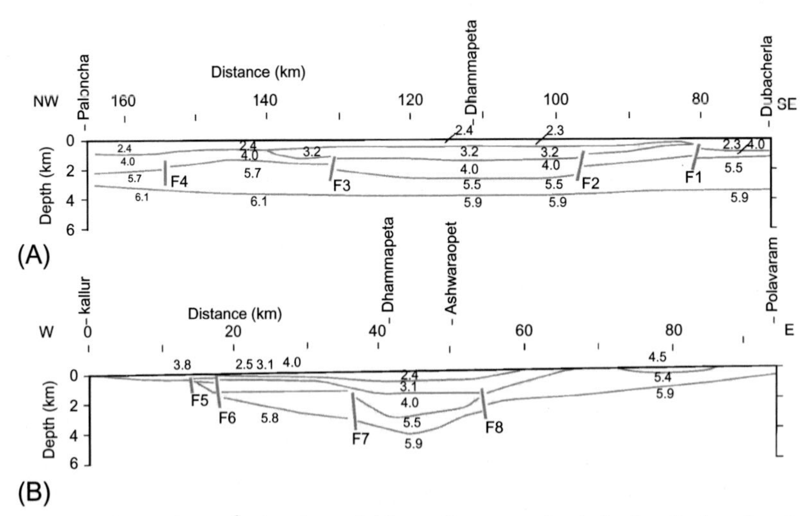

Fig. 4.23 Velocity (km s^{-1})-depth model for sedimentary basin in the Godavari graben along (A) Profile I and (B) Profile II. *(After Rao, G.S.P., Tewari, H.C., Rao, V.K., 2000. Velocity structure in parts of the Gondwana Godavari Graben. J. Geol. Soc. India 56, 373–384.)*

about 2800 m. A comparison between the drilling results of a deep well (Ashwaraopet, at the NE fringe of the graben) and the seismic section reveals that the velocities of 2.4, 3.1 and 4.0 km s^{-1} represent various formations within the lower Gondwana sequence. In this well, the basement to the Gondwana sediments was identified as a metamorphic suite of the eastern ghat. To the east of the Godavari graben, exposed eastern ghat rocks show a velocity of 5.4 km s^{-1}. Though this velocity is similar to the velocity of Proterozoic sediments (5.2–5.6 km s^{-1}) exposed in the Indian Peninsular shield, the geologic evidence does not indicate the presence of these sediments in the Chinthalpudi subbasin. Therefore, the velocity of 5.5–5.7 km s^{-1} in the graben is likely to belong to the eastern ghat group of rocks. These rocks seem to be present till the western margin of the basin, their thickness diminishing to the northeast.

A velocity of 5.8 km s^{-1} is recorded under a weathered layer of 3.8 km s^{-1} on the exposed Archean granites on the western flank of Profile II. The velocity of 5.8–6.1 km s^{-1}, seen in the refraction sections on both the profiles, represents the gneissic basement below the eastern ghat rocks. The depth to this layer is about 4 km in the deepest part of the basin along Profile II and varies between 3 and 4 km along Profile I. While major imprints of the eastern ghat orogeny are visible on Profile II, in the form of faults (F_5 and F_8) on both sides of the Chinthalpudi subbasin, no imprints are seen on Profile I as it runs more or less through the deepest part of the subbasin.

The crust in the Godavari graben has a three-layered velocity-depth configuration (Fig. 4.24) along both the profiles (Kaila et al., 1990). The velocity in the upper crust varies between 6.2 and 6.4 km s^{-1}. A broad dome-shaped feature is seen across the basin on Profile II, the top of the dome being at a depth of about 3.5 km. This feature is, however, not evident in the middle and lower crust. A change in velocity at a depth of 16–20 km indicates transition from the upper to the lower crust. The lower crust consists of two velocity layers, 6.6–6.7 and 6.8–6.9 km s^{-1}, where the latter could represent an underplated material. The commonly found velocity of greater than 7.1 km s^{-1}, which represents the underplated material from the upper mantle in all the rifts, is not seen here. Depth to the Moho boundary varies between 41 and 43 km. At this boundary the velocity increases from 6.9 km s^{-1} in the lower crust to 8.1 km s^{-1} in the upper mantle. Another estimate of the upper mantle velocity, calculated on the basis of the known relationship between heat flow and upper mantle velocity (P_n), has a value of 7.97 km s^{-1} for the main graben and 7.70 km s^{-1} for the coastal basin (Rao, 2002). Thickness of the thermal lithosphere for the two regions, for a Moho temperature of 800°C, is 653 and 503 km, respectively. The lower values of the thermal lithosphere and P_n in the coastal basin, as compared to those in the main graben, indicate its reactivation in the Cainozoic. The seismic results hint that the rift activity did not originate during the Proterozoic and the dome-shaped up warp is an expression of northeast-southwest extension of the lithosphere during the Carboniferous.

The Bouguer gravity anomaly map of the region (Fig. 4.25) shows a large gravity low that represents the Godavari graben and elongated highs on both its shoulders. The central low is attributed mainly to a thick section of the Gondwana sediments (Mishra, 2002). The gravity model (Rao, 2002), based on the seismic data (Fig. 4.26), shows that the gneissic basement (velocity 5.8–6.1 km s^{-1}) has a high density of 2900 kg m^{-3}. The density decreases to 2850 kg m^{-3} in the upper crust and is 2880 kg m^{-3} in the middle crust. The velocity of 6.9 km s^{-1} is considered as an underplated layer, despite the fact that the velocities in such

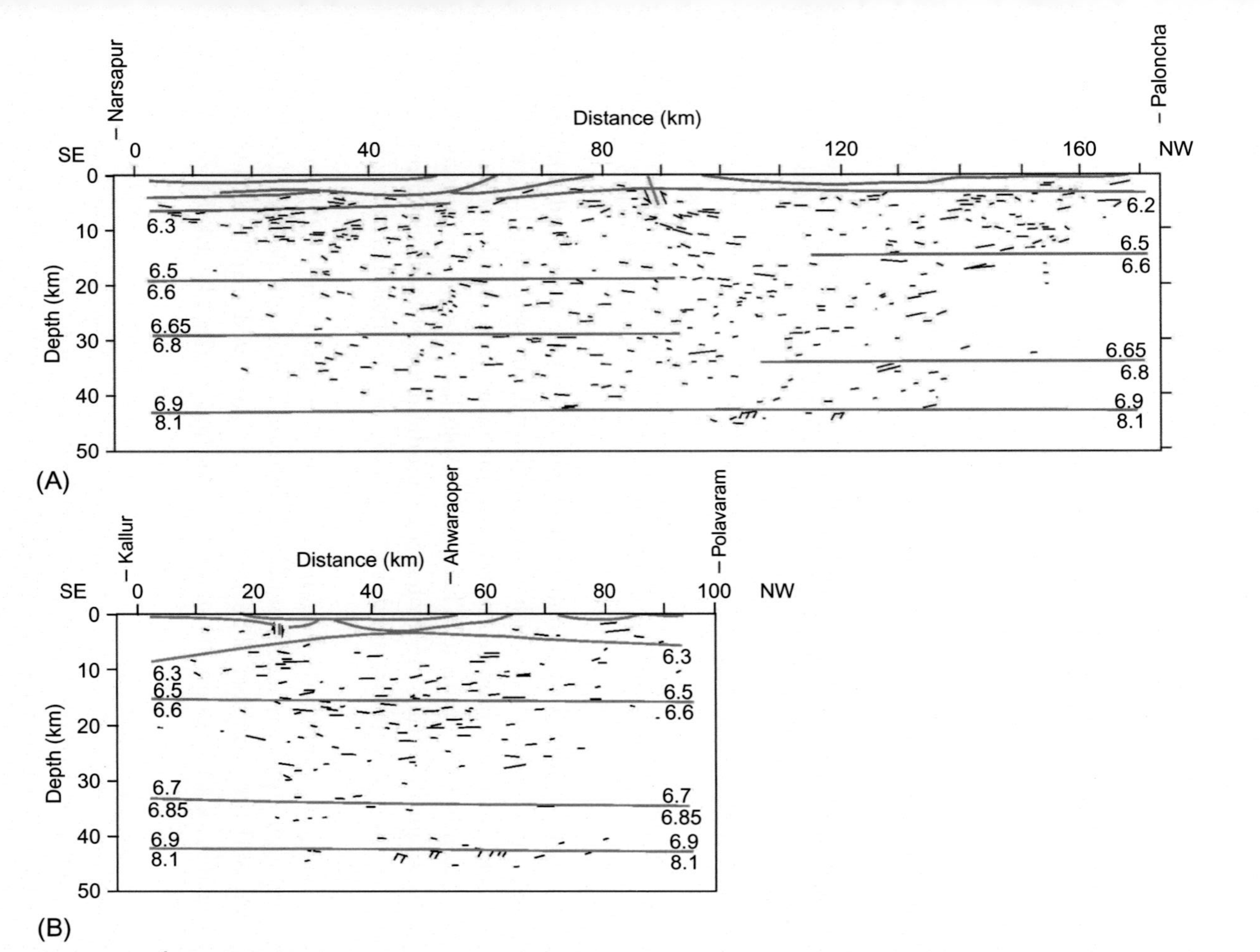

Fig. 4.24 Velocity (km s^{-1})-depth model for crust in the Godavari basin along (A) Profile I and (B) Profile II. *(After Kaila, K.L., Murthy, P.R.K., Rao, V.K., Venkateswarlu, N., 1990. Deep seismic sounding in the Godavari Graben and Godavari (costal) basins, India. Tectonophysics 173, 307–317.)*

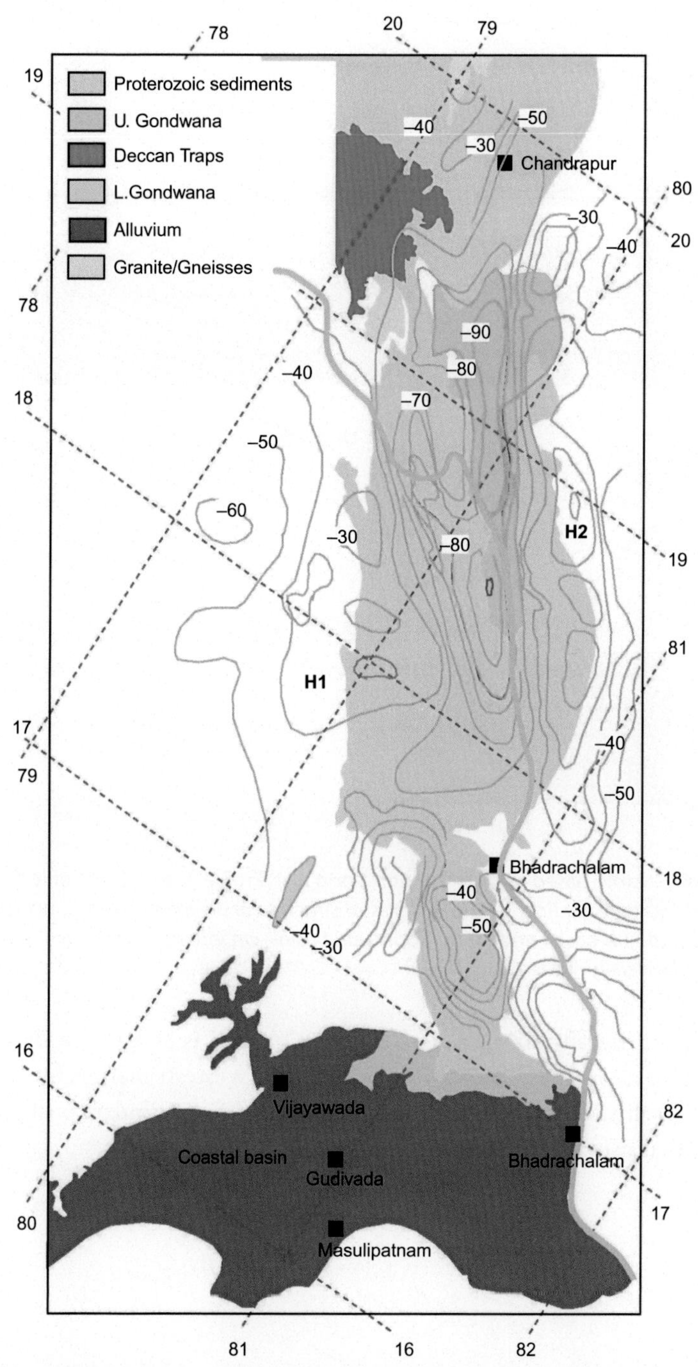

Fig. 4.25 Bouguer gravity anomaly map of the Godavari basin. *(Modified from Mishra, D.C., 2002. Crustal structure of India and its environs based on geophysical studies. Indian Miner. 56, 27–96.)*

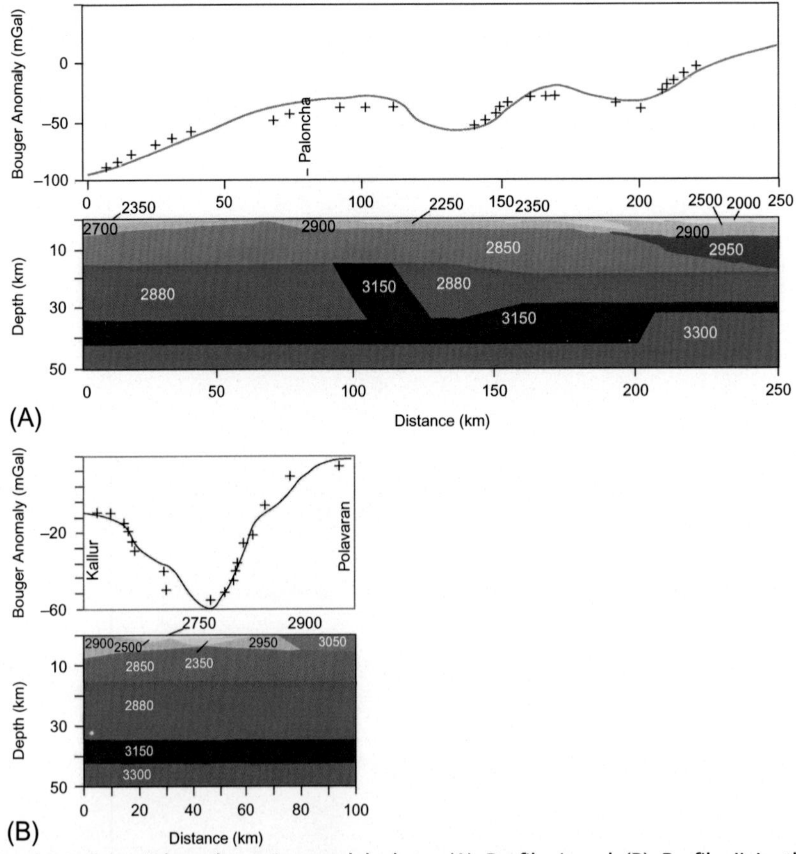

Fig. 4.26 Seismic-based gravity model along (A) Profile I and (B) Profile II in the Godavari basin. *(After Rao, V.K., 2002. Crustal structure and evolution of Godavari graben (Chinthalpudi sub-basin) and Krishna Godavari basin—an integral approach. Ph.D. Thesis, Osmania University, Hyderabad, India.)*

layers are generally in excess of 7.1 km s^{-1}, and is assigned a density of 3150 kg m^{-3}. Extension of this model to the western margin of the eastern ghat belt shows a thrust fault in the ghat region, similar to that in the Indian Peninsular shield (see Chapter 2).

CHAPTER FIVE

Deccan Volcanic Province Near the West Coast

5.1 THE WEST COAST OF INDIA

The west coast of India was tectonically very active during its evolution. Three major Precambrian orogenic trends, namely the north-northwest to south-southeast Dharwar in the southern part, the northeast-southwest Aravalli-Delhi in the northeastern part, and east-northeast to west-southwest Satpura in the central part (Fig. 5.1), converge in this region (Biswas, 1987). The Cambay rift, which is the only rift graben in the western part of India, represents the north-northwest extension of the Dharwar trend. The Aravalli-Delhi trend divides itself into three branches; the Delhi trend takes a westward swing to overlap with the Kutch Rift and the Aravalli trend crosses the Cambay rift to enter the Saurashtra plateau. On its southern side, the Aravalli folding takes an acute eastward swing to merge with the Satpura trend (Biswas and Deshpande, 1983). After breakup of Gondwanaland, development of the structural trends, rift basins and different kinds of igneous intrusions over the west coast of India and adjoining regions were also affected by the major tectonic events during the Mesozoic period. The first of these was the breakup of Africa from the block consisting of India, Madagascar and Seychelles during the middle to late Jurassic period. The second event was the breakup of Madagascar along the west coast of India, during the middle to late Cretaceous, and the last event was the breakup of the Seychelles plateau from the Indian plate, followed by eruption of the Deccan Trap volcanics (Besse and Courtillot, 1988). The second event was probably under the influence of the Marion plume and the last due to interaction between the Reunion plume and the overlying lithospheric plate, during the late Cretaceous (Raval and Veeraswamy, 2003).

On the west coast of India seismic studies for understanding the velocity-depth configuration of the crust were undertaken in three regions: the Koyna region, the Cambay basin and the Saurashtra peninsula. Out of these

Structure and Tectonics of the Indian Continental Crust and Its Adjoining Region
https://doi.org/10.1016/B978-0-12-813685-0.00005-4

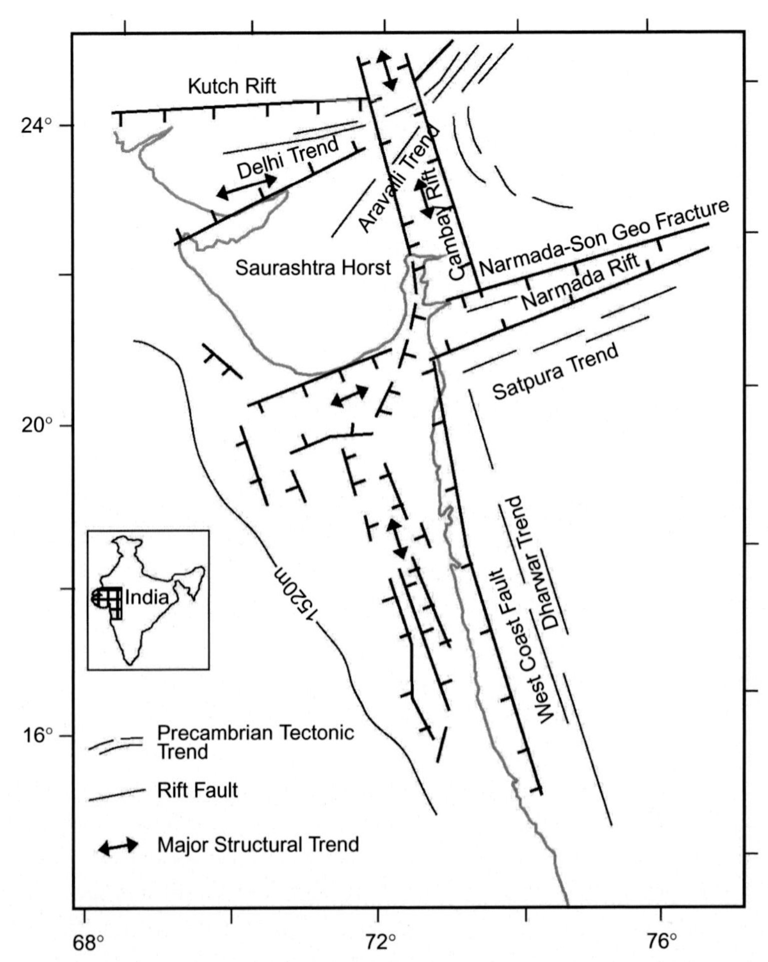

Fig. 5.1 Major Precambrian trends in western India. *(After Biswas, S.K., 1987. Regional tectonic framework, structure and evolution of western marginal basins of India. Tectonophysics, 135, 307–327.)*

three, the Cambay basin has a large cover of the Tertiary and Quaternary sediments overlying the Deccan Trap volcanics, while in most parts of the other two regions these volcanics are exposed. Before proceeding to describe the seismic studies in these areas, it is worthwhile to know a few things about the Deccan Trap volcanics.

5.2 THE DECCAN TRAP

Basaltic lava flows, known as the Deccan Trap, cover about one million cubic kilometers of the western Indian platform. These consist of several flows of mainly tholeiitic lava that extruded from the Reunion mantle plume during passage of the Indian plate over this plume in Cretaceous-early Tertiary time (Morgan, 1981). Thickness of some of the individual volcanic flows is over 40 m. A few of the flows were traced over distances exceeding 100 km. The western Indian volcanic province can be broadly divided into two geomorphologic divisions, namely: (1) the coastal plain, and (2) the plateau region consisting of the western ghat and areas to its further east. A great scarp runs parallel to the west coast in the Deccan Trap covered region. Some workers consider that the scarp is due to a boundary fault downthrown on the western side, while others feel that the edge of this scarp, as also that of the Archean rocks further south, is an erosion feature totally lacking the straight fault controlled edges. There is some evidence of post-trap faulting but no major tilt has been detected that would have altered the inclination of the interface between the volcanics and underlying formations (Auden, 1975).

At about 65 Ma, crustal extension and continental rifting resulted in extrusion of the Deccan Trap volcanics in western India. At that time, the Reunion plume was situated centrally beneath the volcanic outpouring near the west coast of India. Massive volcanic outpouring generated the volcanic lava that overflowed the then existing topography, in places filling the deep valleys. The age of this volcanic extrusion coincides with the Cretaceous-Tertiary boundary (Subbarao, 1988). The extrusion took place during a very short span of geological history, and has been very quick and massive. Various age spans given for the extrusion are: 69–64 Ma (Venkatesan et al., 1986), 68.5–66.5 Ma based on 40Ar/39Ar dates on the western ghat region (Duncan and Pyle, 1988), 65.5 ± 2.5 Ma in central India (Vandamme et al., 1991), 65.6 ± 0.3 Ma based on 187Re/187Os dating for the same region (Allegre et al., 1999). Duration of the bulk extrusion is estimated as 0.5 Ma with the main age at 66 Ma (Courtillot et al., 1988), 67.7–64.4 Ma for alkali volcanics and 66.8 Ma for tholeiitic volcanics (Pande et al., 1988). The 40Ar/39Ar ages determined for biotites from two igneous complexes to the north of the Deccan Volcanic Province is 68.53 ± 0.16 Ma and 68.57 ± 0.08 Ma, respectively, and the one, which intrudes the flood volcanics, is 64.96 ± 0.11 Ma. These indicate that the previous samples represent early and late magmatism with respect to the main pulse of the continental flood volcanism at about 65 Ma. The earlier magmatism shows that about 3.5 million years was the incubation period of a primitive, high -3He mantle plume before the rapid extrusion of the Deccan Trap volcanics (Basu et al., 1993).

Opinions about the reason of outpouring of the Deccan Trap volcanics also vary. While one opinion is that intense volcanism at the time of continental rifting and concurrent igneous activity on the Seychelles and offshore India was a result of rifting above a region of anomalously hot mantle around the Reunion plume (Hooper, 1990), others think that mid-continental rifting associated with the volcanics is the consequence of a mantle plume initiation and not its cause (Richards and Duncan, 1989; White and McKenzie, 1989). Melting in the plume head is considered responsible for the sudden onset and short duration of the continental volcanism. Subsequent decline of the volcanic activity, to a narrow chain of approximately 200-km width, is attributed to melting in the plume tail (Campbell and Griffiths, 1990). Another opinion is that lithosphere pull-apart at the craton boundaries, rather than stretching of the lithosphere, has triggered the extensive continental volcanism (Anderson, 1994).

5.3 THE KOYNA REGION

5.3.1 Geology and Tectonics

The Koyna region (Fig. 5.2), close to the west coast of India, is covered by the Deccan Trap volcanics except in the coastal plains. This region is

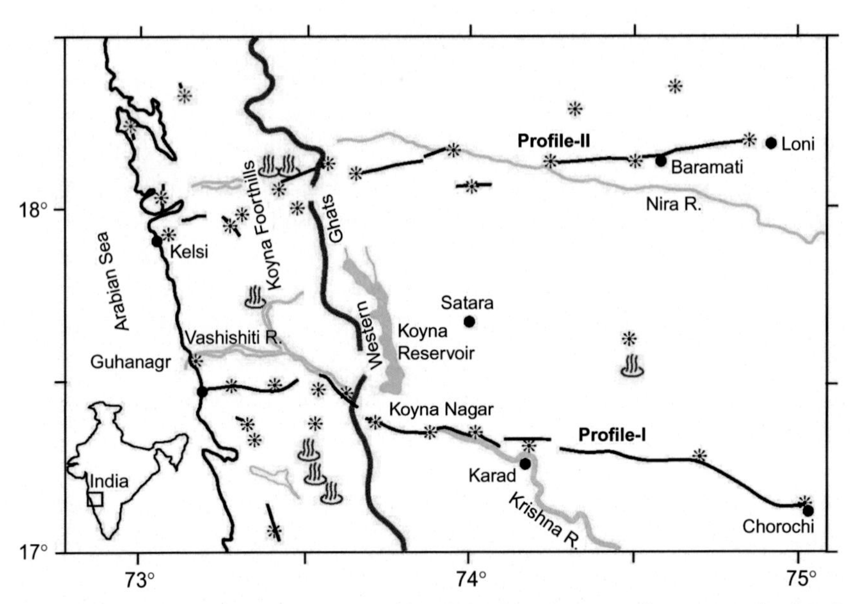

Fig. 5.2 Koyna I and Koyna II seismic profiles plotted on the map of western India. *(Modified from Sharma, S.R., Mall, D.M., 1998. Geothermal and seismic evidence for the fluids in the crust beneath Koyna, India. Curr. Sci. 75 (10), 1070–1074.)*

characterized by a flat top and steplike terraces. The volcanic formations are horizontally disposed with occasional buckling and flexures. In a drill hole in the eastern part of the volcanic covered region, the volcanic formation was drilled down to 338 m and eight flows of tholeiitic volcanics were identified (Gupta and Dwivedy, 1996). In the high plateau, the flows exhibit a very gentle easterly dip. In the low country, the flows show a westerly dip of 30–40 or more. Although dikes are abundant to the north of latitude 180 N, they are very rare in the Koyna region. Hot springs occur along the foothills over a distance of 320 km between latitudes 16040′ and 19035′ N. Three groups of hot springs occur further north. All of them are associated with a south-facing fault scarp of the Deccan volcanics. These hot springs tend to form an en-echelon pattern and are aligned along fractures that are running in a northwest-southeast direction within the volcanics. The hot springs do not occur further south in the exposed Archean basement. Since the springs lie along a fault zone within the basement, the western portion of the Deccan Trap covered area probably has hidden faults and fracture zones.

In December 1967, an earthquake of magnitude 6.3 devastated the Koyna region. It was the first major earthquake recorded in the Indian shield and broke the myth that it is a seismically stable region. Since then a large number of earthquakes, of small and medium magnitude, have been recorded in this region. Seismologists suggest that these earthquakes are related to the west coast fault (Fig. 5.1) running in the north-northwest to south-southeast direction. The focal depth of most of the earthquakes is less than 8 km; however, for some of these it is as much as 30 km. While most of the seismologists agree that these earthquakes are due to induced seismicity from the Koyna reservoir system, some consider them to be of tectonic origin. The loading pattern of the reservoir system appears to have greatly affected the seismicity of the region.

5.3.2 Seismic Results

The seismological data have delineated two parallel epicentral trends in the north northeast-south southwest direction, passing through the two reservoirs (Koyna and Warna) in this region. It is observed that a precursory nucleation process precedes the larger shocks (Gupta et al., 1997). In the presence of a vertical conducting fault a water level change on the order of 1 m in 5 days in the surface loading can propagate 5%–15% of a pore pressure front, corresponding to a pressure of 0.75–2.25 bars, to the hypocentral depth of 6–8 km. These small stress perturbations are sufficient to trigger seismicity on the preexisting, critically stressed faults in the region (Pandey and Chadha, 2003). Continued seismicity at this seismic zone and its triggering by the reservoirs has been explained in terms of southward migration of seismicity from the Koyna reservoir, high filling rate, duration of loading and nucleation process of the moderate earthquakes (Mandal et al., 2000).

Based on data from a large number of earthquakes, the tectonic processes suggested for the seismicity of this region are: (a) back-thrusting of Himalayan collision, (b) northward movement of the Indian shield due to expansion of the Indian Ocean ridges, and (c) the effect of the geothermal regime along the west coast and Narmada-Son belt (Rastogi, 1992).

Velocity configurations based on the earthquake data show (Dube et al., 1973) that the upper crust in the region is 20-km thick and has P- and S-wave velocities of 5.78 and 3.42 km s^{-1}; the lower crust is also about 20 km thick and has P- and S-wave velocities of 6.58 and 3.92 km s^{-1}, respectively. In the upper mantle these velocities are 8.19 and 4.62 km s^{-1} with the Moho discontinuity at about 39 km depth. Seismic tomography of the Deccan volcanic province reveals a thick high-velocity zone that extends to depths of 100–400 km (Iyer et al., 1989). Its western part has a prominent low-velocity region in the depth range of 40–200 km. A coherent colder and rigid lithospheric root extending down to 400 km, decoupled from the underlying weaker layer, underlies this region.

Seismic studies were conducted in this region to provide thickness of the Deccan Trap volcanics, velocity structure of the subtrappean basement and the crust, alignment of possible deep faults and clues to resolve the question regarding the seismicity of the region. For these purposes two profiles, Guhagarh-Chorochi (Koyna I) and Kelsi-Loni (Koyna II), each about 200 km long and separated from the other by about 70 km, were recorded in the east-west direction (Fig. 5.2). These profiles cross the western ghat in a more-or-less perpendicular direction. To determine the depth to the Moho discontinuity in the coastal region, a short north-south profile was also recorded. Refraction and postcritical reflection data collected along all these profiles were in analog form.

Thickness of the Deccan Trap volcanics and the subtrappean topography was determined from seismic refraction data. The velocity in the trap rocks, along the Koyna I profile, varies between 4.7 and 4.9 km s^{-1} and that in its basement between 5.9 and 6.1 km s^{-1}. The trap thickness varies between 400 in the east to 1500 m in the west and is controlled by relief in the pretrappean topography (Kaila et al., 1979b). Along the Koyna II profile the trap velocity varies in the range of 4.8–5.0 km s^{-1} and its thickness is between 700 and 1500 m (Kaila et al., 1981a). The trap rocks directly overlie the Precambrian basement and their maximum thickness is near the west coast. The base of the Deccan Trap has a velocity between 6.0 and 6.15 km s^{-1}.

The pattern of the Deccan Trap thickness along the two profiles indicates that the structural trend varies in a north-south direction in the east and in a north-northwest to south-southeast direction in the west. A flexure, also aligned in the north-northwest to south-southeast direction, is seen in the basement and coincides with the general orientation of the Deccan volcanic scarp (Kaila et al., 1981a).

The depth models of the crust based on the postcritical reflection data along both the profiles show good reflections except in the topmost part of the upper crust (Figs. 5.3 and 5.4). Both the depth models show two blocks separated from each other by a deep fault, where the eastern block has moved up with respect to the western block (Kaila et al., 1979b; 1981a). Two prominent reflections, at depths of about 20 and 27 km, respectively, are seen in these models. The reflection at about 20-km depth probably represents the boundary between the upper and lower crust. Along the Koyna I profile, the Moho boundary is at a depth of about 40 km in the western block and between 36 and 38 km in the eastern block. On the Koyna II profile, the Moho is at a depth of 39 km near the deep fault in the western block and 37 km in the eastern block. Near the west coast the depth to the Moho is 31 km on a short north-south profile.

Reinterpretation of the seismic data along the two profiles in the Koyna region, through the analysis of travel times and relative amplitude modeling

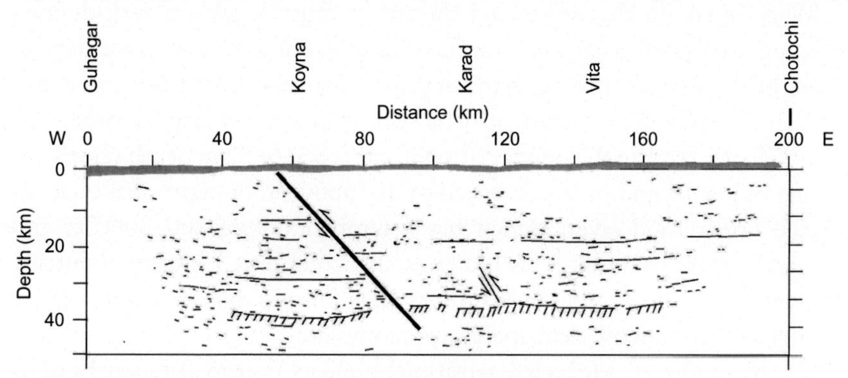

Fig. 5.3 Crustal depth section along the Koyna I profile. *(After Kaila, K.L., Reddy, P.R., Dixit, M.M., Lazarenko, M.A., 1979b. Deep crustal structure at Koyna, Maharashtra, indicated by deep seismic soundings. J. Geol. Soc. India 22, 1–16.)*

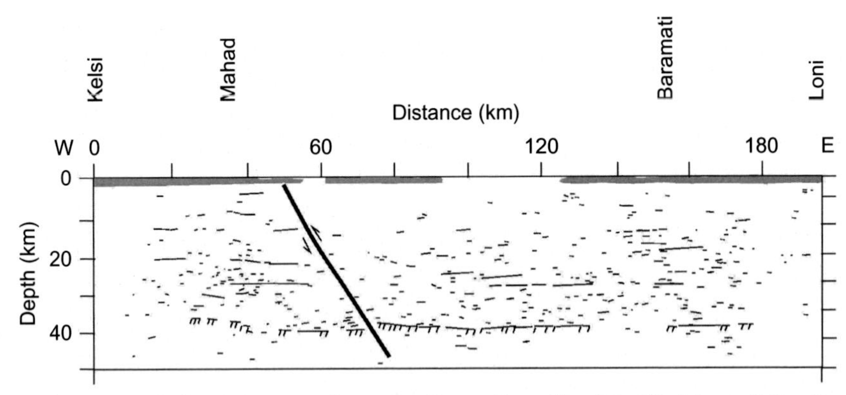

Fig. 5.4 Crustal depth section along the Koyna II profile. *(Modified from Kaila, K.L., Murty, P.R.K., Rao, V.K., Kharetchko, G.E., 1981a. Crustal structure from deep seismic sounding along the Koyna II (Kelsi-Loni) profile in the Deccan Trap area, India. Tectonophysics 73, 365–384.)*

using digitized records, provided a one-dimensional velocity model for the crust and subcrustal lithosphere (Fig. 5.5). The prominent features of this model are: (i) a high-velocity gradient in the 0–5 km depth range, (ii) a low-velocity layer in the upper crust between 6.0 and 11.5 km depth, where the velocity is less by 0.2 km s^{-1} as compared to the layer above it, (iii) a 1 km thick transition boundary at 21–22 km depth that separates the upper crust from the lower crust, (iv) a low-velocity layer in the lower crust between 26 and 28 km depth, where a reduction of 0.4 km s^{-1} is seen in velocity, and (v) at least a 2 km thick transition layer at the base of the lower crust at 35.5–37.5 km depth (Krishna et al., 1989). The two low-velocity layers, in the upper and the lower crust, suggest a well-defined rheological stratification and varying material properties at those depths. Most of the seismic activity in this region is concentrated at 4–5 km depth and an appreciable reduction in it is seen at the larger depths. Since the low-velocity layer in the upper crust with its top at 6–7 km depth represents a layer of lower rigidity, as compared to that above it, the earthquakes occur in the upper rigid layer, which has a temperature of 200–3000°C and relatively more strength. This model is also consistent with the essentially nonseismic nature of the lower crust, which corresponds to a zone of low strength where ductile deformation predominates.

In this model the Moho is a transition boundary instead of a sharp boundary, the transition zone being 2–4 km thick. The velocity in the uppermost mantle is 8.25 km s^{-1}. The upper mantle region has alternating low-velocity

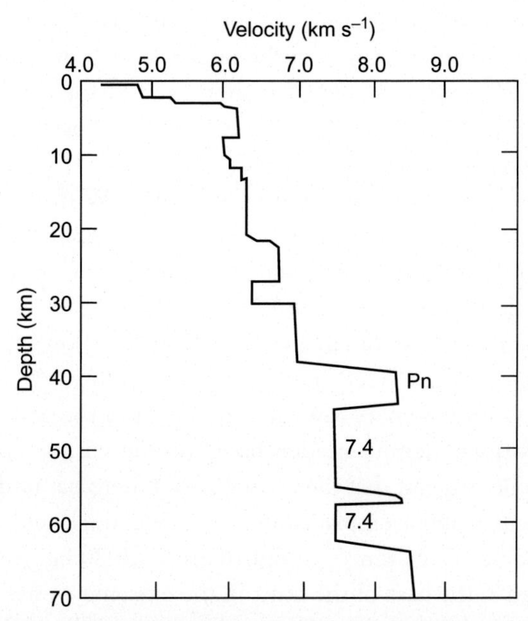

Fig. 5.5 Velocity model of the Koyna region showing lower velocities at different depths in the crust and upper mantle. *(Modified from Krishna, V.G., Kaila, K.L., Reddy, P.R., 1991. Low velocity layers in the subcrustal lithosphere beneath the Deccan Trap region of western India. Phys. Earth Planet. Int. 67, 288–302.)*

layers (velocity decreasing from 8.3 to 7.4 km s^{-1} in each of these), separated by a thin (about 1 km) high-velocity interface at 56–57 km depth. The first of these low-velocity layers, with its top at a depth of about 45 km, is about 10 km thick while the second, with its top at 57-km depth, is about 5-km thick (Krishna et al., 1991). This model shows that the continental subcrustal lithosphere here has a lamellar structure where significant variation of structural and mechanical properties has taken place in the vertical direction. The alternating low-velocity layers that occur at relatively shallow depths below the Moho are likely to be associated with zones of weakness and lower viscosity in the continental lithosphere.

This velocity model is similar to a large number of velocity models for the continental crust where the explanation for the low-velocity layers in the upper crust is either a semicontinuous laccolithic zone of granitic intrusion or a lower density, leading to attenuation of higher velocity in the granitic intrusions, within the surrounding basement rocks. Presence of a number of hot springs in the Koyna region indicates that water, with high pore pressure in the granite, is responsible for the upper crustal low-velocity layer (Krishna et al., 1989). Some other opinions do not agree with the presence of a laccolithic zone of granitic intrusion in the upper crust of the Koyna region, as it demands a temperature increase by about 320°C for a reduction in the velocity in the granite at these depths by 0.2 km s^{-1}. The crust in the Koyna region has a lower than required thermal gradient of 200°C km^{-1}, instead of 40–700°C km^{-1}, and heat flow of 41 mW m^{-2}, instead of 100–200 mW m^{-2}. The lower values of the thermal gradient and heat flow indicate that the reduction in velocity in the upper crust is only due to the occurrence of the fluids (Sharma and Mall, 1998).

The Bouguer gravity anomaly map of the region shows a gravity low in the Koyna region and a steep increase in values towards the west coast (Fig. 5.6). A few more gravity lows are seen towards the east of this region.

Two-dimensional density models based on the seismic data along both the Koyna profiles suggest that the coastal region here has undergone modification through magma emplacement. A 3-km thick and about 40-km wide high-velocity (7.26 km s^{-1}), high-density (3020 kg m^{-3}) anomalous layer is modeled at the base of the crust in the extreme western part of both the profiles, along the coastline. The thickness of this layer decreases gradually to the east, and near the Koyna gravity low the layer ceases to exist.

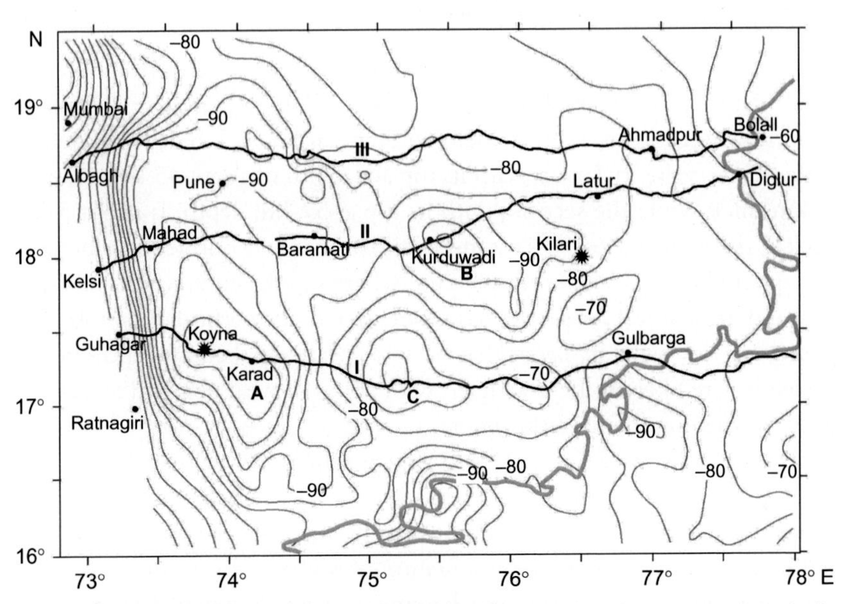

Fig. 5.6 Bouguer gravity anomaly map of Deccan Trap covered region in western India. *(From Singh, A.P., Mall, D.M., 1998. Crustal Accretion beneath Koyna Coastal Region, (India) and late Cretaceous Geodynamics. Tectonophysics 290, 285–297.)*

This anomalous layer is a result of igneous accretion at the base of the crust, an imprint of magmatism caused by the Reunion plume when the northward migrating Indian plate passed over it (Singh and Mall, 1998). Another density model (Fig. 5.7) is similar to this model, but shows the density in the high-density lower crustal layer as 3100 kg m^{-3} and existence of an additional layer of lower than normal density (3200 kg m^{-3}) in the upper mantle between the west coast and the western Ghat (Tewari et al., 2001). The low-density upper mantle layer continues to the east of the Koyna gravity low. Further east the mantle density is 3300 kg m^{-3}. On Profile I a high-density layer (2850 kg m^{-3}) exists at the base of the upper crust but this layer is not seen on Profile II. The gravity lows to the east of the seismic profiles are modeled as low-density (2620 kg m^{-3}) near-surface layers.

5.4 THE CAMBAY BASIN

Three pericontinental rift basins (Kutch, Cambay and Narmada) originated in the western part of India after breakup of Gondwanaland. Northward movement of the Indian subcontinent, between the early Jurassic and Tertiary period, was responsible for the origin of these rift basins (Biswas, 1987). At the close of the Mesozoic time intersecting sets of tensional faults developed along ancient basement trends in the northwest Indian platform. The trends of these faults follow the three main Precambrian orogenic trends in western India and bound these basins. The movements responsible for creation of these faults were also accompanied by large-scale Deccan volcanic activity, and the volcanics that were produced due to this activity constitute the floor of the Tertiary sediments of the Cambay basin (Raju, 1968). The evolution of this basin can be deciphered with some degree of certainty only above the volcanic floor of the basin, as its geological history earlier than the volcanic extrusion is hidden under a thick volcanic cover.

5.4.1 Geology and Tectonics

The Cambay basin is a narrow, buried graben that trends in a north-northwest to south-southeast direction and enters the Arabian Sea at the Gulf of Cambay (Fig. 5.8). It was formed due to the northward extension of the Dharwar trend, which runs parallel to the Indian west coast into the Indian shield (Fig. 5.1). Tectonically it is an intracratonic rift or graben between the Saurashtra horst, consisting of the Deccan Trap volcanics, in the west and the Aravalli hills comprising the igneous and metamorphic rocks of

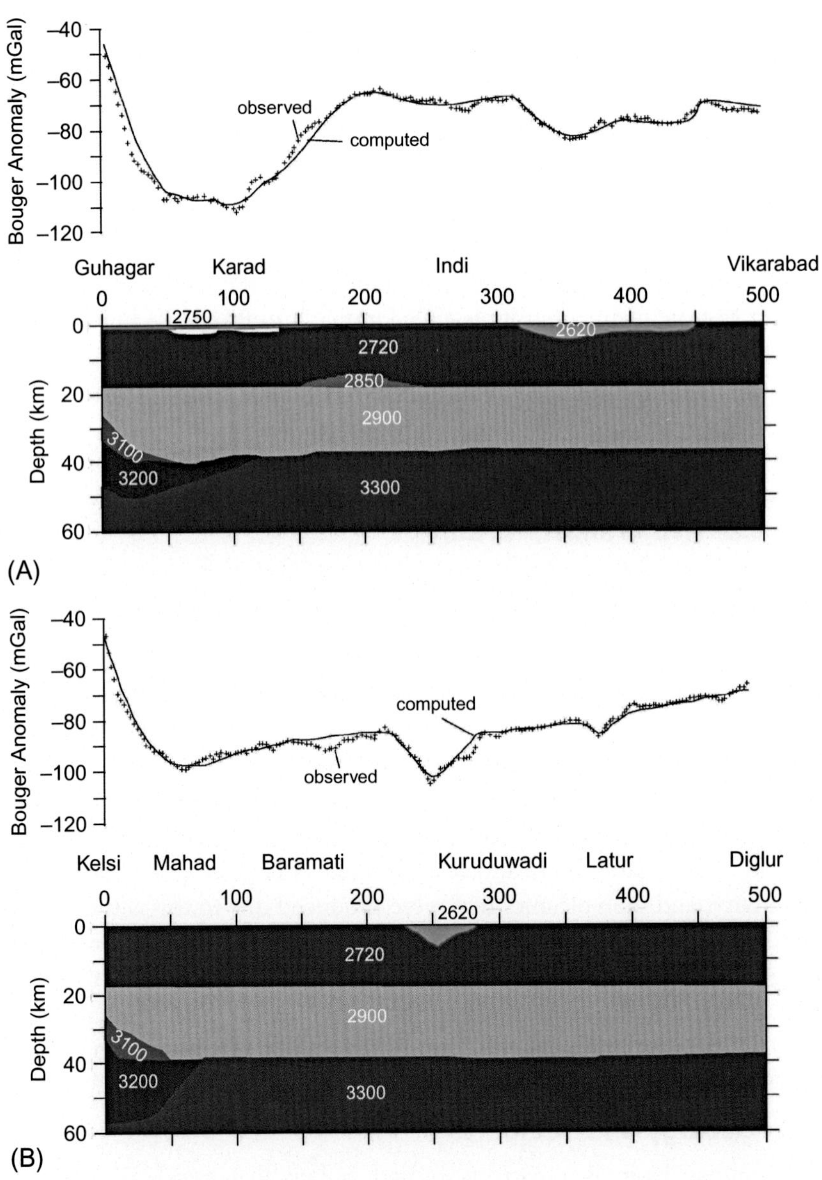

Fig. 5.7 Density models in the Koyna region and to its east along: (A) Profile I, (B) Profile II. Densities are in kilograms per cubic meter. *(After Tiwari, V.M., Vyaghreswara Rao, M.M.B.S., Mishra, D.C., 2001. Density inhomogeneities beneath Deccan Volcanic Province, India as derived from gravity data. J. Geodyn. 31, 1–17.)*

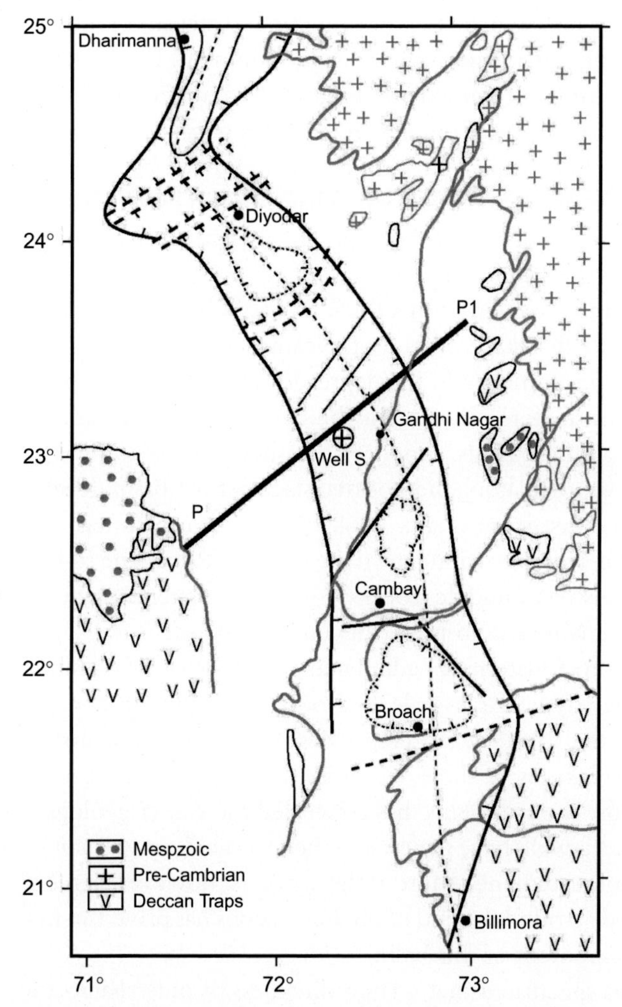

Fig. 5.8 Location map of the Cambay basin showing its tectonics. The deep seismic line is plotted with dashes and dots. Generalized sedimentary depth section across the line PP_1 is shown in Fig. 5.10. *(Modified from Tewari, H.C., Dixit, M.M., Sarkar, D., 1995. Relationship of the Cambay rift basin to the Deccan volcanism. J. Geodyn. 20, 85–95.)*

Precambrian age in the northeast. To the east of the basin, most of the exposures consist of the Mesozoic sediments (lower Cretaceous, upper and middle Jurassic) and Precambrian. Exposures to the west are the Mesozoic sediments and the Deccan Traps.

The Cambay basin is bounded by steep faults on its eastern and western margins, the former being more severely faulted than the latter. Faulting of the Deccan Trap, which also controlled the structure in the overlying sediments, has played a major role in its development. Even within the basin a number of structural elements were generated, differentiating it into uplifts and depressions, some of them being very pronounced. The basin is broadly divided into four major structural blocks, bounded by basement faults that have large throws. These blocks are characterized by different fold and fault trends and basement depths. To a certain extent the pattern of sedimentation also fits into the block scheme.

The basin is covered with a thick layer of Quaternary and Tertiary sediments. In the first stage of its postvolcanic subsidence, early Tertiary sediments were accumulated in the depressions adjoining the uplifts within the basin. In this stage reactivation at the basin margins, along the northwest-southeast basement faults, produced an uneven relief in the volcanic floor. The negative, oscillatory and positive stages of the basin development then followed the first stage (Raju, 1968). Maximum thickness of various elements of the Tertiary sediments in the basin is: Paleocene to Lower Eocene ~1000 m, lower to middle Eocene ~1800 m, upper Eocene to Oligocene ~600 m and Miocene to post-Miocene ~1200 m.

Tertiary to Quaternary sedimentary geology of this basin is quite well known due to the large-scale hydrocarbon exploration program in this region but the pre-Deccan Trap (Mesozoic) geology is not well known, as a large thickness of these volcanics, lying at depths of a few thousand meters below the sediments, has concealed the earlier geologic formations. Very few deep wells have penetrated the Deccan Trap volcanics and reached the Mesozoic sediments. Most of the workers, however, agree that the volcanic episode around the end of the Cretaceous has played an important role in the development of this basin.

It is also speculated that a large thickness of older Mesozoic sediments exists under the volcanics in some parts of the Cambay rift basin. In the northernmost part, a few deep wells on the basin margins were drilled through the Deccan Trap volcanics. In some of these wells Mesozoic (Cretaceous) sediments were encountered beneath the volcanics.

Three deep wells located in the southern part of the basin have either reached or crossed the Deccan Traps (Fig. 5.9). Well A touched the Archean basement at 3063 m after penetrating 518 m of the Quaternary, 865 m of the Tertiary, 1550 m of the Deccan Traps and 130 m of the Cretaceous sediments. Well B, drilled down to 5481 m, encountered 580 m of the Quaternary, 4092 m thick Tertiary and reached the Deccan Trap at 5472 m depth. Well C encountered 526 m of the Quaternary, 1976 m of the Tertiary and

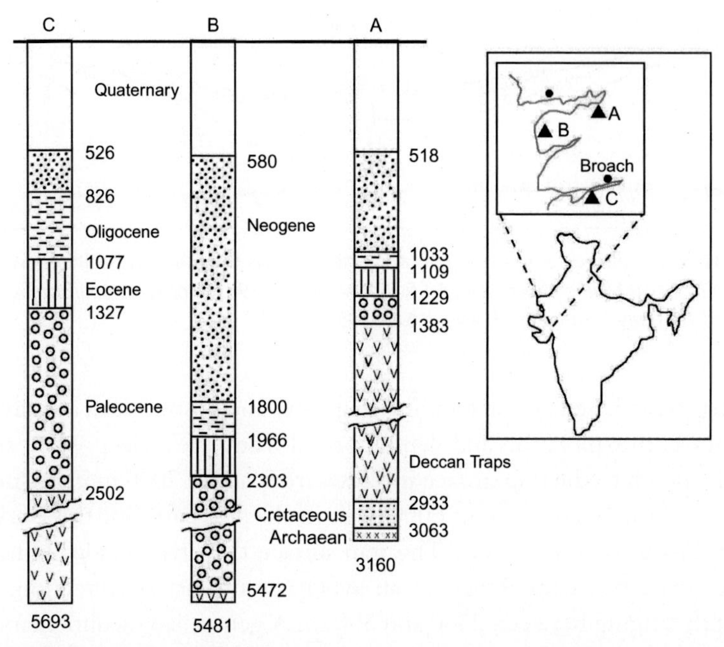

Fig. 5.9 Deep well lithology in the southern part of the Cambay basin. *(Modified from Roy, T.K., 1991. Structural styles in southern Cambay basin, India and role of Narmada geofracture in formation of giant hydrocarbon accumulation. ONGC Bull 27, 15–56.)*

3191 m of the Deccan Trap volcanics. It was abandoned at a depth of 5693 m, as the base of Deccan Traps could not be reached in this well. Paleogeologic reconstruction of the unconformity surface of the Deccan Traps suggests that a great thickness of volcanic flows is present in the Narmada fault zone and nearby areas within the Cambay basin, indicating that this fault is a conduit through which a huge quantity of lava flowed and spread on either side of the fault (Roy, 1991).

5.4.2 Seismic Results

Due to the hydrocarbon potential of the Cambay basin, seismic studies here have continued since the early fifties. Because of these studies the sedimentary basin configuration in the basin is well known, at least down to the top of the Deccan Trap surface. The geometry of the subtrappean sediments and their basement is, however, not well known because of the large thickness of Quaternary and Tertiary sediments that cover this surface and also because penetration of the trap surface by the seismic method has not been very successful at these depths. A structural map of the Deccan Trap surface in the

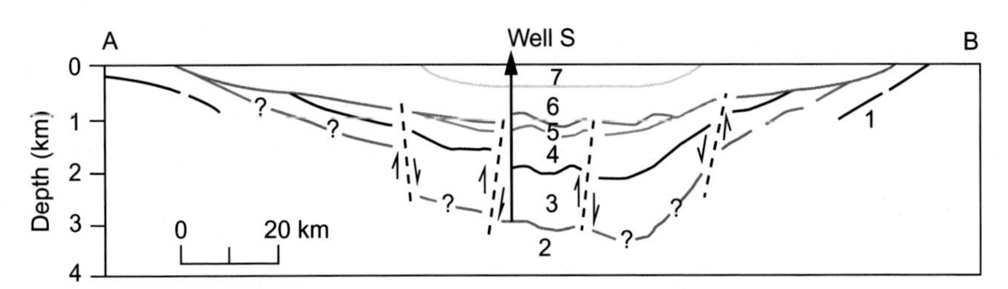

Fig. 5.10 Generalized sedimentary depth section across the Cambay basin along Profile PP1 of Fig. 5.8. *(Modified from Raju, A.T.R., Srinivasan, S., 1983. More hydrocarbon from well explored Cambay basin. Petrol. Asia J. 6, 25–35.)*

Cambay basin based on the synthesis of gravity and seismic data, acquired for hydrocarbon exploration and depth control from a few deep wells, shows that the depth to the trap surface increases from about 2000 m in the northern part of the basin to about 6000 m in its deepest part near the Narmada River (Awasthi et al., 1971). The trap surface then rises gradually further south of the Narmada River, in an average southeasterly direction, with its depth ranging between 2500 and 500 m. A generalized sedimentary section across the basin (Fig. 5.10) shows that the Paleocene-Eocene sediments constitute the largest element of the sedimentary thickness above the Deccan Trap (Raju and Srinivasan, 1983).

Seismic refraction and postcritical reflection studies along the basin, to delineate the basement and determine the velocity-depth configuration of the crust, were carried out in two phases. In the first phase analog data were collected in the southern part of the basin (south of 22030′) during the mid-1970s. In the second phase digital data were collected in the northern part of the basin in the mid-1980s. The two profiles together (Fig. 5.8) cover the basin in its entire length and provide reliable information on its basement and crustal features.

Results of these studies divide the Cambay basin into several subbasins (Fig. 5.11), which are separated from each other by basement ridges or faults (Kaila et al., 1981b, 1990b). Various ridges further divide some of these subbasins. The velocity-depth configuration in most of these subbasins shows four sedimentary layers with velocities of 1.8–2.1(Quaternary), 3.2–3.3 (Tertiary), 4.3–4.75 (Deccan Trap) and 5.8–6.0 km s^{-1} (Archean/Proterozoic basement) respectively. This configuration holds good for the entire basin except the extreme northern and southern parts. In the extreme north the basement directly lies under the Quaternary sediments. In the south an additional velocity of 2.5–2.6 km s^{-1} is delineated below 1.8–2.1 km s^{-1},

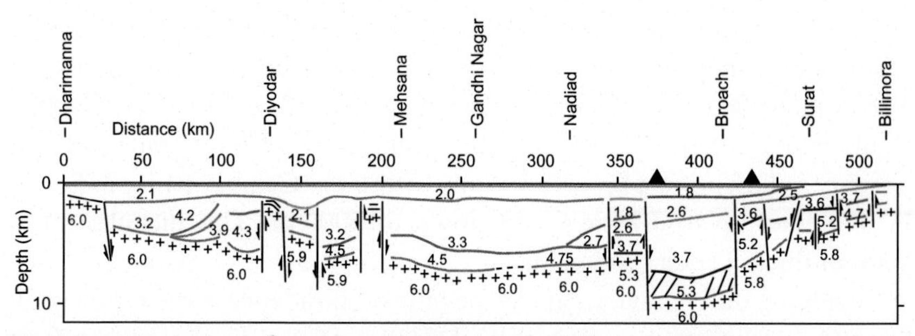

Fig. 5.11 Velocity (km s^{-1})-depth model for the sedimentary part of the Cambay basin along the seismic profile of Fig. 5.8. *(Modified from Dixit, M.M., Tewari, H.C., Visweswara Rao, V., 2010. Twodimensional velocity model of the crust beneath south Cambay basin, India from refraction and wide-angle reflection data. Geophy. J. Int. 181, 635–652.)*

along with velocities of 3.6–3.7 (Tertiary) and 5.2–5.3 km s^{-1} (Deccan Trap). In some of these subbasins another layer of velocity 3.9–4.0 km s^{-1} (Mesozoic) is seen below the Deccan Traps.

Thickness of the sediments (velocity 2.1 km s^{-1}) at the extreme northern end of the profile is about 350 m and increases to 1000 m to the northwest of a fault in the basement. Here the sediments directly lie over the basement of velocity 6.0 km s^{-1}. To the southeast of this fault this thickness is 2900 m and the layer with velocity of 3.2 km s^{-1} appears between the layers of velocities 2.1 and 6.0 km s^{-1}. Another layer of velocity 4.2 km s^{-1} is present below the layer of velocity 3.2 km s^{-1} to the further south. Comparison with the lithology of a deep well in the region shows that the layer of velocity 4.2 km s^{-1} corresponds to the Deccan Traps. Two deep wells in this area, one close to the profile and another on the basin margin, show that the Mesozoic sediments are present under the Deccan Traps of thickness between 475 and 570 m. Assuming an average thickness of 500 m for the trap layer and the velocity in the underlying Mesozoic sediments as 3.9 km s^{-1}, the maximum depth to the basement is about 5000 m in this subbasin. This subbasin terminates against the basement uplift at the Diyodar ridge, where the depth to the basement varies between 1300 and 2000 m. To the southeast of this ridge, a more or less similar velocity configuration is seen in most of the subbasins. However, the Mesozoic sediments have not been delineated due to absence of any deep wells that penetrate the Deccan Traps. Maximum depth to the basement between the Diyodar ridge and another ridge close to Mehsana is about 6000 m. This subbasin terminates against the latter ridge, where the maximum depth to the basement is 2200 m. To the southeast of this ridge, in the Gandhinagar subbasin, maximum depth to the basement is about 7700 m. A deep well on the fringe of this subbasin encountered the Deccan Traps at a depth of about 2400 m. Maximum depth to the basement in the Broach subbasin is about 8000 m (Kaila et al., 1981b, 1990b). Combined thickness of the Quaternary and Tertiary sediments in a few depressions within the Cambay basin, indicated by velocities up to 3.7 km s^{-1}, is down to 5000 m. This includes up to 4000 m thick Tertiary sediments.

There are different views about the time of subsidence of the Cambay rift basin. According to one view the subsidence of the rift was concomitant with the Deccan Trap volcanism (Raju et al., 1971). Another view is that the rift originated in the early Cretaceous and a major part of its subsidence was caused by reactivation of the west coast fault due to the India-Asia collision during the early Tertiary (Biswas, 1987). A third view is that the part of the Cambay basin, which was occupied by the extension of the Proterozoic Aravalli/Delhi system until the late Cretaceous, subsided only during the early Tertiary (Raju and Srinivasan, 1983). Yet another opinion, based on the large thickness of the volcanic traps in the basin (estimated to be between 1000 and 3200 m) and their rapid extrusion, is that the rift either originated or its main prevolcanic subsidence took place in the late Cretaceous (Tewari et al., 1995).

It is believed by several workers that an upwelling thermal body (Reunion plume) was present in the vicinity of the Cambay basin during the early Cretaceous. This indicates that ideal conditions existed at that time for magmatic extrusion in this region. Since the duration of the volcanic extrusion was short, withdrawal of the magma from the neighborhood of the Reunion plume head axis must have been rapid. This led to deflation of the plume, subsequent loading of the then-existing surface of the Cambay basin by the volcanics and its further subsidence. The relatively lower thickness of the volcanics immediately outside the basin supports this hypothesis. During the volcanic extrusion surface elevations in various parts of this rift basin were different and lower than the surroundings. These were filled by lava and resulted in a difference of the volcanic thickness in different parts of the basin. Older sediments, which are encountered under the traps in some of the deep wells drilled within the basin, were present before the rifting took place. Large postvolcanic subsidence of this rift basin is evidenced by the presence of up to 4000-m thick Tertiary sediments that were deposited in various phases of the basin development. The subsidence curves (Fig. 5.12) for some of the deep wells in the basin show that the largest postvolcanic subsidence of the Cambay basin was in the first phase (65–50 Ma,

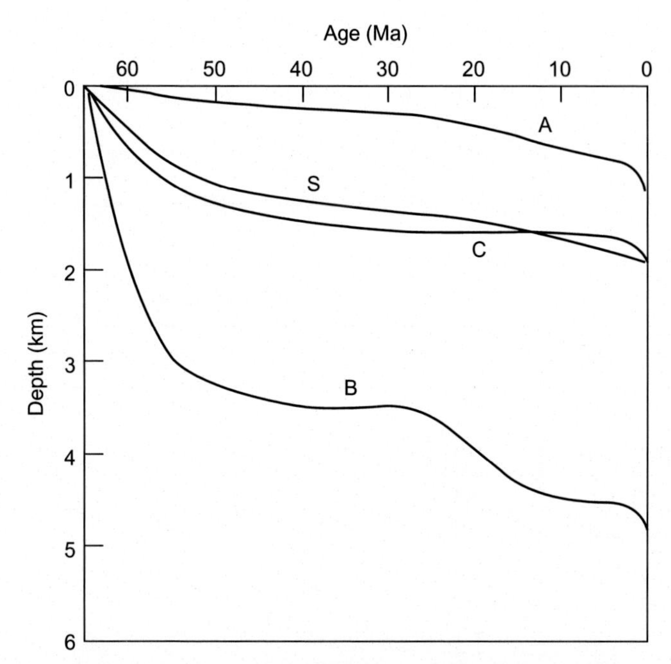

Fig. 5.12 Subsidence curves for some of the deep wells (A, B, C of Fig. 5.9 and S of Fig. 5.8) in the Cambay basin. *(After Tewari, H.C., Dixit, M.M., Sarkar, D., 1995. Relationship of the Cambay rift basin to the Deccan volcanism. J. Geodyn. 20, 85–95.)*

i.e., within 15 million years of the volcanic episode at about 65 Ma). Later phases of the subsidence are probably due to the India-Asia collision and relative thermal cycles associated with different rifting phases of the basin (Tewari and Rao, 1995).

The velocity-depth model of the crust along the north Cambay basin profile (Fig. 5.13A) shows four layers of different velocities (Kaila et al., 1990b). Within the Archean/Proterozoic basement the velocity increases from 6.0 km s^{-1} at a depth of about 10 km to 6.2–6.3 km s^{-1} at about 12 km. A 2–3 km thick low-velocity zone, of velocity 5.5–5.8 km s^{-1}, lies at this depth. This zone is at the base of the upper crust. After the low-velocity zone the velocity in the lower crust is 6.6 km s^{-1} and increases to 6.7–6.9 km s^{-1} at a depth of 23–25 km. A velocity jump to 7.2–7.4 km s^{-1}, representing the underplated crust, is seen here. Another velocity jump to 8.0–8.1 km s^{-1}, which represents the base of the crust (Moho), is seen at 31–33 km depth.

The southern part of the north Cambay seismic profile overlaps the northern part of the south Cambay profile (SP 31 of the north profile is

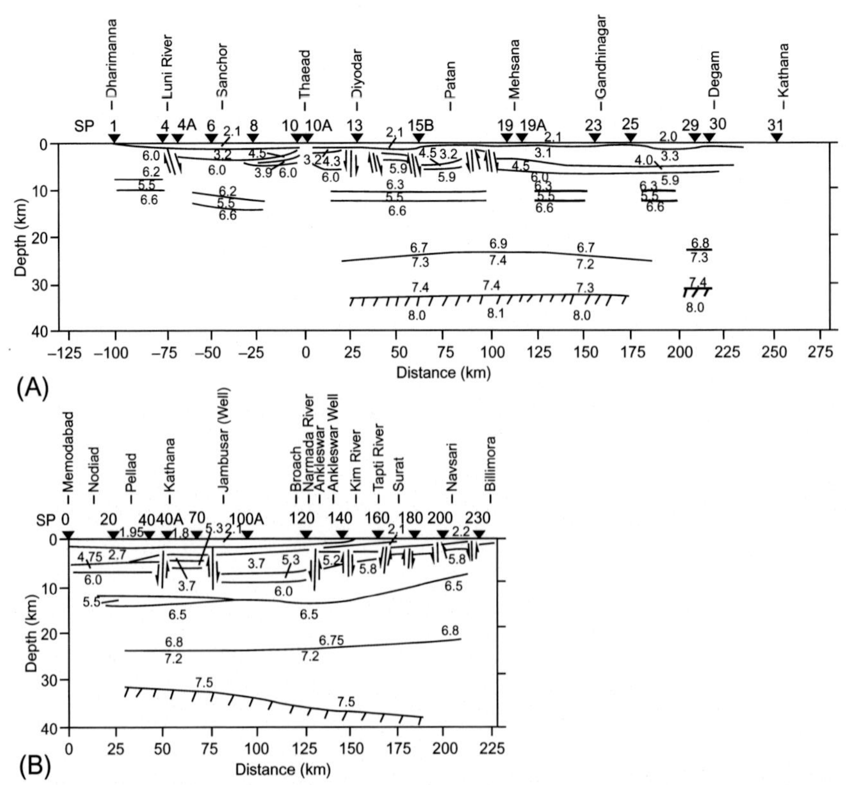

Fig. 5.13 Velocity (km s^{-1})-depth model for the crust of the Cambay basin: (A) in northern part; (b) in southern part. *((A) After Kaila, K.L., Tewari, H.C., Krishna, V.G., Dixit, M.M., Sarkar, D., Reddy, M.S., (1990b). Deep Seismic sounding studies in the North Cambay and Sanchor basins, India. Geophys. J. Int. 103, 621–637; (B) Modified from Dixit, M.M., Tewari, H.C., Visweswara Rao, V., 2010. Twodimensional velocity model of the crust beneath south Cambay basin, India from refraction and wide-angle reflection data. Geophy. J. Int. 181, 635–652.)*

the same as SP 40A of the south profile). Digitization, reprocessing and reinterpretation of the earlier recorded analog data of the southern seismic profile (Fig. 5.13B) shows that the upper crustal low-velocity zone (velocity 5.5 km s^{-1}) present in the northern part of the basin tapers and merges with the top of the lower crust (6.5–6.8 km s^{-1}) to the south of Jambusar, close to SP 100A (Dixit et al., 2010). Other crustal velocities are more or less the same as in the northern part. The crust is 33–34 km thick up to about SP 70 (Jambusar) and deepens to the south till it reaches a thickness of about 37 km at SP 180 (Surat), beyond which it cannot be delineated. This is contrary to the earlier results (Kaila et al., 1981b), which showed a 38–40 km thick crust in the central part (Broach region around SP120) that becomes shallower to about 23 km at the southern end of the profile (Billimora).

The main features of the crustal model of the Cambay basin are: the upper crustal low-velocity zone, higher than normal velocity in the lower crust, and a crust thinner than that in the Indian shield region. The contributing factor for generation of the upper crustal low-velocity zone is the high heat flow, 75–93 mW m^{-2}, in the basin (Gupta, 1981), which is also related to the tectonic processes that occurred here during the Cainozoic period. The low-velocity zone is also associated with the presence of free fluids, either due to metamorphism at lower crustal depths or partial melting at upper crustal depths, both of which could be caused by the underlying thermal anomaly, the Reunion plume (Raval, 1989). Higher than normal velocity in the lowermost part of the crust is due to underplating by the upper mantle magma, which was generated during passage of the Indian plate over the Reunion plume, after its breakup from Gondwanaland (Kaila et al., 1990b). Proximity of the basin to the axis of this plume and tension produced due to it led to thinning of the crust. The large thickness of the Tertiary sediments and the relatively thinner crust, as indicated by a shallow Moho (31–37 km) within the Cambay basin, indicate several stages of mantle upwarping and basin subsidence from the late Cretaceous to the Recent time (Dixit et al., 2010).

The Bouguer gravity anomaly map of the Cambay basin shows a high gravity anomaly within the basin relative to the gravity values outside it (Fig. 5.14). This anomaly, on the order of 50–60 mGa1, is located in the central part of the basin and extends to about 100 km in the northwest-southeast direction. To the northwest of this gravity high the gravity values within the basin are either comparable to or less than those outside it. An anomaly of this order within the basin, despite the presence of a large thickness of low-density sediments, is rather unusual and appears to be related to deep-seated features.

Various explanations were provided for this gravity high before the availability of the crustal seismic model. It was attributed to a possible Moho upwarp (Negi, 1951), mantle depth becoming shallower in parts of the basin (Sen Gupta, 1967), thickening of the volcanic layer at the base of sediments (Rao, 1968), and a shallow (10–20 km) high-density (3000 kg m^{-3}) intrusion in the crust (Verma et al., 1968). However, the gravity model based on the seismic configuration shows that a uniform thickness of the crust within the Cambay basin is not able to explain this gravity high. Various crustal configurations, where the crust inside the basin is thinner than that outside it

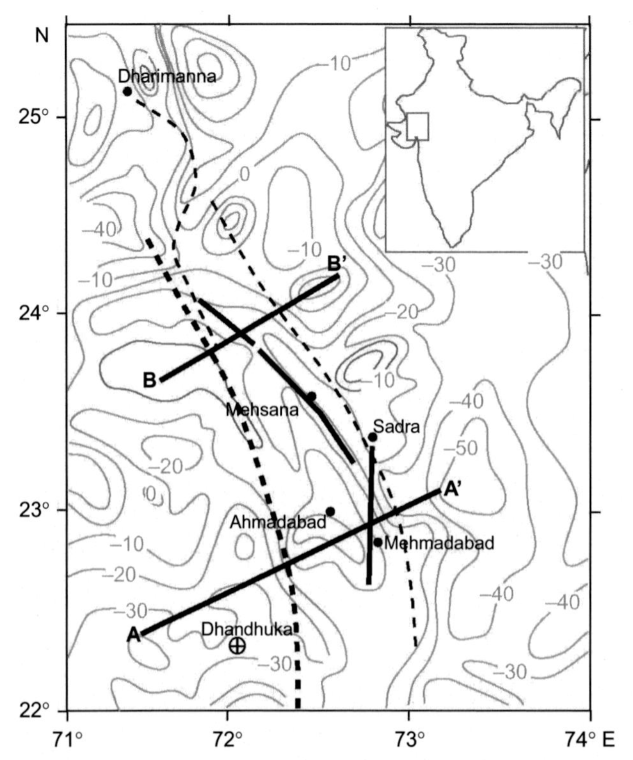

Fig. 5.14 Bouguer gravity anomaly map of the Cambay basin. Model along the line AA′ is presented in Fig. 5.15. *(After Tewari, H.C., Dixit, M.M., Sarkar, D., Kaila, K.L., 1991. A crustal density model across the Cambay basin, India, and its relationship with the Aravallis. Tectonophysics 194, 123–130.)*

accompanied by high-density (3150 kg m^{-3}) underplating, represented by a high velocity of 7.2–7.3 km s^{-1}, can explain the gravity high. These two factors together are responsible for the gravity high within the basin (Tewari et al., 1991). Fig. 5.15 shows one of the possible models controlled by the seismic data.

The shallow Moho and the high-velocity lower crust (7.2–7.5 km s^{-1} layer), which has higher density than the normal crust, appears to have adequately compensated for the low-density sedimentary accumulation in the Cambay basin. The basin, therefore, represents a typical example of rift basins associated with crustal thinning. The relatively thin crust (31–37 km) beneath the Cambay basin, as compared to the thicker crust (35–42 km) in the Saurashtra and Aravalli/Delhi trend and across the Narmada-Son lineament, suggests that the crustal thinning is related to several stages of mantle upwarp and basin subsidence.

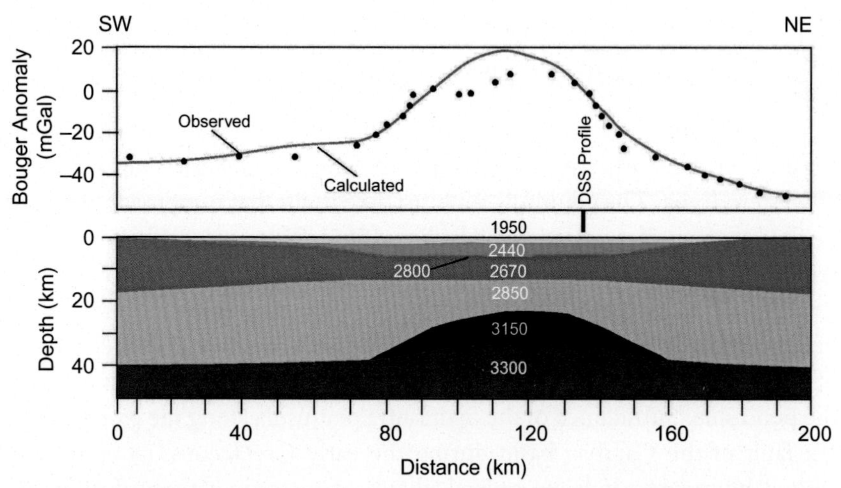

Fig. 5.15 Density model based on the seismic results along profile AA′ of Fig 5.14 across the Cambay basin. Densities are in kilograms per cubic meter. *(Modified from Tewari, H.C., Dixit, M.M., Sarkar, D., Kaila, K.L., 1991. A crustal density model across the Cambay basin, India, and its relationship with the Aravallis. Tectonophysics 194, 123–130.)*

It is now well established that the Reunion plume has affected the western part of India when the Indian plate passed over it during its northward movement after its breakup from Gondwanaland. Seismic tomography using P-wave arrival times (Kennett and Widiyantoro, 1999) reveals an approximately cylindrical region of lower velocity, at a depth of about 80 km in the upper mantle to the north of the Cambay basin and the present exposures of the Deccan Traps. This anomaly interrupts the characteristically high seismic velocities in the lithosphere beneath the peninsular India and extends from a shallow mantle depth down to a more extensive low-velocity zone beneath 200 km depth. The center of the low-velocity anomaly lies just below the earliest alkaline magmatism to the north of the Cambay basin, which precedes the main Deccan Trap eruptions by at least 3 million years. The association of this anomaly with the Deccan Traps is intriguing, since the heat flow values in the Cambay basin are high. It is therefore felt that the seismic anomaly may represent the signature of the initial conduit through which plume material forced its way through the lithosphere beneath the Cambay rift. The low-velocity zone beneath the lithosphere might then represent the remains of the initial plume head, which would have moved with India over the 65 Ma since the extrusion of the Deccan Traps.

5.5 THE SAURASHTRA PENINSULA

The Saurashtra peninsula (Fig. 5.16), and the Kutch region to its north, is at the northern boundary of the exposed Deccan Trap volcanics in western India. The peninsula is a horst-like uplift that was probably caused by the Deccan volcanic activity (Valdiya, 1984). The geologic history of this region is obscure before the volcanic extrusion, though the Mesozoic (Jurassic-Cretaceous) sediments exposed in the northeastern part of the peninsula are expected to be present below the trap volcanics. A few exposures of the Tertiary and Quaternary sediments can be seen in the coastal belt of the peninsula. Subsidence of the Saurashtra peninsula along the eastern margin fault of the Cambay basin during the early Cretaceous, its uplift, and faulted boundaries indicate several phases of tectonic activity during and after the Cretaceous period (Biswas, 1987).

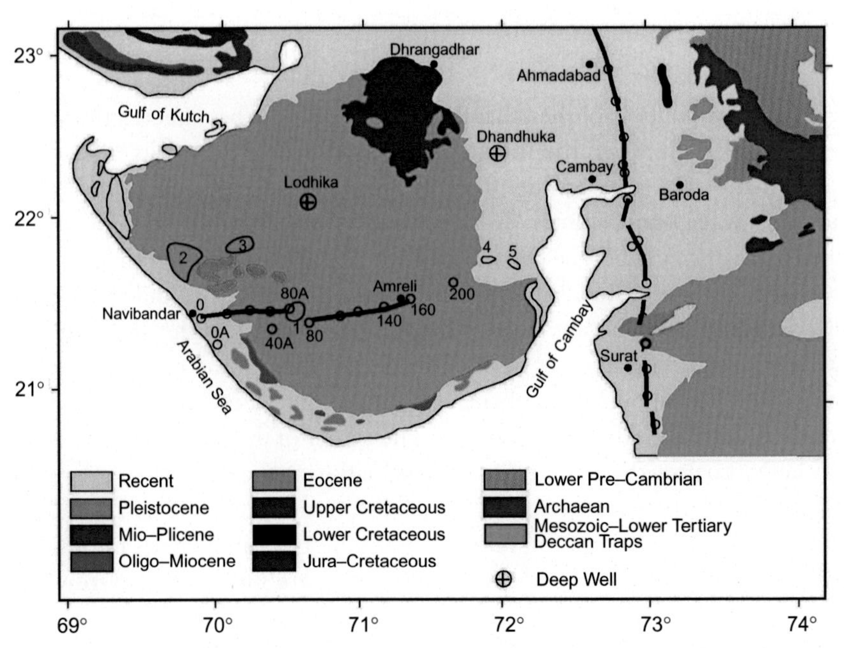

Fig. 5.16 Seismic profile in the Saurashtra peninsula (east-west) plotted on its geological map the numbers (1–3) indicate volcanic plugs of Girnar, Barda, Alech and (4, 5) Chogat, Chamardi. The north-south profile is along the Cambay basin, to the east of the Saurashtra peninsula. *(Modified from Rao, G.S.P., Tewari, H.C., 2005. The seismic structure of the Saurashtra crust in northwestern India and its relationship with the Reunion plume. Geophys. J. Int. 160, 319–331.)*

5.5.1 Geology and Tectonics

A number of volcanic plugs in the west (Girnar, Barda, Alech) and in the southeast (Chogat, Chamardi) are present in the Saurashtra peninsula (Fig. 5.16). The Girnar plug has acidic and alkaline to mafic/ultramafic rocks (Chandra, 1999). The Barda and Alech plugs are predominantly acidic in composition (Merh, 1995) while the Chogat and Chamardi plugs are basically alkaline (Karanth and Sant, 1995). Dike swarms oriented in the east-west and northeast-southwest directions, representing Deccan volcanic activity, are also found in the central and southeast Saurashtra. These swarms follow dominant structural trends of this region, as their orientation is almost the same as that of igneous intrusions and alkaline complexes of ~65 Ma (Mishra et al., 2001).

Two deep wells were drilled in the Saurashtra peninsula, at Lodhika and Dhanduka (Fig. 5.16), to find the stratigraphic configuration of the region below the trap surface. Results obtained by drilling of these wells have vastly improved knowledge of this peninsula. The Lodhika well is situated in the north-central part of the peninsula and the Dhanduka well at its eastern fringe. The geological sequence obtained from these deep wells (Fig. 5.17) shows several flows of the volcanic basalts and indicates that the Cretaceous/Jurassic sediments, volcanic tuff and possibly another sedimentary layer were deposited over a crystalline basement of Precambrian age prior to extrusion of the Deccan Trap (Singh et al., 1997). It is not possible to find whether the earlier flows of the volcanic basalt belong to the Deccan Traps or some earlier phases of volcanism, as the age determination has not been done.

5.5.2 Seismic Results

Analog seismic refraction and postcritical reflection data were acquired along an east-west profile (Navibandar-Amreli) in the western part of the Saurashtra peninsula, in the late 1970s (Fig. 5.16). The purpose of acquiring these data was to find the geometry of the basement of the Deccan Traps, determine the crustal configuration, and understand the evolution and tectonic history of the peninsula. Initial results of the seismic studies along this profile (Kaila et al. 1980) show a 400-m thick layer of Quaternary/Tertiary sediments at the western end of the profile (Navibandar) and another layer with a thickness of about 200 m and velocity 3.85 km s^{-1} (Tertiary) close to Junagarh. Except for these low-velocity exposures, a layer with a velocity varying between 5.20 and 5.25 km s^{-1}, representing the exposed Deccan Trap, covers the entire segment. The trap layer is also present under the

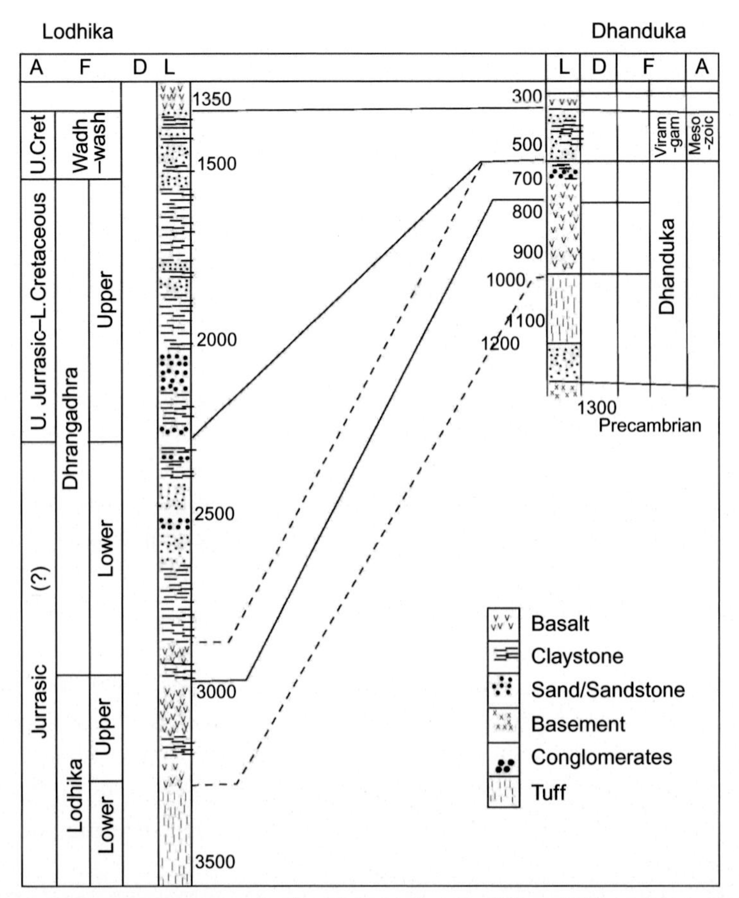

Fig. 5.17 Geological sequences in deep bore wells at Lodhika and Dhanduka in the Saurashtra peninsula shown in Fig. 5.16. *(Modified from Singh, D., Alat, C.A., Singh, R.N., Gupta, V.P., 1997. Source rock characteristics and hydrocarbon generating potential of Mesozoic sediments in Lodhika area, Saurashtra basin, Gujarat, India. In: Proc. Second International Petroleum Conference and Exhibition, PETROTECH-97, New Delhi, Oil and Natural Gas Corporation Limited (ONGC), Dehradun, India, pp. 205–207.)*

layers of velocity 2.80 and 3.85 km s^{-1}. A low-velocity layer, attributed to the Mesozoic sediments, is assumed to be present below the Deccan Trap volcanics in the entire profile. The basement to the Deccan Trap/Mesozoic sediments is a layer of velocity 5.80–6.00 km s^{-1}. This layer represents the Precambrian basement in the region. The depth to this layer varies between 1400 and 2600 m along the profile.

This model is, however, speculative as no evidence was available to show a particular thickness of the Deccan Traps above the assumed presence of the

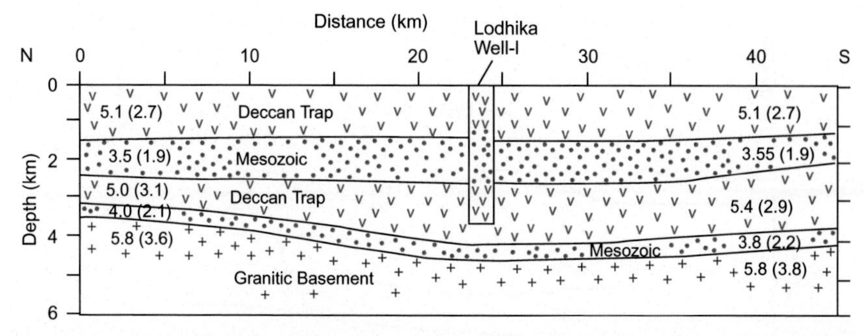

Fig. 5.18 Velocity-depth configuration down to the basement in the vicinity of the Lodhika deep well in Saurashtra showing the P- and S- (in brackets) wave velocities in km s^{-1}. *(Modified from Dixit, M.M., Satyavani, N., Sarkar, D., Khare, P., Reddy P R., 2000. Velocity inversion in the Lodhika area, Saurashtra peninsula, western India. First Break 18, 499–504.)*

Mesozoic sediments. A more reliable model of the sedimentary thickness in Saurashtra is available after drilling results of the Lodhika deep well. The P- and S-wave velocities along a small north-south seismic profile, recorded close to the deep well, were correlated with the geological horizons obtained in the well. This has led to substantial improvement in understanding the subtrap geometry down to a depth of about 3.50 km. This velocity-depth model (Fig. 5.18) shows the thickness of the exposed Deccan Traps as about 1500 m and its P-wave velocity as 5.1 km s^{-1}. The Mesozoic sediments below the trap layer, with a thickness of about 900 m, have a velocity of 3.5–3.55 km s^{-1}. The second trap layer of velocity 5.0–5.4 km s^{-1} is about 600-m thick to the north but thickens considerably to 2000 m in the middle and to the south of the deep well. Another layer, of possibly Mesozoic sediments (400–500 m thick) with velocity of 3.8–4.0 km s^{-1} lies below this layer and above the granitic basement of velocity 5.8 km s^{-1} (Dixit et al., 2000). The initial model of the crust along the Navibandar-Amreli profile (Kaila et al., 1980) was revised in view of later geological and geophysical results that became available for the western Indian region. This model (Fig. 5.19), based on reinterpretation of the digitized data through a dynamic forward modeling technique (Rao and Tewari, 2005), shows that the upper crust consists of three layers of velocities 5.80–6.00, 6.10–6.15 and 6.35–6.40 km s^{-1}, respectively. The boundary between the upper and lower crust is at a depth of about 16 km close to the western part of the profile and at about 13-km depth near the eastern end. The lower crust consists of two layers: the upper layer has a velocity of 6.70–6.75 km s^{-1} and the lower layer 7.10–7.20 km s^{-1}. The depth to the latter varies between

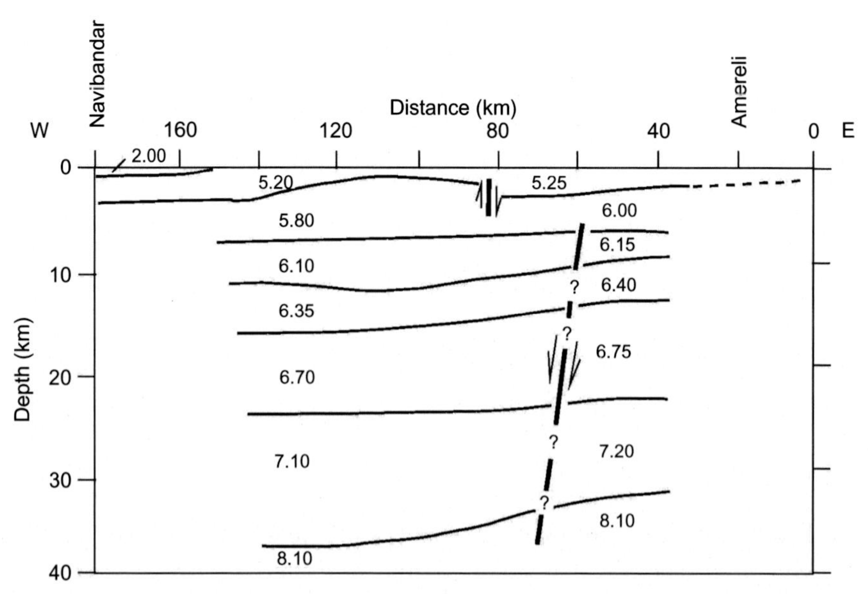

Fig. 5.19 Velocity (km s^{-1})-depth model of the crust along the seismic profile in Saurashtra. *(After Rao, G.S.P., Tewari, H.C., 2005. The seismic structure of the Saurashtra crust in northwestern India and its relationship with the Reunion plume. Geophys. J. Int. 160, 319–331.)*

24 and 22 km from west to east. The crust, with an upper mantle velocity of 8.10 km s^{-1} below the Moho boundary, is about 36 km thick in the west. A steep updip brings it to a thickness of about 33 km almost in the middle of the profile. From there the updip is gentle to about 32-km thickness in the east.

The reduction in thickness towards the east is not only at the Moho boundary but is seen in all the crustal boundaries. Thickness of the crust in the western part of Saurashtra is close to that in the Indian shield, while in the eastern part it is close to that in the Cambay basin. The zone of sharp reduction in the crustal thickness represents a deep fault, thrown up towards the east. The fault comes close to the surface at 20–30 km west from the eastern end of the profile (Amreli). This deep fault has probably acted as a lava feeder from the lower crust/upper mantle to the near surface. The large number of dikes and volcanic plugs in the Saurashtra peninsula and the high velocity in the upper crust are consistent with this possibility.

The main features of the Bouguer gravity anomaly map of the Saurashtra peninsula (Fig. 5.20) are a series of northeast-southwest trending gravity highs (D, E, F) of 40–60 mGal amplitude in the southeastern part and of

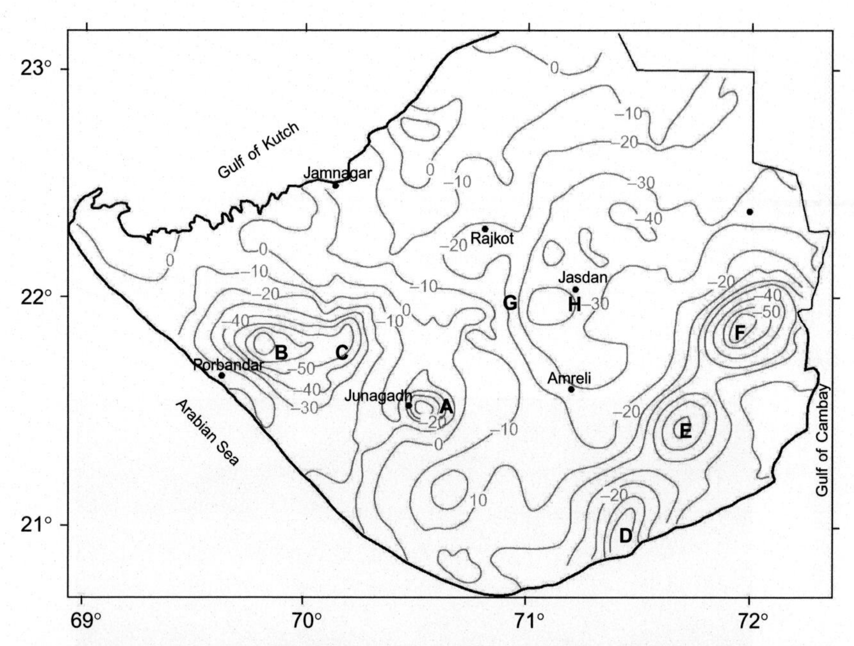

Fig. 5.20 Bouguer gravity anomaly map of the Saurashtra peninsula. *(From Gravity Map Series of India, 1975. National Geophysical Research Institute, Hyderabad.)*

almost the same amplitude (A, B, C) in the western part. The highs in the western part correspond to the Barda and Alech volcanic plugs (Mishra et al., 2001) and are related to their deep roots (Chandrasekhar et al., 2002). The bulk density of the plugs in western Saurashtra is 2850–2900 kg m^{-3}. A broad low (H) with a steep gravity gradient (G) to its west is observed in the central part of Saurashtra.

The two-dimensional gravity model (Fig. 5.21) based on the seismic data across the Saurashtra peninsula in the east-west direction indicates that the base of the Deccan Traps (density 2740 kg m^{-3}) in the extreme western part has a higher density (2850 kg m^{-3}) than that of the granitic layer, down to a depth of about 10 km (Rao and Tewari, 2005). This indicates that the upper part of the crust here is not granitic but volcanic and conforms to the idea of deep-rooted high-density volcanic plugs in that part.

To the east the density at the base of the Deccan Trap is 2670 kg m^{-3} but the high density of 2850 kg m^{-3} is present below it until the lower density replaces it. To the east of Amreli, high density (2820 kg m^{-3}) is indicated again below the density of 2670 kg m^{-3}, probably due to the effect of volcanic plugs of southeast Saurashtra.

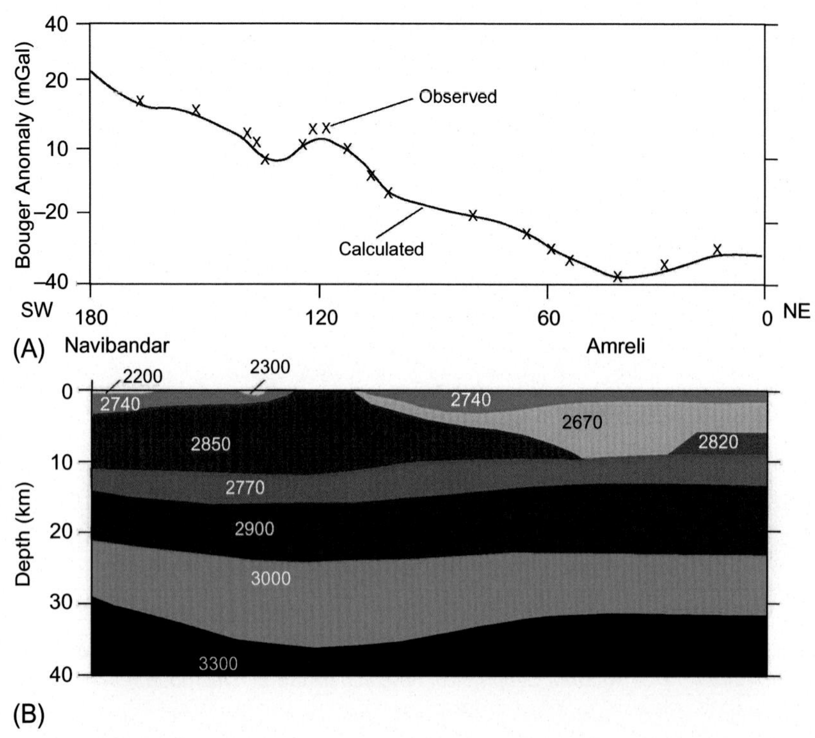

Fig. 5.21 Two-dimensional density model along the seismic profile and to its east in the Saurashtra peninsula. (A) Observed and computed Bouger gravity values. (B) Depth model obtained through combined analysis of seismic and gravity data. Density values in 'B' are in kg m^{-3}. *(Modified from Rao, G.S.P., Tewari, H.C., 2005. The seismic structure of the Saurashtra crust in northwestern India and its relationship with the Reunion plume. Geophys. J. Int. 160, 319–331.)*

5.6 REASONS FOR THINNER AND UNDERPLATED CRUST

The deep fault, interpreted by the seismic studies in the Saurashtra peninsula, is also seen in the gravity anomaly picture as the gravity gradient (G of Fig. 5.20), as well as in the satellite data (Mishra et al., 2001) as a pair of long north-south trending lineaments. It is in line with extension of the Aravalli trend in the Saurashtra peninsula and divides the crust of the peninsula into two distinct parts. To the east of this fault the crustal configuration is more or less similar to that in the Cambay basin. In the Cambay basin, the velocity-depth model shows an upper crust down to 13 km depth, the Moho at 32–33 km depth and an 8–10 km thick layer of velocity 7.3 km s^{-1} at 23–25 km depth. Configuration of the upper crust, however, is somewhat different in the two regions: the low-velocity layer (5.50 km s^{-1}) that lies at the base of the Precambrian basement in the Cambay basin is absent in Saurashtra. The thickness of the crust in the eastern, the Cambay basin and even in the region of the Vindhyan exposure to the immediate east of the Aravalli formation (see Chapter 3) is on the order of 32–35 km as against the depth of 37–40 km in the western Saurashtra and beyond the eastern margin fault of the Cambay basin. From the seismic results and subsequent gravity models (Tewari et al., 2009), it is clear that the crust in the eastern part of the Saurashtra and Cambay basin, between the two trends of the Aravalli system (one that crosses the Cambay basin and Saurashtra and the other that turns to the east and merges with the Satpura trend) and the Narmada Fault running to the south of the Saurashtra peninsula, is uplifted by as much as 4–6 km as compared to the regions outside these trends. This uplift took place either after deposition of the Mesozoic sediments or was concomitant with the rise of the Reunion plume prior to the extrusion of the Deccan volcanic, as the region was close to the axis of the plume when the Indian plate moved over the plume during the late Cretaceous. Rise of the Reunion plume was accompanied by a phase of tectonic activity that caused uplift of the crust in a large part of western India, including the western part of the Koyna profiles, the eastern part of the Saurashtra peninsula, and the Cambay basin. The upper crustal structure of the Cambay basin was probably modified during its subsidence in the early Tertiary, after the extrusion of the Deccan volcanics (Rao and Tewari, 2005).

The high-velocity (7.1–7.30 km s^{-1}) layer at the base of the crust is a common feature in the western part of India. This 10–15 km thick layer is seen in the Cambay basin, Saurashtra peninsula and also in a large part of the Narmada zone, up to at least 600 km east of the Cambay basin (see Chapter 6). It is also seen at the western end of the Koyna profiles. Passage of the western part of India over the Reunion plume resulted in partial melting initiated by the arrival of a hot plume, which thinned and conductively heated up the lithosphere. Rise of the plume head through the lithosphere causes rapid eruption of the continental volcanics (Renne and Basu, 1991). Large plume heads also lead to extrusion of the magmatic material from the plume head to the base of the crust. The head of the Reunion plume had a diameter of 1000–2000 km (White and McKenzie, 1989), and it led to underplating in a large part of the crust in western India. Further rise of the plume through the lithosphere caused rapid eruption of the Deccan Trap volcanics.

CHAPTER SIX

Central Indian Region

The central part of the Indian subcontinent consists mainly of four tectonic units (Fig. 6.1). These are: (1) the Bundelkhand craton (Late ArcheanArchean to Neoproterozoic), (2) the Satpura mobile belt (Palaeoproterozoic to Mesoproterozoic), (3) Kotri-Dongargarh mobile belt (middle ArcheanArchean to Mesoproterozoic), and (4) the Bastar craton (Neoproterozoic). Other important tectonic elements of the region are the Narmada-Son lineament and the central Indian suture.

The Dharwar craton lies to the west and south of these tectonic units. Tectonic history of the Bundelkhand craton includes Late ArcheanArchean compression. The Archean gneiss is exposed in a large part of this craton and includes migmatitic gneiss, older migmatitic enclaves, basement relics as well as paragneiss. The Archean assemblage has banded, streaky and augen gneisses, amphibolites and pillow volcanics. This assemblage experienced amphibolites-facies metamorphism (Sarkar et al., 1981).

6.1 THE NARMADA REGION

The Narmada region follows the east-west course of the Narmada River. It consists of the Narmada-Son lineament and some other features, including the Satpura mobile belt, and lies between the Bundelkhand craton in the north and Dharwar and Bastar cratons in the south. The lineament appears to form the northern boundary of the Satpura mobile belt. Tectonics of the central Indian region are greatly influenced by this almost east-west trending lineament, a megageofracture extending eastward to about 1300 km from the west coast of India. The central Indian suture, which also has a more or less similar trend to the Narmada-Son lineament, forms the southern boundary of the Narmada region. Ongoing seismic activity in the region indicates that movements against certain faults are still continuing (Nayak, 1990). The Jabalpur earthquake on May 21, 1997 ($M=5.8$) caused widespread damage in the region.

Structure and Tectonics of the Indian Continental Crust and Its Adjoining Region
https://doi.org/10.1016/B978-0-12-813685-0.00006-6

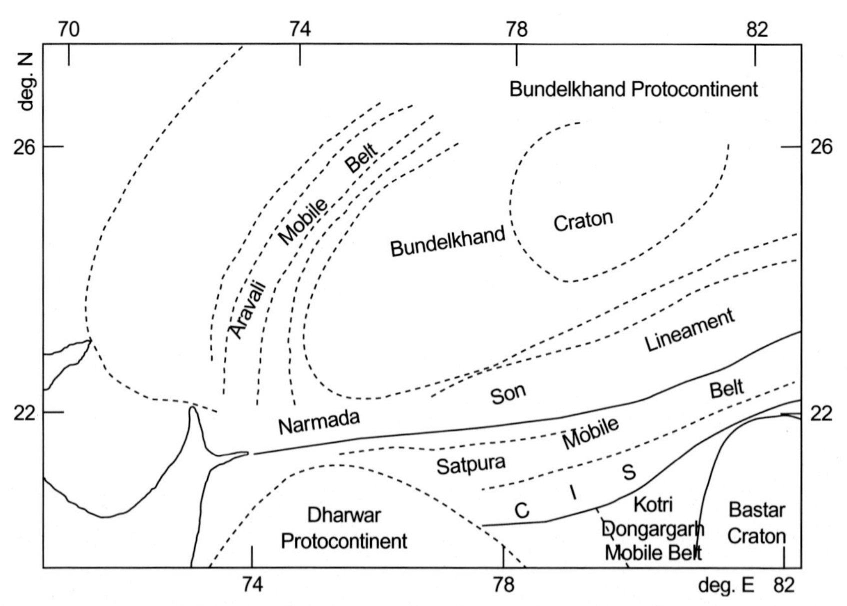

Fig. 6.1 Tectonic framework of central India. *CIS*, central Indian suture. *(After Kumar, P., 2002. Seismic Structure of the Continental Crust in the Central Indian Region and its Tectonic Implications, Thesis submitted to Osmania University, Hyderabad (unpublished)).*

6.1.1 Geology and Tectonics

The Bundelkhand craton, to the north of the Narmada region, has undergone multiple phases of granitic plutonism of the Late Archean age. It consists of the Late Archean tonalites and Paleoproterozoic granitoids intruded by the younger basic dikes and quartz veins. The Proterozoic rocks (Bijawar) rest unconformably over the granitoid rock and are exposed in an elongated basin to the south of Bundelkhand granite. The intracratonic Meso-Neoproterozoic Vindhyan basin, which lies to the further south of the craton, is bounded on the southeast by the basement ridge of the Narmada-Son lineament and laps onto the Bundelkhand craton in the northwest. The Satpura mobile belt lies to the immediate south of the Narmada region and extends in an east-northeast to west-southwest direction. It is part of a major mobile belt that originated as a result of typical compression. The Deccan Trap volcanics cover a large part of the Bundelkhand craton. A north-south compression of several blocks against a crustal mobile belt is indicated in the western part of the craton where metasedimentary and metavolcanic sequences are exposed (Naqvi and Rogers, 1987). The Bastar craton, which lies to the southeast of the Narmada region, has a continental shelf depositional environment although the sediments rest unconformably on the older basement gneiss and granite. This craton consists mostly of the granite and unmetamorphosed Proterozoic sediments. The Kotri-Dongargarh mobile belt consists of the gneiss, acid/basic volcanics and Proterozoic sediments. The Dharwar protocontinent lies to the southwest of the Narmada region and consists of a central ancient craton of gneissic complex with greenstone and younger Proterozoic fold belts. It is warped on the south by a granulitic mobile belt and contains intracratonic sedimentary basins. Dikes of dolerite, related largely to the Proterozoic mafic/magmatic activity, traverse the gneissic complex.

The Narmada Rift is the last of the three pericontinental rifts that originated in the western part of India after the breakup of Gondwanaland, i.e., after the Kutch and Cambay rift basins (see Chapter 5). It is warped, uplifted and rift faulted. This rifting influenced the deposition and folding of Meso-Neoproterozoic Vindhyan sediments to the north. Some of the opinions expressed about the origin and nature of the Narmada region are: (a) a zone of persistent weakness, (b) a swell, (c) a tectonic rift reactivated since Precambrian time, (d) a failed arm. It is a zone of weakness that separates the regions of Meso-Neoproterozoic deposits to the north from the Gondwana (Permo-Carboniferous-lower Cretaceous) deposits to the south. The Gondwana sediments, which have a maximum thickness of 6–7 km, are exposed to the south of the Narmada region (West, 1962).

The hills of the Vindhyan and Satpura region are the uplifted shoulders of the Narmada Rift zone. Most of the area is completely filled and covered by the Deccan Trap volcanic flows (Fig. 6.2). Almost horizontal volcanic piles, varying in thickness between 100 and 800 m, cover the earlier topography. At a few places the trap flows are eroded and the pretrappean topography is visible. The volcanics become thinner from the west coast to the central part of India. Evidence of sedimentation during the Jurassic period is also seen. Geological evidence suggests that various zones of fracture, belonging to early Precambrian, Cretaceous and post-Deccan Trap periods, are close to this region. Faulting in the Narmada region is generally of step type without any tilt of the blocks. These faults have successive downthrow toward the south of the Narmada River. This is a zone of tectonic truncation of regional structural trends and is bounded by the Narmada north and south faults (Acharya et al., 1998). Other important faults in the region are the Barwani-Sukta fault and the Tapti fault.

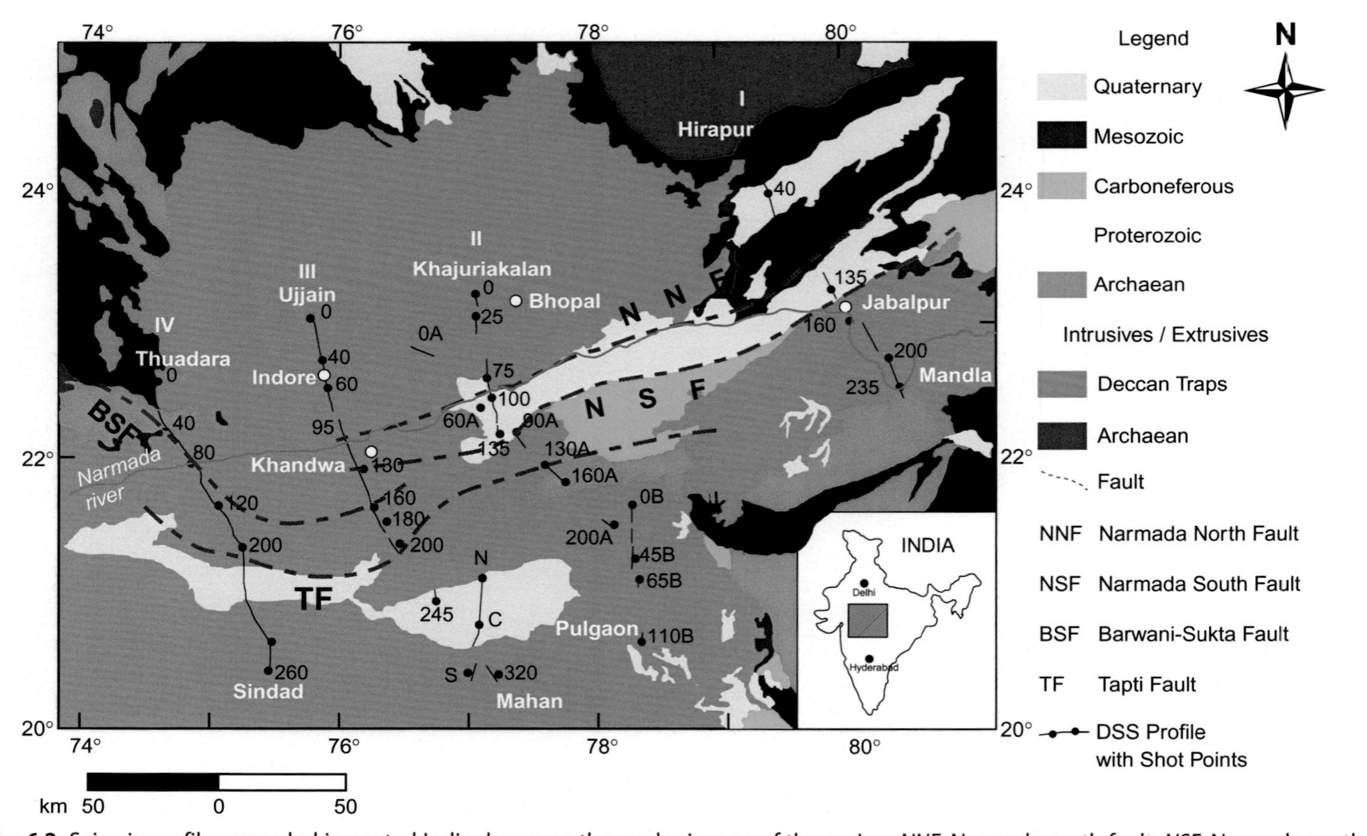

Fig. 6.2 Seismic profiles recorded in central India drawn on the geologic map of the region. *NNF*, Narmada north fault; *NSF*, Narmada south fault; *BSF*, Barwani-Sukta fault; *TF*, Tapti fault; *I*, Hirapur-Mandla; *II*, Khajuriakalan-Pulgoan; *III*, Ujjain-Mahan; *IV*, Thuadara-Sindad. *(After Kumar, P., 2002. Seismic Structure of the Continental Crust in the Central Indian Region and its Tectonic Implications, Thesis submitted to Osmania University, Hyderabad (unpublished)).*

The Narmada north and south faults were initiated earlier than the late Archean. The Narmada north fault is traceable in the east-northeast to west-northwest direction in a large part of this region (Nair et al., 1995). Its western continuation is concealed under alluvium of the Narmada valley. The last reactivation of this fault was as a thrust heading toward south, due to which the older Mahakoshal (Paleoproterozoic) group of rocks are seen overriding the basal Vindhyan rocks. No major reactivation of this fault has taken place in the post-Vindhyan period. The Narmada south fault has a general east-west trend and marks the southern boundary of the Mahakoshal belt. In the eastern part, a rectilinear narrow zone of intensive shearing within the gneiss marks this fault. Activity along this fault probably started again during or after the breakup of Gondwanaland. There is no geological evidence for activation of the north fault at that time. In the Satpura region, a few exposures of the Gondwana sediments are present to the north of this fault but absent to its south, indicating that its reactivation continued even after the deposition of the Gondwana sediments. Occasional seismic activity along this fault indicates that it is still active. This fault has controlled the lower Gondwana sedimentation to its south by synchronous sinking, similar to that of the Vindhyan basin by the Narmada north fault (Jain et al., 1995). The Barwani-Sukta fault extends from the northwest to southeast and its trend represents the basement grain. This fault is marked by shearing and brecciation of the basalt, alignment of the hot springs, presence of the earthquake epicenters, and steeper lava flows to the south of the fault. Some dike swarms are also present along this fault zone. The Tapti fault extends in a curvilinear manner with a general east-west trend and marks the southern boundary of the Satpura belt and northern boundary of the alluvium deposits in the Tapti River basin. Further east, the Tapti fault trends in the northeast direction and enters the trap-covered area of the Satpura region and enters the Gondwana terrane.

6.1.2 Seismic Results

Knowledge of the crustal configuration of the Narmada region is important for determining the tectonic configuration of the central Indian region. The Narmada region is also seismically active, as suggested by occurrence of at least six earthquakes of magnitude greater than 5.0 in the last 100 years, the largest being the Jabalpur earthquake of magnitude 5.8 on 21 May 1997. Geological importance of this region is indicated by the fact that a large amount of geophysical data has been acquired in this region. These include study of the velocity-depth configuration of the crust along four seismic profiles, each 200–250 km long, across this zone in the N-S direction (Fig. 6.2). Analog refraction and postcritical reflection seismic data were acquired along these profiles in the late 1970s and early 1980s. The profiles are: Hirapur-Mandla (Profile I), Khajuriakalan-Pulgaon (Profile II), Ujjain-Mahan (Profile III), and Thuadara-Sindad (Profile IV). These profiles are almost orthogonal to the Narmada region. The first results of the study along these profiles were based on interpretation of the analog seismic data. Since continued seismic activity in the Narmada region prompted a need to correctly decipher the velocity-depth configuration of the crust, as velocity is the most important parameter to constrain the focal depth of the earthquakes, the analog data have been digitized and reinterpreted using synthetic seismogram modeling and inversion techniques. This attempt also helped in generating a unified crustal model to understand the seismotectonics of the region.

The velocity-depth model along Profile I (Fig. 6.3) shows that from the northern end of the profile to about 40 km distance (Upper Vindhyan exposures) the velocity of 4.2–4.5 km s^{-1} directly overlies the basement (velocity 5.9 km s^{-1}) followed by a velocity of 6.5 km s^{-1} that belongs to its basement (Murty et al., 2004). Between this point and the north Narmada fault, a three-layered configuration is seen where the velocities are 4.5–4.7 km s^{-1}

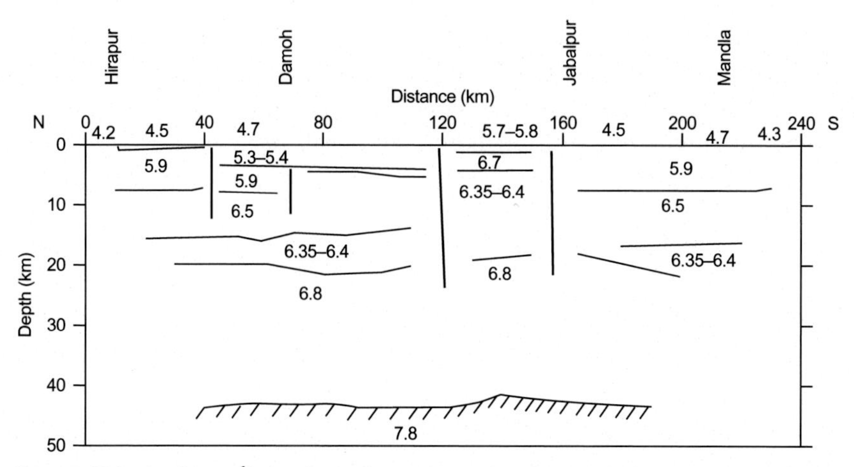

Fig. 6.3 Velocity (km s^{-1})-depth configuration across the Narmada region along Profile I of Fig. 6.2. *(Modified from Murty, A.S.N., Tewari, H.C., Reddy, P.R., 2004. 2-D crustal velocity structure along Hirapur-Mandla profile in Central India: an update. Pure Appl. Geophys. 161, 165–184).*

(upper Vindhyan), 5.3–5.4 km s^{-1} (lower Vindhyan) and 5.9 km s^{-1} (basement), respectively. Maximum thickness of the Vindhyan sediments is about 5 km. Between the Narmada north and south faults the basement, with velocity of 5.7–5.8 km s^{-1}, is at a depth of about 300 m. The region between the Narmada south fault and southern end of the profile is covered with the Deccan Trap volcanics (about 800 m thick) of velocity 4.5–4.7 km s^{-1} that lie over the basement rocks of velocity 5.9 km s^{-1}. The crust consists of four layers with seismic velocities of 5.7–5.9, 6.5–6.7, 6.35–6.50, and 6.8 km s^{-1}. The top of the layer with a velocity of 6.5–6.7 km s^{-1} in the upper crust lies between 2- and 8-km depth. Between the two Narmada faults it is at less than 2-km depth. Its thickness is also less (~3 km) between these faults as compared to that beyond the faults (6–8 km). High velocity of this order is anomalous for the upper crust; in crustal seismic studies the velocity of 6.5–6.7 km s^{-1} is associated with the lower crustal rocks. Normally the velocity in the upper crust does not exceed a value of 6.45 km s^{-1} (Murty et al., 1998). Existence of a layer with velocity of 6.35–6.40 km s^{-1} below the velocity of 6.5–6.7 km s^{-1} in the Narmada region indicates that the high-velocity zone is an intrusion within the upper crust. The upper boundaries of other layers are at approximately 17- and 22-km depth. The velocity in the upper mantle is 7.8 km s^{-1} and the thickness of the crust varies between 40 and 44 km (Murty et al., 1998).

The postcritical seismic reflection data analyses of this profile through ray trace inversion indicates that the Narmada north and south faults are deep penetrating and responsible for intrusion of the upper mantle material in the upper crust. The lateral and vertical inhomogeneities in the crust have caused instabilities in the crust and played an important role in reactivation of the Narmada south fault during the 1997 Jabalpur earthquake (Murty et al., 2008).

Amplitude normalized seismic records, plotted after digitizing the analog data (Fig. 6.4), along this profile show some prominent wave groups of large amplitudes below the Moho (crust-mantle boundary) reflection. The apparent velocities of these wave groups are slightly higher than those of the Moho reflection, indicating that these belong to layers within the upper mantle (Murty et al., 2005). The upper mantle velocity model based on these phases shows a lamellar structure with two alternate low-velocity layers of velocity 7.2 km s^{-1}, separated by a 6-km thick layer of velocity 8.1 km s^{-1} (Fig. 6.5). These reflection phases are from depths of 49, 51, 57, and 60 km, respectively. This type of change in velocity with depth indicates that the subcrustal lithosphere in the central Indian region has varying structural

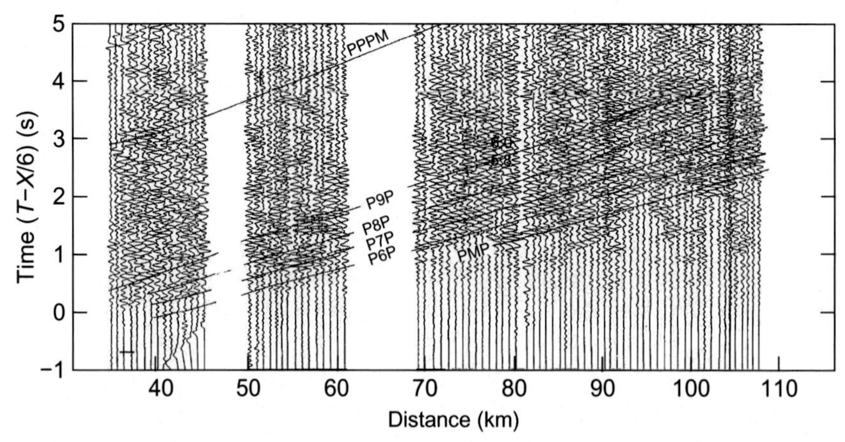

Fig. 6.4 Seismograms indicating the existence of subcrustal low-velocity layers for a shot point. PMP is the reflection from the Moho boundary. Reflections P6P to P9P are from the upper mantle. *(After Murty, A.S.N., Tewari, H.C., Kumar, P., Reddy, P.R., 2005. Sub-crustal low velocity layers in central India and their implications. PAGEOPH 162, 2409–2431).*

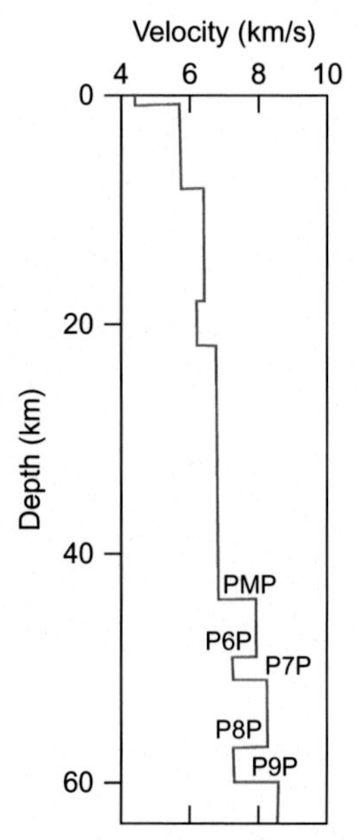

Fig. 6.5 Velocity-depth configuration along Profile I showing subcrustal low-velocity layers in central India. *(After Murty, A.S.N., Tewari, H.C., Kumar, P., Reddy, P.R., 2005. Sub-crustal low velocity layers in central India and their implications. PAGEOPH 162, 2409–2431).*

and mechanical properties. Similar layers in the upper mantle, at slightly different depth, are also present along Profile IV (Sridhar et al., 2007).

The low-velocity layers, occurring at relatively shallow depths below the Moho, are likely to be associated with the zones of weakness and lower viscosity in the upper mantle, suggesting its continued mobility and a possible thermal source as in the Koyna region (see Chapter 5). However, the thermal source is not identified here. The similarity in the structure between the Koyna region on the west coast of India and the Narmada region suggests that a common process, effective from central India to the west coast, might be responsible for the upper mantle low-velocity zones (Murty et al., 2005).

The exposures along Profiles II and III mostly consist of the Deccan Trap volcanics. The exposed Quaternary/Tertiary sediments of the Tapti River (maximum thickness 300 m), which lie over the Deccan Traps, have a velocity of 3.5 km s^{-1} on Profile II and 3.2 km s^{-1} on Profile III. The Deccan Trap volcanics, with a velocity of 4.4–4.5 km s^{-1} and a maximum thickness of about 800 m, directly overlie the basement of velocity 6.0–6.1 km s^{-1} on Profile II (Fig. 6.6) and 5.8–6.1 on profile III (Fig. 6.7). A larger depth to the basement is indicated at a few places in the northern part of the profiles. Between the Deccan Traps and the basement in the extreme north, and at the southern end of Profile II lies another layer with velocity of 5.2 km s^{-1}. This layer is also present at the extreme southern end of Profile III. Between the Narmada north and south faults the basement is shallower as compared to that beyond these faults (Kumar et al., 2000).

The velocity-depth models of the crust along Profiles II and III (Figs. 6.6 and 6.7) are similar to that along Profile I. The high-velocity layer seen in the upper crust on Profile I, and representing a lower crustal intrusion, is seen on these profiles as well. On Profile II this layer (velocity 6.5–6.6 km s^{-1}) is at 7–10 km depth, except in the region bounded by the Narmada north and south faults, where it is at a depth of about 5 km. This layer is also thinner (4.5 km) between the two Narmada faults as compared to that beyond them ($\sim$6 km). The velocity of 6.25–6.30 km s^{-1}, typical for the upper crust, is seen below this layer at a depth of about 8 km between the Narmada faults and 16 km beyond these faults. The velocities 6.7 km s^{-1}, at 20–23 km depth, and 7.2 km s^{-1} at a depth of about 30 km represent the other boundaries within the crust. The Moho, with upper mantle velocity of 8.0–8.1 km s^{-1}, is at a depth of 38–42 km.

On Profile III the layer of high velocity (6.5–6.6 km s^{-1}) in the upper crust is present at a depth of about 2 km between the Narmada faults as

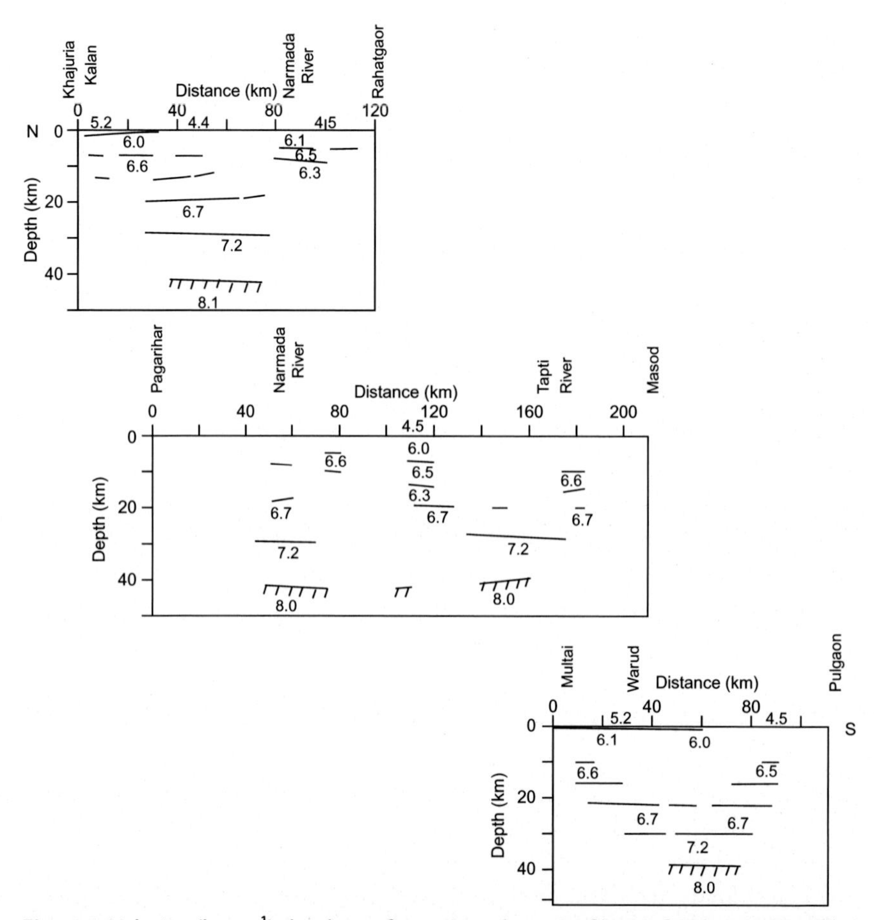

Fig. 6.6 Velocity (km s^{-1})-depth configuration along Profile II of Fig. 6.2. *(Modified from Kumar, P., 2002. Seismic Structure of the Continental Crust in the Central Indian Region and its Tectonic Implications, Thesis submitted to Osmania University, Hyderabad (unpublished)).*

compared to its depth of 9–10 km beyond these faults and is thinner (4–4.5 km) between the faults than beyond them (6–11 km). In the lower crust, the top of the layer with velocity 6.8 km s^{-1} is at 20–24 km depth and that of velocity 7.2 km s^{-1} is at a depth of about 30 km. The Moho, with an upper mantle velocity 8.0 km s^{-1}, is at a depth of 38–44 km (Tewari and Kumar, 2003). A major difference in the crustal structure along Profiles II and III, as compared to that along Profile I, is the presence of a layer with velocity 7.2 km s^{-1} at the base of the crust on Profiles II and III and its absence on

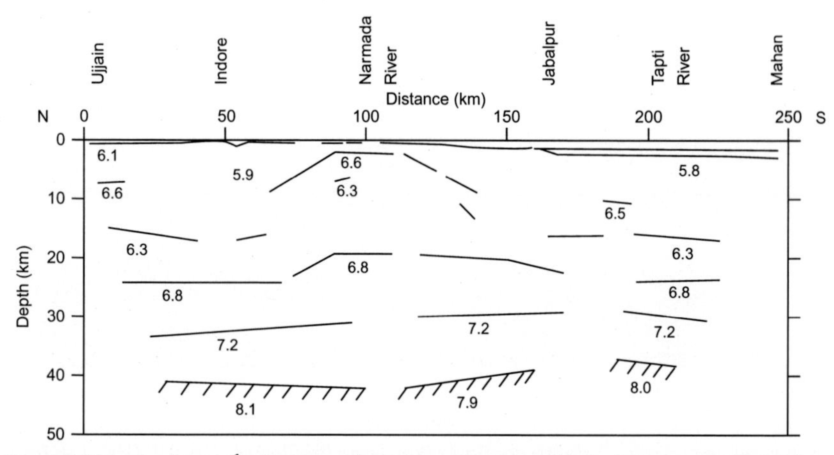

Fig. 6.7 Velocity (km s^{-1})-depth configuration along Profile III of Fig. 6.2. *(Modified from Kumar, P., 2002. Seismic Structure of the Continental Crust in the Central Indian Region and its Tectonic Implications, Thesis submitted to Osmania University, Hyderabad (unpublished)).*

Profile I. This layer, which represents an underplated crust, is either absent or very thin on Profile I and has not been modeled on it.

Profile IV also lies mostly over the Deccan Trap volcanics, except at its extreme northern end where it lies over the exposed Archean and Proterozoic sediments. The upper crustal velocity-depth configuration along this profile is different from other profiles in the Narmada region. A sedimentary graben below the exposed Deccan Traps is indicated on this profile between the Narmada and Tapti rivers. Within this graben the maximum depth to the basement (velocity 5.9–6.2 km s^{-1}) is 5000–5500 m (Fig. 6.8). This graben contains 1000–2800 m thick low-velocity (3.2–3.6 km s^{-1}) sediments under a thick (maximum 2500 m) cover of the Deccan Traps. The Barwani-Sukta and Tapti faults bound this graben (Sridhar and Tewari, 2001). The crustal structure along this profile is more or less similar to that along the Profiles II and III, except that the high-velocity layer in the upper crust is not seen here. The high-velocity layer in the lower crust (7.0 km s^{-1}) is at a depth of about 27 km. The thickness of the crust varies between 38 and 42 km (Sridhar et al., 2007).

The seismic results show the presence of low-velocity sediments under the Deccan Traps on Profile IV, but do not provide any evidence for occurrence of these sediments along Profiles I, II, and III. It is, however, possible that these sediments, being very thin, could not be delineated in the seismic

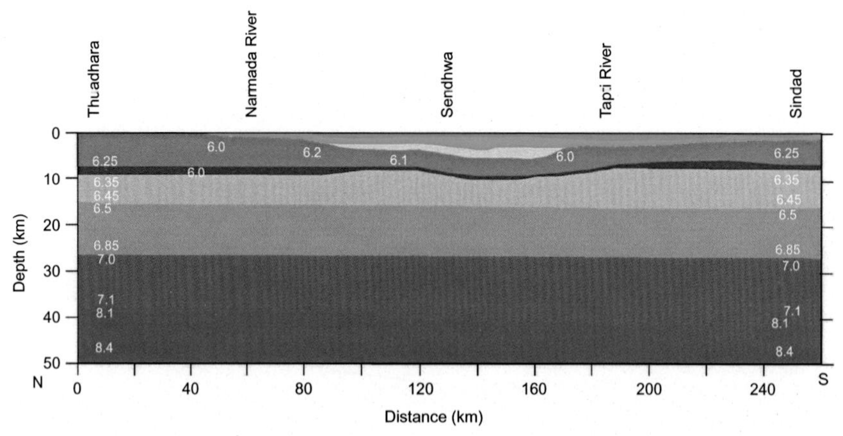

Fig. 6.8 Velocity (km s^{-1})-depth configuration along Profile IV of Fig. 6.2. *(Modified from Sridhar, A.R., Tewari, H.C., Vijaya Rao, V., Satyavani, N., Thakur, N.K., 2007. Crustal velocity structure of the Narmada-Son Lineament along the Thuadara-Sendhwa-Sindad profile in the NW part of central India and its geodynamic implications, J. Geol. Soc. India 69, 1147–1160).*

studies as the seismic parameters were fixed for the much thicker crustal layers. A few gravity and magnetic studies, however, show the presence of Vindhyan sediments under the traps in some parts of central India. The possibility for existence of the Gondwana/Mesozoic sediments in the southern part of Profiles II and III seems to be strong. The basement depth in the southern part of Profile III is about 2500 m. Here, the exposed Deccan Trap layer (velocity 4.4–4.5 km s^{-1}) has a maximum thickness of about 1000 m and another 2000–2500 m thick layer of velocity 5.2 km s^{-1} exists under it. Since the Deccan Trap volcanics have a number of flows, the second velocity layer may represent another flow of the volcanics. It is, however, unlikely that these would be 2000–2500 m thick, as this region is about 800 km away from the main source of volcanic eruption near the west coast of India. Various drill holes in central India indicate that the thickness of the Deccan Traps in this region does not exceed 600 m. The magnetotelluric studies, a little to the east of Profile III, have also shown two distinct layers of Deccan Traps with the total thickness varying between 300–1000 m on the flanks and about 2000 m in the center and also the presence of 300–2000 m thick Gondwana sediments below the traps (Rao et al., 1995). Comparison of the seismic results with the magnetotelluric results indicate that both the 4.4–4.5 and 5.2 km s^{-1} velocity layers belong to the Deccan

Traps, and the sediments delineated on Profile IV are likely to extend below the traps toward south.

The seismic models across the Narmada region show that the northwest-southeast trending Barwani-Sukta fault is the northern limit of the sediments that exist below the traps. This fault also divides the upper crust of the region in two parts. While to the west of the fault exists a sedimentary graben under the Deccan Traps, the region to its east is basement uplift between the Narmada north and south faults. The absence of the shallow high-velocity (6.5–6.7 km s^{-1}) layer in the upper crust seen along Profiles I, II, and III also appears to be caused by this fault. The Barwani-Sukta fault, however, does not affect the lower crust.

The contour maps depicting the basement configuration (Fig. 6.9) and top of the upper crustal high-velocity layer (Fig. 6.10) along Profiles I, II, and III show that both these layers are at shallower depth between the two Narmada faults as compared to those outside these faults and an east-west trending horst exists in large parts of central India (Kumar et al., 2000). This horst is very prominent at least till the base of the upper crust.

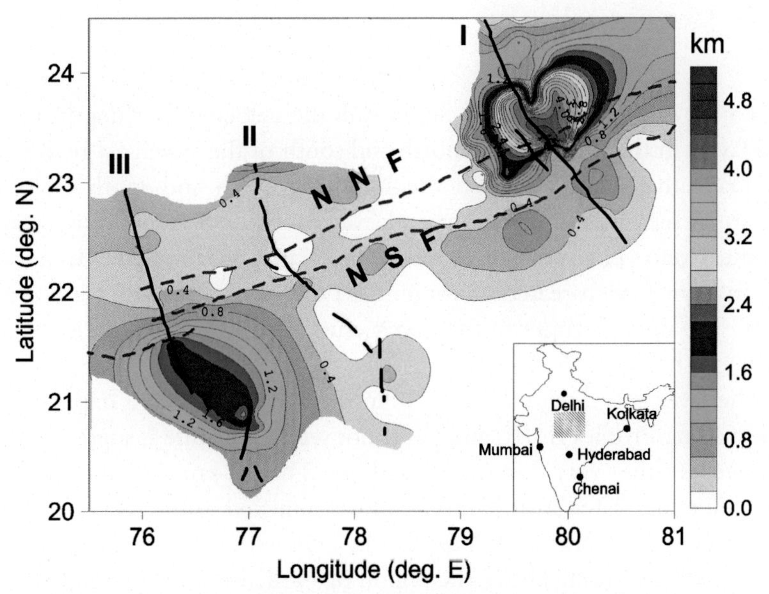

Fig. 6.9 Basement depth contour map of the Narmada region. *NNF*, Narmada north fault; *NSF*, Narmada south fault (Kumar et al., 2000).

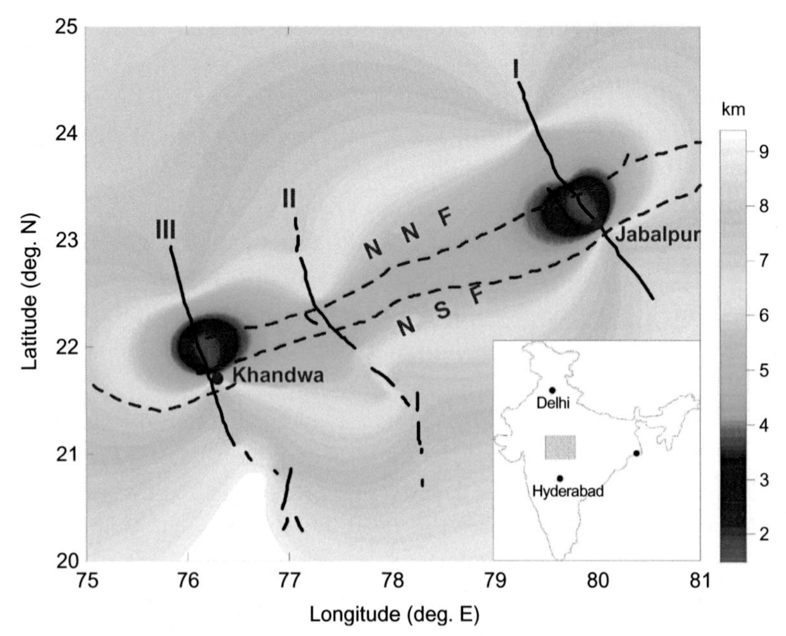

Fig. 6.10 Contour of top of upper crustal high-velocity layer in the Narmada region. *(After Kumar, P., Tewari, H.C., Khandekar, G., 2000. A high velocity layer at shallow crustal depths in the Narmada region. Geophys. J. Int. 142, 95–107).*

To the east of the Barwani-Sukta fault the velocities and depths of various layers in the crust, to the north and south of the Narmada region, are essentially the same. Between the Narmada north and south faults the upwarp is seen at the basement and even in the other intracrustal layers and the Moho. This indicates that both the Narmada faults have deeper origin, involving deep-seated tectonics, and the Narmada region is a crustal uplift. The uplift is larger in the upper crust as compared to the lower part of the crust.

The layer of velocity 7.2 km s^{-1} in the lower crust, at a depth of 28–33 km along Profiles II, III, and IV, is not seen on Profile I, suggesting that this layer is either very thin or absent there. The 7.2 km s^{-1} velocity represents the underplated mantle layer and is connected to the igneous activity concurrent with extrusion of the Deccan Trap volcanics near the west coast of India, due to rifting above the region of anomalously hot mantle around the Reunion plume. A similar layer has also been seen at the base of the crust in the Cambay basin, Saurashtra Peninsula and on the western end of the

Koyna region, thus supporting a large diameter for the plume head (see Chapter 5).

The Bouguer gravity anomaly map (Fig. 6.11) of the central Indian region (Verma and Banerjee, 1992) shows three prominent lows (L1, L2, L3). L1 is just north of Jabalpur on Profile I; L2 is close to Profile II and L3 on Profile III. The gravity low L2 is associated with thick (down to about 5 km) Gondwana sediments that are exposed in that part. A large gravity high is present to the south of Jabalpur. Smaller gravity highs are seen over the Vindhyan sediments in the northern part of Profile I and also in the southern part of Profile III.

To fill in the gaps in the seismic models, two-dimensional gravity models along Profiles I, II, and III were derived (Fig. 6.12). The Bundelkhand craton, to the north of Profile I, has low gravity values. The high-velocity layer in the upper crust along Profile I has a density of 2820 kg m^{-3} in the northern part and 2860 kg m^{-3} in its southern part. Since the high velocity in the

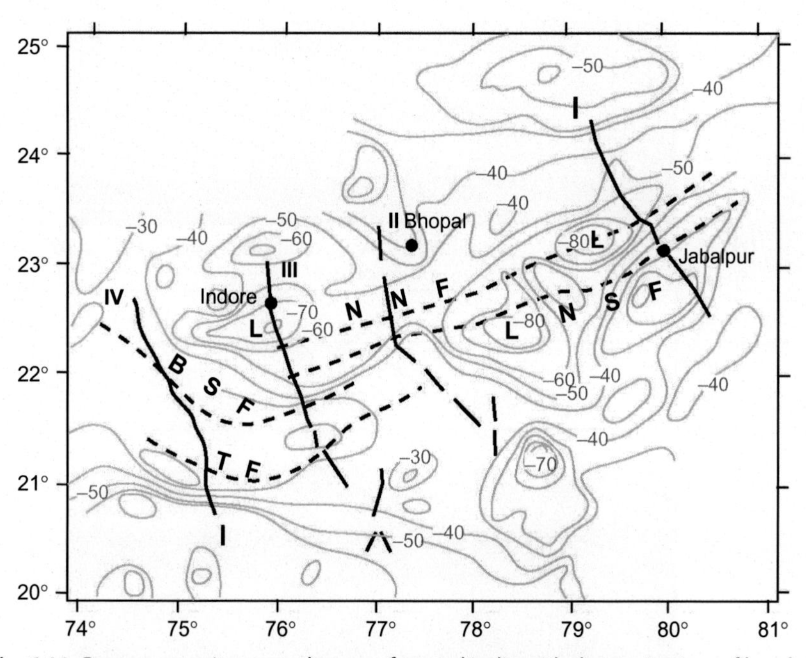

Fig. 6.11 Bouguer gravity anomaly map of central India with deep seismic profiles plotted on it. *(After Kumar, P., 2002. Seismic Structure of the Continental Crust in the Central Indian Region and its Tectonic Implications, Thesis submitted to Osmania University, Hyderabad (unpublished)).*

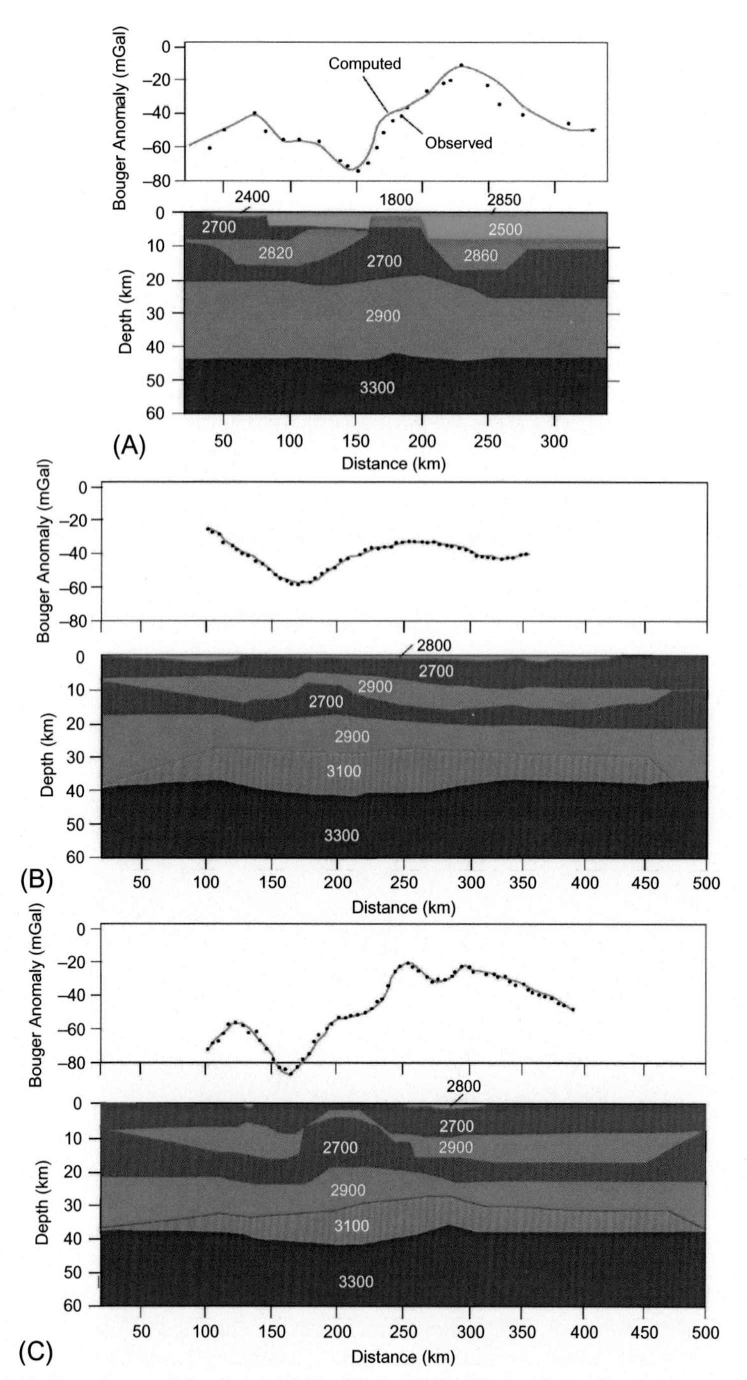

Fig. 6.12 Density models along (A) Profile I, (B) Profile II, (C) Profile III. Density is in kg m^{-3}. *(Modified from Kumar, P., Tewari, H.C., Khandekar, G., 2000. A high velocity layer at shallow crustal depths in the Narmada region. Geophys. J. Int. 142, 95–107).*

upper crust is associated with material that has intruded from the lower crust/upper mantle, a higher density is considered in this layer than that normally associated with the upper crust. The gravity lows that lie to the north of the seismic profiles are probably due to either a thinner volcanic layer or presence of the subtrappean Vindhyan sediments and/or absence of the high-density upper crustal layer. In the absence of the seismic results the gravity high over the exposed Vindhyan sediments was interpreted either due to underlying volcanics, which form a part of the Palaeoproterozoic Bijawar formation (Verma and Banerjee, 1992), or extension of the high-density rocks from the Bundelkhand craton up to the Narmada region (Mishra and Kumar, 1998). The seismic-based gravity model along Profile I shows that this high is due to presence of a high-density body below the Archean basement. Due to large thickness (~5 km) and lower density of the Vindhyan sediments (2500 kg m^{-3}), as compared to that of the basement (2700 kg m^{-3}), the gravity high in the Vindhyan region is not as prominent as the southern gravity high. If the basement density replaces the density of 2500 kg m^{-3}, the shape of the gravity high would be different (Murty et al., 2004). The gravity low L1 is due to large thickness of the Vindhyan sediments and thinning of the high-velocity/high-density body in the upper crust between the Narmada north and south faults as compared to its thickness beyond these faults. The gravity high south of Jabalpur was explained because of a 5-km thick high-density causative source at 12-km depth (Mishra, 1992). However, the seismic results show that it is due to an intrusive magmatic body in the upper crust (Mall et al., 1991). This intrusive body is about 9-km thick and lies between 8- and 17-km depth.

The seismic and magnetotelluric studies (Gokaran et al., 1995) show that the relatively higher seismic velocities beneath the Vindhyan sediments are associated with a conductive crust and a conductor is located in the upper crust in the region of gravity high to the south of Jabalpur. Geomagnetic depth studies, overlapping Profile I, mapped an 8–10 km thick conductive anomaly as an anticline structure at 10–12 km depth to the south of Jabalpur and related it to a cap enclosing the magmatic body (Arora et al., 1995).

The gravity models along Profiles II and III (Fig. 6.12) consider densities similar to those on Profile I for various layers in the crust. The only difference is that these models show the presence of the underplated layer with a density of 3100 kg m^{-3}, which is not seen in the model along Profile I. The gravity low L3 has been caused by lesser thickness of the high-velocity/high-density body between the Narmada north and south faults, on Profile III, as compared to a larger thickness beyond these faults (Tewari et al., 2001). Total magnetic intensity values in the region show several short wavelengths but large amplitude anomalies. The heat-flow values in the northern part are 40–70 mW m^{-2} and rise sharply to about 100 mW m^{-2} to the south of the Narmada-Son lineament.

While the deep seismic studies that show a Moho upwarp across the width of the Narmada-Son lineament along Profile I, the teleseismic receiver function studies (Rai et al., 2005) show a Moho downwarp to about 50 km to the south of Jabalpur, against an average depth of 40 km elsewhere along the profile. The crust beneath the lineament zone also has a higher V_p/V_s ratio of 1.84 compared to about 1.73 in other parts of the profile. The high V_p/V_s ratio suggests a high-density mafic to ultramafic lower crustal composition that compensates the crustal root. Presence of such a mass in the deep crust may lead to gravitationally induced stresses in the lower crust that contribute to the failure of the rock along the preexisting fault leading to the earthquake in the deep crust.

6.1.3 Tectonic Significance of Seismic Results

Seismic studies along the profiles crossing the Narmada region show that to the east of the Barwani-Sukta fault the crust is upwarped between the Narmada north and south faults. It is believed that the Moho upwarp under the Narmada region is related to vertical mass transfer, from the upper mantle into the crust, between the late Archean and Paleoproterozoic period (Ghosh, 1975). To the south of the Narmada region the crust is thick and contains linear and extensive basic intrusive bodies in its upper part. These bodies were placed there due to deep-seated phenomena of asthenospheric upwelling related to a mantle plume or hotspot trace. This activity appears to have been transmitted in time from central to the east coast of India and provided the necessary thermomechanical framework for separation of India from Gondwanaland during the late Jurassic or the early Cretaceous (Mishra, 1992). It is also suggested that rifting or replacement of the crust due to emplacement of mantle-derived rocks at shallow crustal levels followed reduction in the Moho depth (Venkat Rao et al., 1990). The Narmada north and south faults seem to have acted as fissure zones through which molten magma erupted and got emplaced on either side of the Narmada region. Intense crustal movements during this process disturbed the Moho configuration (Sivaji and Agrawal, 1995). Singh and Meissner (1995) suggest that the Deccan Trap eruption in central India is associated with huge magmatic intrusion at the base of the crust.

Major reactivation of the Narmada north fault has not taken place after the deposition of the Vindhyan sediments (Acharya et al., 1998). It has also been mentioned in earlier discussion (Section 6.1.2) that the upper crustal intrusive mafic body has smaller thickness between the two Narmada faults, as compared to those beyond the Narmada faults, and the upper crust is uplifted. These indicate that the postintrusion activities on two sides of the Narmada region are independent of each other; meaning that the two Narmada faults were active at different geological times. Major activity along the north fault probably occurred during the Proterozoic time while that along the south fault started again during or after the breakup of Gondwanaland. Occasional seismic activity observed in the neighborhood of the south fault is evidence that it is still active. Smaller thickness of the intrusive layer between the two Narmada faults is an indication that the uplift observed here is older than the activity along the south fault.

The seismic velocity-depth models of the crust along the seismic profiles indicate that the Barwani-Sukta fault played a major role in development of the crust in this region. To the east of this fault, the Narmada north and south faults divide the upper crust in a horst-graben structure. The gravity models show that the low gravity axis in central India, except for the low L2, corresponds to the zone of Narmada uplift that would normally have given rise to a gravity high. This gravity axis divides the region into two distinct parts, north and south of the Narmada region Mafic intrusion in the upper crust, represented by the high-velocity/high-density body, seems to be an important element in shaping the present structural trends of the Narmada region to the east of the Barwani-Sukta fault. Its lower thickness at the uplift, as compared to that beyond the uplift, enhances the gravity low caused by the presence of the sediments in the Vindhyan basin. The displacement in this layer is consistently observed along Profiles I, II, and III, which indicates that both the north and south faults extend at least up to Profile III. This displacement is probably a result of repeated reactivation of the Narmada fault system. These models also reveal that major crustal disturbances are confined down to the upper crust. The arguments may be: (a) either the Narmada north and south faults are prominent only down to the upper crust and die out with depth due to ductility of the lower crust, or (b) relief on the lower crust and the Moho was subsequently erased by the ductile flow, whereas the structures were preserved in the cooler crust above (Kumar, 2002).

Since high-velocity/high-density layers in the upper crust correspond to the rocks of lower crustal or upper mantle origin, the upper crustal body is likely to represent a pre-Archean granulites/amphibolites enclave in the Archean crust. Granulites are present within the exposed gneiss of central India. During this Proterozoic tectonic activity, the Archean crust bounded by the faults got uplifted, as a result of which the high-velocity body was subjected to a large displacement. The region remained tectonically inactive since the end of the Precambrian until about 300 Ma (the beginning of the Gondwana sedimentation) when activity started again. The only major known tectonic activity after this period is marked by the Deccan Trap volcanism, which might have resulted in reactivation of the fault system. Another consequence of the Deccan Trap volcanism is the presence of the underplated layer above the Moho. Absence of the high-velocity/high-density layer in the upper crust to the west of the Barwani-Sukta fault suggests that this fault has controlled the supracrustal engulfment of the granulite/amphibolite bodies within the upper crust, indicating that this fault may be of Proterozoic age.

A cartoon of the present-day crustal and lithospheric structure of the Narmada region (Fig. 6.13) shows that this structure developed in at least

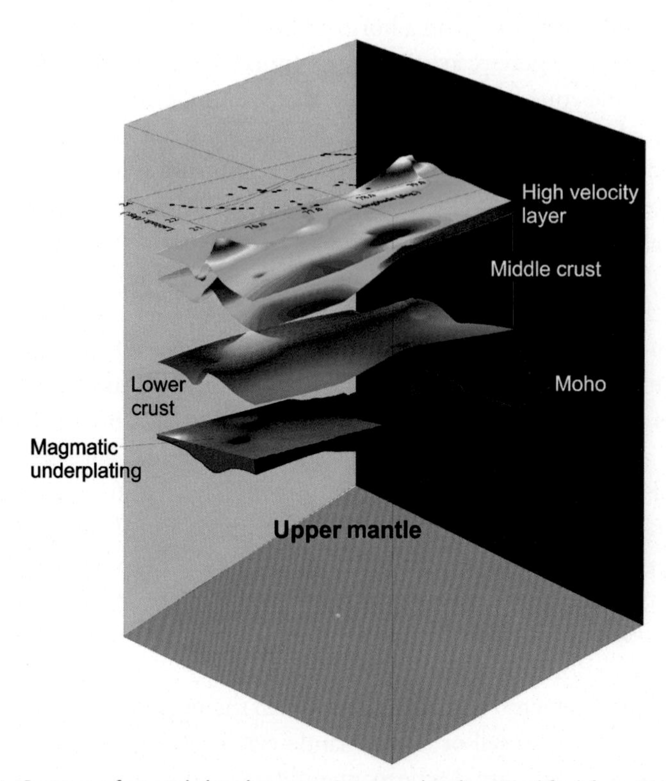

Fig. 6.13 Cartoon of crustal development in central India. *(Modified from Tewari, H.C., Murty, A.S.N., Kumar, P., Sridhar A.R., 2001. A tectonic model of the Narmada zone. Curr. Sci. 20, 273–877).*

three major tectonic phases. During Phase I, in the Archean-Proterozoic, the north Narmada fault was active. In the earlier part of this phase high-density intrusive material was engulfed in the upper crust. The mafic intrusion in the upper crust was limited to the region east of the Barwani-Sukta fault. In Phase II, activity along the south fault started again during the Jurassic-Cretaceous period. Phase III coincided with the passage of India over the Reunion plume during the late Cretaceous. The plume head at that time was situated close to the west coast of India, to the south of the Narmada region. Due to the tectonic disturbances, material from the plume head got emplaced at the base of the crust and extended to the east to a large distance. This caused underplating in large parts of the crust. Subsequent extrusion from the plume head led to the Deccan Trap activity and its deposition on the then existing topography. Around the same time a minor eruption might also have taken place from the upper crustal mafic body to the south of Narmada zone (Tewari et al., 2001).

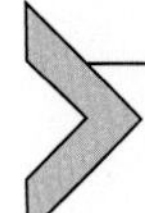

6.2 DEEP REFLECTION STUDIES ACROSS THE CENTRAL INDIAN SUTURE

Another important geologic feature in central India is the central Indian suture (CIS), the boundary between the Satpura mobile belt to the north and Kotri-Dongargarh mobile belt to the south (Fig. 6.1). This is a megashear zone extending for hundreds of kilometers across the region. On the basis of petrological, structural, chemical and chronological data, this suture has been identified as a major Paleoproterozoic ductile shear zone separating two distinct tectono-magmatic terranes of the Bundelkhand protocontinent in the north and the Dharwar protocontinent in the south (Yedekar et al., 1990). The two-pyroxene granulites occurring in a narrow tectonic slice along the southern margin of the former craton are inferred to be the metamorphic equivalents of oceanic basalt. Three volcano-sedimentary successions, i.e., the Nandgaon group, the Khairagarh group and the Dongargarh granitoid, occur to the south of the suture. The Nandgaon volcanics form the base of the Dongargarh rocks. The Khairagarh group of rocks overlie the Nandgaon group with an angular unconformity (Fig. 6.14).

To study the seismic signatures of this suture a deep reflection profile (Seoni-Kalimati) was recorded across it (Fig. 6.14). This northwest-southeast profile is about 150 km long and passes through the Deccan Trap, central Indian gneissic complex (Paleoproterozoic to middle Archean),

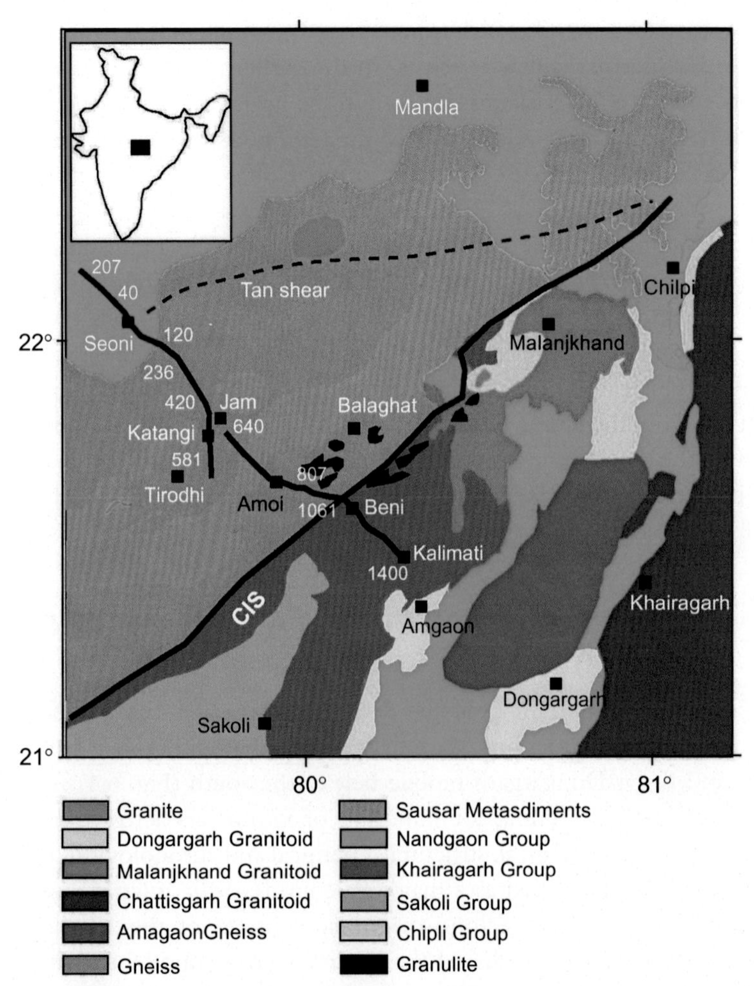

Fig. 6.14 Location map of the deep reflection profile across the central Indian suture zone in central India. *(After Reddy, P.R., Murty, P.R.K., Rao, I.B.P., Khare, P., Kesava Rao, G., Mall, D.M., Koteswara Rao, P., Raju, S., Sridhar, V., Reddy, M.S., 1995. Deep crustal seismic reflection fabric pattern in Central India-preliminary interpretation. In: Sinha-Roy, S., Gupta, K.R. (Eds.), Continental Crust of NW and Central India. Geol. Soc. India, Mem., vol. 31, pp. 537–544).*

Dongargarh group (late Archean) and ends in the Chattisgarh granitoid (Neoproterozoic).

The reflectivity along this profile is good except in its northwest part (Fig. 6.15), which is covered by the Deccan Trap volcanics. Here only a

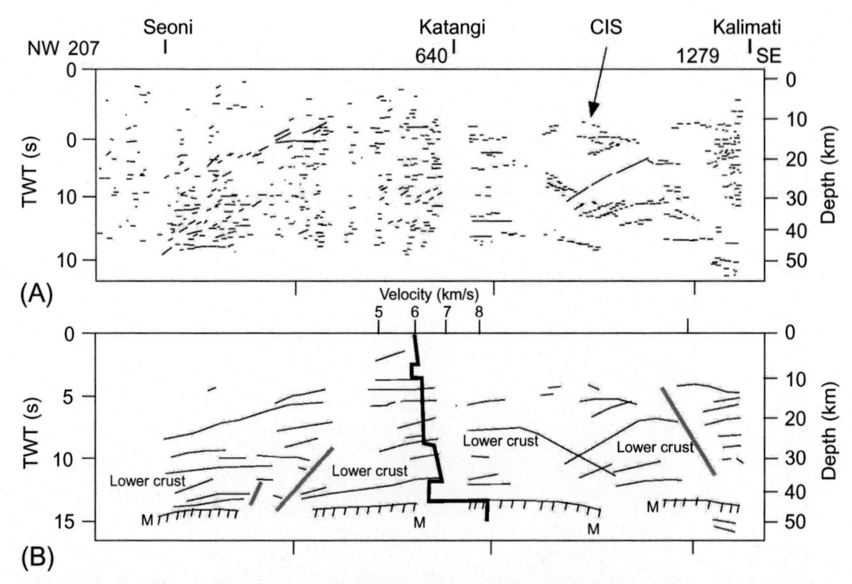

Fig. 6.15 (A) Line drawing of the reflectivity pattern and (b) structural trends across the central Indian suture zone. One-dimensional velocity-depth distribution along the profile is plotted on (B). *(After Reddy, P.R., Murty, P.R.K., Rao, I.B.P., Khare, P., Kesava Rao, G., Mall, D.M., Koteswara Rao, P., Raju, S., Sridhar, V., Reddy, M.S., 1995. Deep crustal seismic reflection fabric pattern in Central India-preliminary interpretation. In: Sinha-Roy, S., Gupta, K.R. (Eds.), Continental Crust of NW and Central India. Geol. Soc. India, Mem., vol. 31, pp. 537–544).*

few reflections are seen in the upper crust, down to 4.5 s TWT—about 15-km depth. Along the rest of the profile the reflections are short and dense, especially in the middle and lower crust, and the reflectivity is good from about 5 s TWT down to 14 s TWT (42–44 km depth). Structural trends in the crust are broad and show a gentle dip. Termination of reflectivity at about 14 s TWT represents the crust-mantle boundary (Moho). The reflectivity pattern along the profile shows a distinct reversal of the dip direction on the two sides of this suture; to the northwest of it the reflections dip to the southeast, and to its southeast the dips are to the northwest (Reddy et al., 1995).

The two adjacent seismic domains dipping toward each other represent a suture between the two crustal blocks. The suture itself is not imaged as a sharp boundary, indicating that it probably has a near vertical orientation due to the disturbed character of the crust in a 20–30 km-wide zone. A much clearer picture of the reflectivity along the profile (Fig. 6.16) has

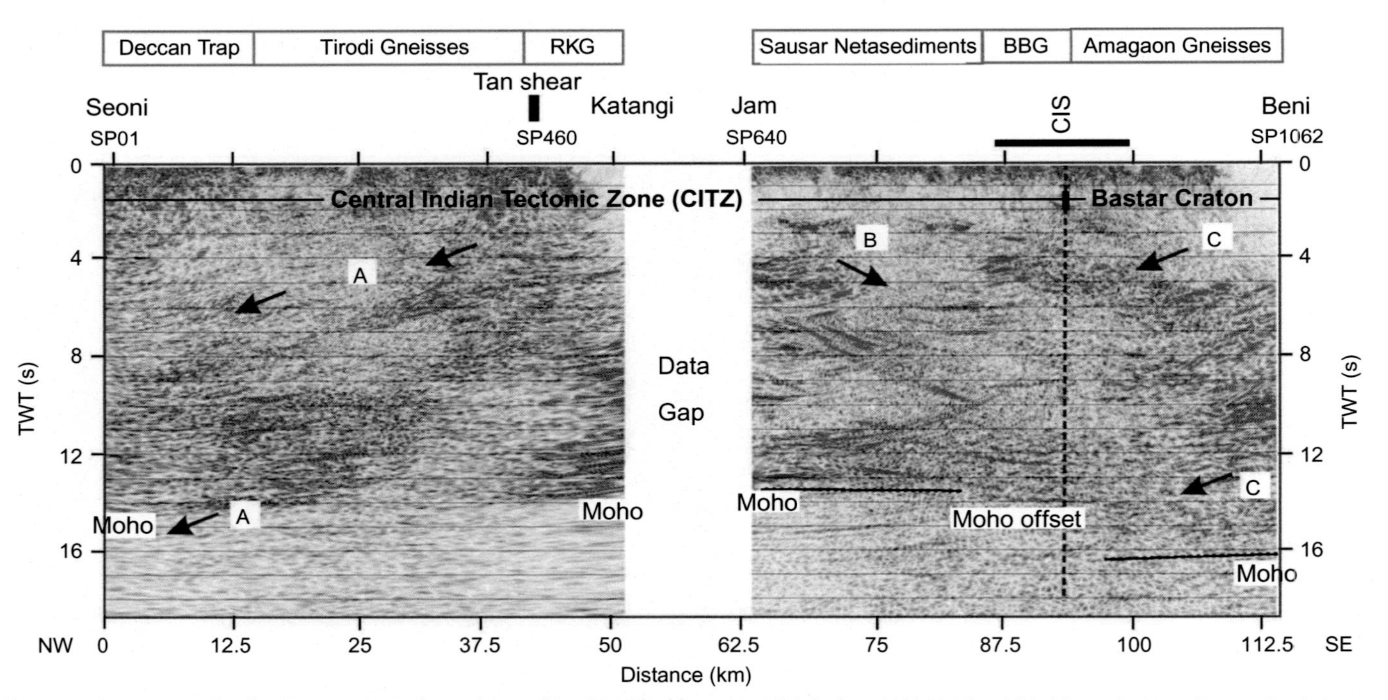

Fig. 6.16 Reprocessed reflectivity section along the profile. *(Modified from Mandal, B., Sen, M.K., Vaidya, V.V., Mann, J., 2014. Deep seismic image enhancement with the common reflection surface (CRS) stack method: evidence from the Aravalli–Delhi fold belt of northwestern India. Geophys. J. Int. 196, 902–917).*

come up after reprocessing the data (Mandal et al., 2014). The main features of the crust close to the suture are: an oppositely dipping reflection fabric, a Moho offset, a positive-negative gravity anomaly. Combined with the geological anomaly, these indicate that the suture is a collision zone developed due to the interaction of the Bastar and Bundelkhand cratons. Strong bands of reflectors, with a predominant dip toward the northwest, covering the entire crustal column in the first 65 km of the northwestern portion of the profile (between Seoni and Katangi) create a dome type structure with the apex around 30 km northwest of the suture. The reflection Moho, taken as the depth to the deepest set of reflections, varies in depth between 41 and 46 km and is imaged sporadically across the profile, with the largest amplitude occurring in the northwest.

The velocity-depth model of the region shows two low-velocity zones, one in the upper crust and the other in the lower crust just above the Moho. In the central Indian region, the Moho is dipping and the root of the crust is well defined. In most of the other older cratons around the world, the Moho is either missing and/or is flat, as it is likely to be equilibrated in the period of postorogenic extension. In very few of the older belts, such as in the Baltic shield, the roots of the crust are as clearly defined as in the central Indian region (Reddy et al., 2000).

The reflectivity northwest and southeast of the central Indian suture, in a zone of approximately 30-km width, is not very good. These data record the presence of a Mesoproterozoic collision between two microcontinents, with the Satpura mobile belt in the north being thrust over the Bastar craton in the south (Mall et al., 2008). Based on the reflectivity pattern of the region, Mall et al. (2008) propose a tectonic model that explains the key elements of the region, the main features of the crust and their relationship to the suture (Fig. 6.17). This model shows thrusting of the Satpura mobile belt over the Bastar craton and the presence of granulites on the surface due to obduction of the lower crust during the collision of the two crustal blocks.

The Bouguer gravity anomaly map of the region shows scattered lows and highs between −50 and −80 mGal. 2.5D forward modeling of the gravity data along the Seoni-Rajnandgaon profile is consistent with the seismic results. High-frequency anomalies have been modeled as near-surface intrusive bodies in the upper crust (Fig. 6.18). The model also shows that all the crustal layers of the Bastar craton consistently dip toward the north near the suture zone, as if the crust of the Bastar craton was subducted beneath

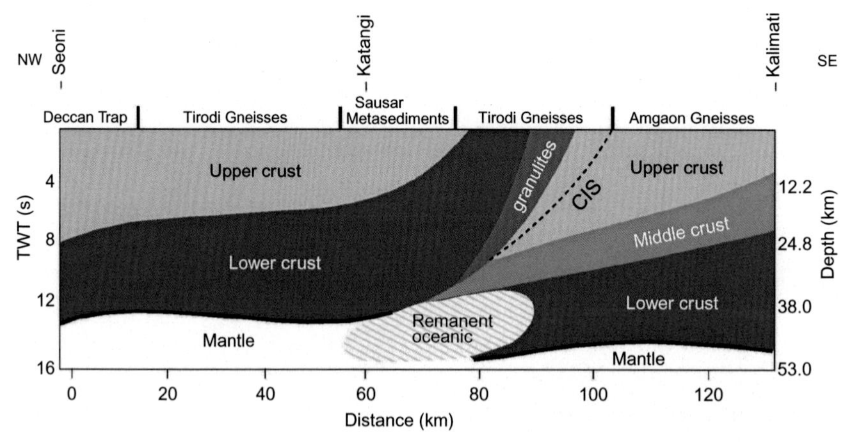

Fig. 6.17 The crustal division in the area of CIS and the elements of the tectonic model. *(After Mall, D.M., Reddy, P.R., Mooney, W.D., 2008. Collision tectonics of the Central Indian Suture zone as inferred from a deep seismic sounding study. Tectonophysics 460, 116–123).*

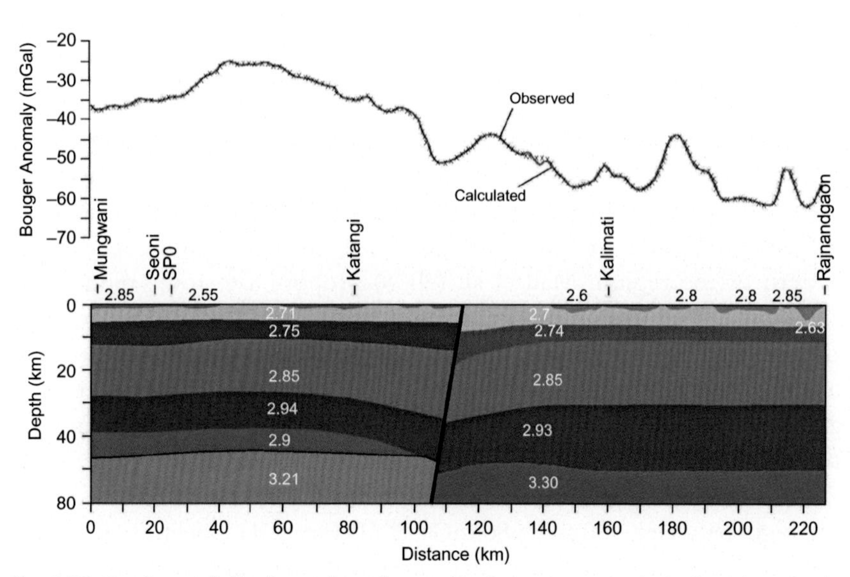

Fig. 6.18 Gravity model in the region of central Indian suture zone based on the seismic data. The densities shown are in g cm^{-3}. *(After Nageswara Rao, B., Kumar, N., Singh, A.P., Prabhakar Rao, M.R.K., Mall, D.M., Singh, B., 2011. Crustal density structure across the Central Indian Shear Zone from gravity data, J. Asian Earth Sci. 42, 341–353).*

the suture. A low-velocity (6.4 km s^{-1}) layer at the base of the crust corresponding to density of 2900 kg m^{-3} and relatively lower density (3210 kg m^{-3}) subcrustal mantle may be the imprint of thermal remobilization beneath the central Indian tectonic zone. This indicates a mechanically strong lower crust and relatively weak subcrustal lithosphere where the requisite mass deficiency is balanced by a lower density (3210 kg m^{-3}) in the subcrustal mantle beneath the Satpura mobile belt. It has been hypothesized that through underplating and fractionation of basaltic magma at the base of the crust, the ultramafic root of the crust became stronger (Nageswara Rao et al., 2011).

The synthesis of seismic results of the profile across the central Indian suture and Profile I (Hirapur-Mandla) of the Narmada region with geologic, gravity, magnetic and heat-flow results, along a 100-km wide corridor, is able to explain various geophysical anomalies existing in the central Indian region and shows that the geometry of the Narmada region and the central Indian suture is consistent with the results obtained through the seismic and gravity data (Divakara Rao et al., 1998).

Based on the available geological and geochronological data, Mandal et al. (2014) suggest a model for tectonic evolution of the CIS. They suggest that an oceanic crust existed between the Bundelkhand and Bastar cratons during 2.2–1.8 Ga (Fig. 6.19A). Compressional forces that developed in the region led to the subduction of the oceanic crust of the Bastar craton toward the north. The active subduction system culminated at 1.5 Ga with a continent-continent collision (Fig. 6.19B). The subduction process is manifested by the north-dipping reflection band in the seismic section between Seoni and Katangi (Fig. 6.19C). The rifting took place in the southern part with the opening of the ocean, and the Sausar sediments were deposited in the northern part of the Bastar craton during the early Neoproterozoic (Fig. 6.19D). During the formation of the Rodinia supercontinent, the compressional forces that were developed over the Indian shield brought together the Bundelkhand craton in the north and the Dharwar protocontinent in the south with the northward subduction of the Bastar craton and the formation of the Sausar orogeny (Fig. 6.19E) during the early Neoproterozoic. The thickened crust of the Sausar orogenic belt collapsed under its own weight due to delamination (Fig. 6.19F). During this process, middle-lower crustal granulites were brought to the surface. This has resulted in the thinning of the crust in the Sausar belt.

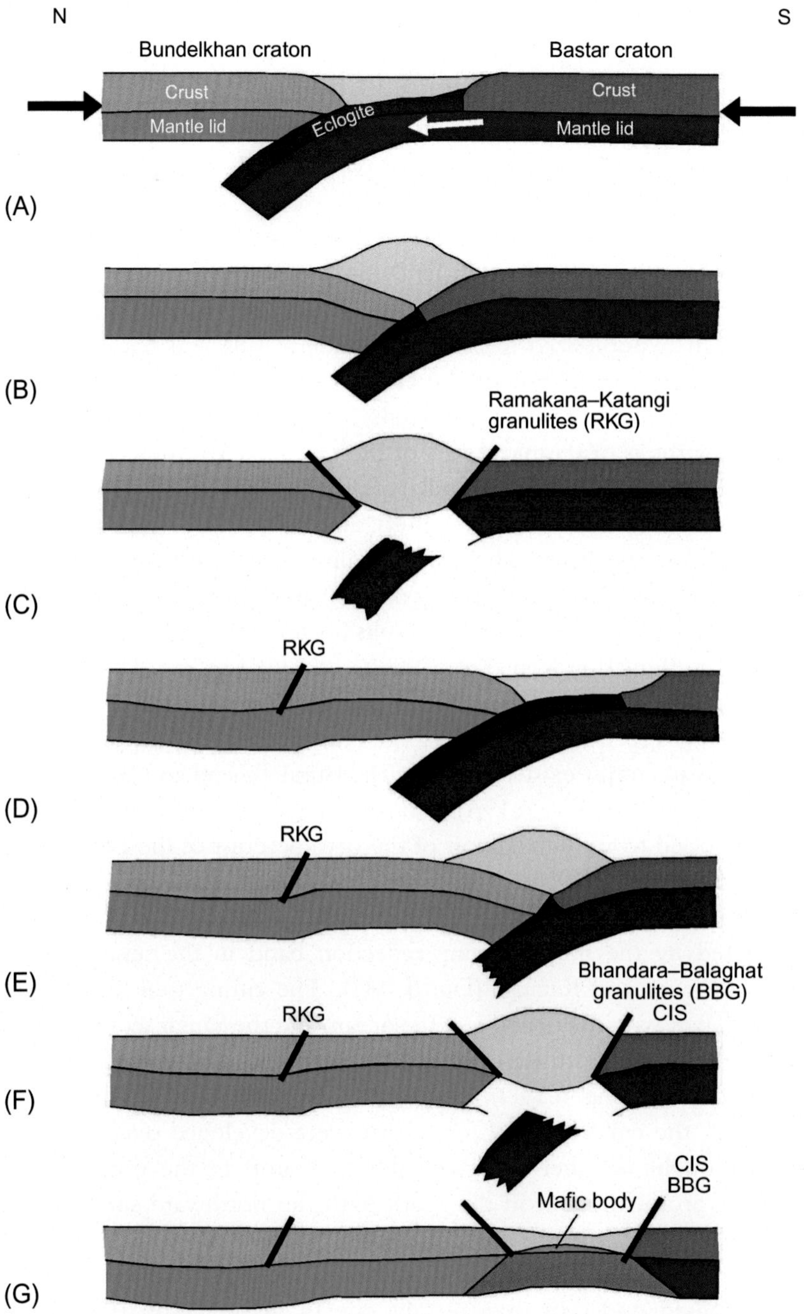

Fig. 6.19 Schematic diagram showing tectonic evolution of the CIS, central India. (A–C) The earlier collisional episode during 1.5–1.6 Ga, (D–F) the Sausar orogeny during 1.1–1.0 Ga, and (G) the present-day crustal features. *(Modified from Mandal, B., Sen, M.K., Rao, V., 2013. New seismic images of the Central Indian Suture Zone and their tectonic implications. Tectonics 32, 908–921. https://doi.org/10.1002/tect.20055.)*

Northwest Himalayan, Tibet, and Karakoram Regions

The Himalaya and the Karakoram mountains and the adjacent Tibetan plateau are the Earth's largest region of elevated topography. The 200–300 km wide mountain chain is 2500-km long from the northwest to the southeast. Understanding of the deep crustal structure in most of the Himalayan region is primarily based on geologic and tectonic modeling, which is derived from geology, a small amount of regional gravity/magnetic data, geoelectrical studies and interpretation of the seismicity. The collision of the Indian plate with the Eurasian plate at about 45 Ma and subsequent thrusting of the former under the latter are responsible for the origin of the Himalaya. The convergence and thrusting in the Himalaya are still continuing, as is evident from the seismicity associated with this region.

Most of the young orogens, like the Himalaya, show crustal roots that are in isostatic equilibrium with their topography. But despite its large area, the seismic data available in the Himalayan region is negligible and does not permit proper estimation of the seismological parameters of this orogenic belt. During the last 30 years sparse but more reliable information, through seismic and related studies, has been obtained in the eastern part of the higher Himalaya and Tibet (Hirn et al., 1984a; Sapin and Hirn, 1997). Geophysical studies carried out under various collaborative programs between China, Nepal, France, the United States and other western countries has increased our knowledge of the crustal structure in the eastern Himalayan region. The INDEPTH program, particularly the results along INDEPTH-1, has helped in better understanding of the collision boundary between the Indian and Eurasian plates (Hauck et al., 1998; DeCelles et al., 2002).

In recent times several new concepts have emerged concerning the crustal and lithospheric structure in the northwestern Himalaya (Fig 7.1), particularly across the Indus-Tsangpo suture zone, which is believed to be the boundary between the Indian and Eurasian plates. A significant change in the anisotropy across this suture suggests a change in the direction of deformation at large depth (Hirn et al., 1995). In the Indian plate, where the collision effects are small, orientation of the anisotropy is parallel to

Structure and Tectonics of the Indian Continental Crust and Its Adjoining Region
https://doi.org/10.1016/B978-0-12-813685-0.00007-8

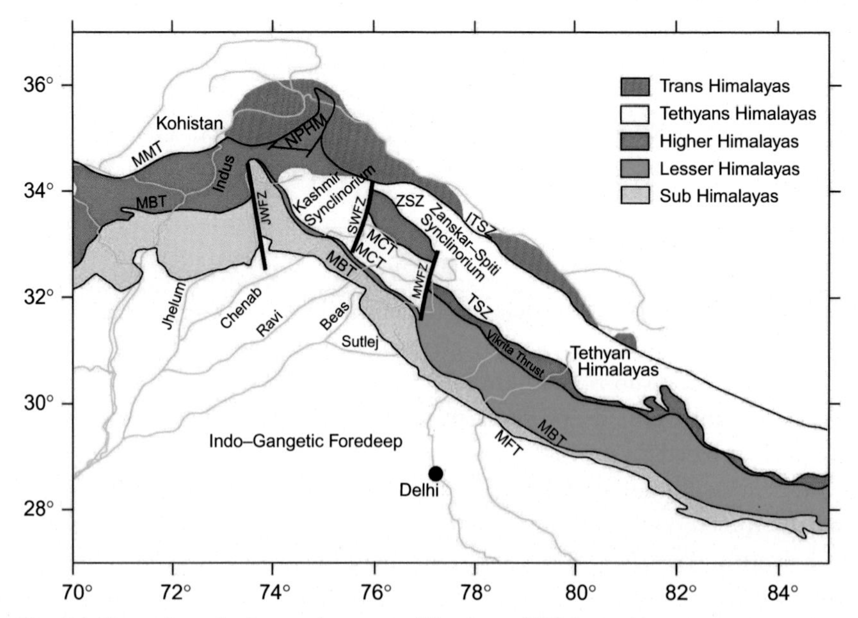

Fig. 7.1 Tectonic units in northwestern Himalaya. *ITSZ*, Indus-Tsangpo suture zone; *MMT*, main mantle thrust; *ZSZ*, Zanskar suture zone; *TSZ*, Tethyan shear zone; *MCT*, main central thrust; *MBT*, main boundary thrust; *MFT*, main frontal thrust.

direction of the plate motion, i.e., to the northeast, while within the Tethyan zone they are aligned with the northwest differentiating motion between India and Tibet (part of the Eurasian plate). To the north of the Indus-Tsangpo suture a ductile mantle flow, without any signature of the Indian lithosphere, is indicated. A suggestion has also been made that before the India-Asia collision the Indian plate was subducting to the north but the polarity of subduction changed after the collision, subsequent to which the Eurasian mantle has subducted to the south, thus limiting the present Indian lithosphere to the Indus-Tsangpo suture (Willet and Beaumont, 1995).

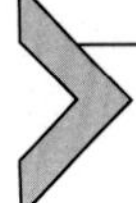

7.1 GEOLOGY AND TECTONICS OF THE NORTHWEST HIMALAYA

The 200–300-km wide Himalayan mountain chain is 2500-km long from the northwest to the southeast and consists of six litho-tectonic zones from the north to the south (Fig 7.1). These units and their rock types are: (i) the trans-Himalayan zone batholiths, to the north of the Indus-Tsangpo suture zone, which are interpreted as the Andean-type northern margin of the Tethyan Himalaya; (ii) the Indus-Tsangpo suture zone, which represents the line of collision of India and Asia, and is composed of the marine sediments, ophiolite, arc volcanics and melange; (iii) the Tethyan Himalaya, which are Cambrian to Paleocene sediments deposited on the Indian continental terrace; (iv) the higher Himalaya, which is the regionally metamorphosed Indian continental crust of mainly Proterozoic age and leucogranites; (v) the lesser Himalaya that is represented by non- or weakly metamorphosed Indian continental crust that ranges in age from the Proterozoic to Paleozoic and Paleocene foreland basin sediments; and (vi) the Sub-Himalaya, which consists of the foreland basin sediments eroded from the rising orogen and deposited in the peripheral foreland basin in front of the mountain belt. These units are separated from each other by various detachment thrusts and faults (Le Fort, 1996).

In the large part of the central Himalaya, metamorphism is associated with the suturing and later thrusting of the Indian plate under the Eurasian plate and is placed at 50–10 Ma (Harrison et al., 1997). The process of thrusting created crustal shortening, which is accommodated by several thrusts and faults. These thrusts extend beneath the Indo-Gangetic foredeep, the Sub-Himalaya and the lesser Himalaya with very shallow dip, and separate the sedimentary wedge from the basement thrust also known as the detachment (Seeber and Armbruster, 1981). Formation of the northwest Himalayan Syntexis began at about 55 Ma, when the edge of the Indian passive margin thrusted under the Eurasian active margin (Guillot et al., 2003). Several workers have discussed the geometry and evolutionary model of the northwest Himalayan Syntexis. At least three mechanisms have been put forward to explain consequences of the collision and missing oceanic crust. The first mechanism describes the thrusting of the Indian continental crust under the Tibetan crust (Powell et al., 1988). The second is the extrusion or escape tectonics mechanism (Molnar and Tapponnier, 1975) in which the Indian plate is seen as an indenter that squeezed the Tibetan block out of its way. The third mechanism proposes that thrusting and folding of the sediments of the Indian passive margin together with the deformation of the Tibetan crust accommodated a large part of the crustal shortening (Dewey et al., 1989).

Different phases of deformation of the northwest Himalaya include: (a) the northeast directed movements of the region, (b) main compressional phase affecting the sedimentary series of the Tethys Himalayan deformation that propagated gradually from the Indus-Tsangpo suture zone to the southwest, (c) north to north-northwest thrusting under the Tethys that eventually forms the higher Himalayan crystalline, (d) activation of thrusting along the main central thrust, (e) dome formation of the higher Himalayan crystalline sequence, and (f) high-angle brittle ductile normal faulting.

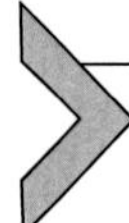

7.2 SEISMOTECTONICS OF THE HIMALAYAN AND TIBET REGIONS

Four devastating earthquakes, of magnitude larger than 8, have occurred in the Himalayan region during the last century. Earthquakes of magnitude 7 and above frequently take place in a narrow band just north of the higher Himalaya (Ni and Barazangi, 1984). These earthquakes have ruptured the decollement ahead of the higher Himalaya up to the frontal structures in the Sub-Himalaya and to its further south. Most of the strain is accumulated at the tip of the dislocation, which plastically absorbs only a fraction across the Himalayan ranges (Kind et al., 2002). The topographic distribution in the Himalayan region is the consequence of an interseismic deformation. Most of the Himalayan earthquakes nucleate at the ramp-flat transition of the lesser Himalaya and relax the elastic strain accumulated around the locking zone during the interseismic period. These earthquakes, of magnitude 6–7, activate only a section of the flat beneath the lesser Himalaya. The whole flat, up to the main frontal thrust, would be activated only during the events of magnitude larger than 8. Though there is no evidence of surface rupture along the main frontal thrust, it is assumed that the growth of this thrust is due to the large earthquakes (magnitude larger than 8). These major earthquakes account for most of the India-south Tibet convergence and therefore would be much more efficient in transferring the slip as compared to the smaller events.

Seismological studies suggest that the Indian mantle lithosphere underthrusts the Tibet plateau horizontally and deep into the plateau (Ni and Barazangi, 1984; Kosarev et al., 1999). Receiver function data, on the other hand, suggests that the conversion boundary is at a moderate angle (Bilham et al., 1997). Results of the teleseismic array, spanning 500 km from the lesser Himalaya to central Tibet, suggest three zones with different parameters (Hirn et al., 1995). Significant changes are seen in the anisotropy across the Indus-Tsangpo suture. In the Indian plate the anisotropy orientation is to the northeast, parallel to the ancient plate motion, while within the Tethyan Himalaya these are aligned with the northwest differential motion between India and Tibet, consistent with stacking of the respective lithospheres. To the north of the suture zone the orientations are inconsistent, indicating a ductile mantle flow without any signature of the Indian lithosphere.

Intermediate period Rayleigh and Love waves propagating across Tibet indicate marked anisotropy within the middle to lower crust, consistent with thinning of the middle crust by about 30%. The orientations of crustal anisotropy are not consistent with the shear wave splitting fast polarization directions, implying distinct motions in the crust and mantle (DeCelles et al., 2002). The middle to lower crust appears to have thinned more than the upper crust, consistent with deformation of a mechanically weak layer of flows as if confined to a channel (Shapiro et al., 2004).

7.3 CRUSTAL SEISMIC STUDIES IN THE EASTERN HIMALAYAN REGION OF NEPAL AND TIBET

Seismic refraction studies in the eastern Himalayan region, in Nepal and southern Tibet, were undertaken by the Sino-French groups during and after the 1980s. Results of these studies show that the boundary between the upper and lower crust is at about 28-km depth to the north of the higher Himalaya (Hirn et al., 1984a; Sapin and Hirn, 1997). The Moho dips, in a stepwise fashion, from about 35-km depth in the foreland basins of India to about 55-km depth between the main central thrust and the higher Himalaya. In the region between the higher Himalaya and southern Tibet it is at about 75-km depth. To the south of the higher Himalaya the deeper segments of the Moho reflection (~55-km depth) underlie the southern shallow segments (~35-km depth). Evidence of northerly dips is seen toward the south of the higher Himalaya. The crust-mantle transition to the north of the Himalaya is also a result of interbedding of the crust and mantle material. The depth interval between the deeper reflectors indicates that only the lower crustal layers would have underplated each other. Since the superposition of corresponding upper crustal layers along large thrusts observed on the surface could not be confirmed, it is not possible to support the concept of evolution of a crust of double than normal thickness by doubling of both the upper and lower crustal units. The average velocity in the crust should have been higher because of its larger thickness but is on the lower side, of the order of 6.25–6.30 km s^{-1}, similar to normal values in the crust that is twice thinner. This supports a model of Himalayan convergence in which the Indian crust has undergone phase transformation. Mostly upper crustal type material remains to form the thickened crust above the Moho (Hirn et al., 1997). The upper mantle velocity is high at 8.7 km s^{-1}.

A north-south wide-angle fan profile across the Lhasa block shows several sharp changes in the crustal thickness (Hirn et al., 1984b; Sapin et al., 1985;

Hirn, 1988). The deep Moho (~75 km) of the Himalayan region is traced at more or less the same depth until it is interrupted at the southern boundary of the Lhasa block, where it rises sharply to a depth of about 50 km before going down again to about 70-km depth within 50 km to the north; indicating that at the Moho level the Lhasa block is limited to the south by a sharp boundary. The coincidence of the Moho step with the major structural and mechanical boundary at the edge of the Gangdese batholith points a vertical contact of the crustal segment, suggesting that the plate boundary is related to the postsuture strike-slip motion and may represent one of the shear zones for eastward expulsion of lithospheric segments in the continuing motion of India toward Asia.

The north of the Lhasa block is also limited by a suture, where a sharp step of the same polarity and slightly less than 20 km in depth is seen at the Moho. The verticality of the contacts indicates postsuture strike-slip activity here also. The departure of the Moho boundary from the horizontal along the suture suggests that the crustal thickening did not occur through only underplating of the Indian crust under Tibet. Vertical steps of the Moho, coinciding with surface lineaments above them, indicate horizontal strike-slip on major east-west lithospheric features due to expulsion of Tibet to the east. Strong heterogeneity and lack of continuity of major interfaces across the crustal blocks suggest thickening by imbrications of both the upper and lower crust (Hirn, 1988). That the crust beneath Tibet is nowhere less than 50-km thick, at least 65 km but less than 80-km thick in most of the areas, has also been confirmed by study of dispersion of the surface seismic waves (Molnar, 1988).

Wide-angle reflection-refraction profiles at crustal scale along a 700-km transect of northeastern Tibet (Jiang et al., 2006) indicates that the Moho depth increases toward the center of the plateau. These occur with different styles across the different block boundaries, as shown also by the changes resolved in the internal architecture and layer velocities in the crust. The study reveals an abnormally felsic average composition for the whole crust and its lower half along the entire length of this transect. In the northeastern part, the Moho depth varies between 55 and 65 km, and shows an inverse correlation with the basement topography. In the middle part, the crust is 70-km thick. This may have resulted from a thin-layer inclusion marking the northward tectonic superposition of one crust on the other. The corresponding upper part could have been transported northwards to thicken also the upper crust of the present northern block. In the southern part, the crust is 75-km thick and its lower half has a crust with average velocities characteristic of a felsic composition. Together with the felsic composition of the lower part of the crust, a Moho offset suggests imbrication of the crust, in the middle of the transect, in the lithosphere in the southern part between a south-dipping Moho above and a north-dipping Moho below.

P- and S-wave attenuation studies in the thickened crust of Tibet (Galve et al., 2006) suggest that the composition, mineralogy, temperature and hydration conditions of the lower half of this crust are related to its evolution. The material presently in the thickened crust has a felsic composition and upper to middle crustal lithology. The temperature conditions suggest that basic material underlying it may be eclogitic and may not appear above the seismic Moho. To the further north/northeast the felsic material in the lower half of the crust appears as hot and dry. Its burial may have occurred earlier or may have been moderate in the postcollision phase.

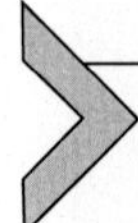

7.4 PROJECT INDEPTH STUDIES IN THE HIGH HIMALAYA AND TIBET

A multidisciplinary collaborative program (INDEPTH) was undertaken in the early 1990s wherein several institutions from a number of countries took part to study the Earth's crust and lithosphere in the Himalaya and Tibet plateau, which is the type example of a continent-continent collision (Fig. 7.2). At least three phases of this program are completed at this point. Various types of geophysical studies, viz. seismic reflection, postcritical reflection, broadband earthquake recording, and magnetotelluric profiling were undertaken under these programs. The INDEPTH-I profile is in the region that comprises Palaeozoic and Mesozoic miogeoclinal strata that were deposited on the north-facing continental margins of India and were subsequently folded and thrust southward during the collision between India and Asia. Other phases of the program are to the further north of this.

The broadband recordings indicate a substantial south-to-north variation in the crust beneath south Tibet. A midcrustal low-velocity zone is revealed in the northern part of INDEPTH-1. Postcritical angle reflection studies reveal that growth of the Tibet plateau is due to oblique stepwise rise, continuous thickening and widespread viscous flow of the crust and mantle. It is also due to time-dependent, localized shear between coherent lithospheric blocks (Tapponier et al., 2001). There is a 10–15-km variation in crustal thickness across the Tibet plateau. This is due to the independent

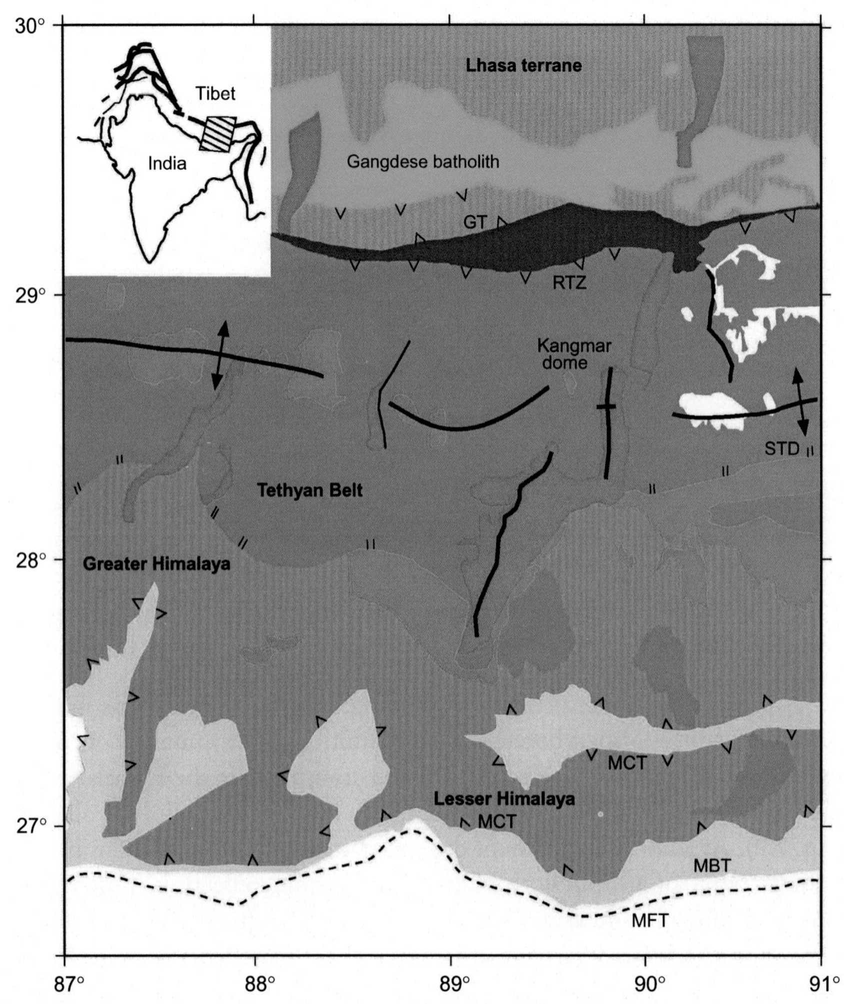

Fig. 7.2 Map showing the location of INDEPTH profiles as *thick lines. (Modified from Hauck, M.L., Nelson, K.D., Brown, L.D., Zhao, W., Ross, A.R., 1998. Crustal structure of the Himalayan Orogen from project INDEPTH deep seismic reflection profiles at ~90°E. Tectonics 17, 481–500.)*

adjustments in the Moho to provide isostatic compensation for lateral density contrasts in the upper mantle (Zhao et al., 2001).

Seismic reflection studies along the INDEPTH-1 profile (Zhao et al., 1993; Hauck et al., 1998; DeCelles et al., 2002) suggest that a reflection in the middle crust, dipping gently to the north, is a very prominent crustal boundary. It dips from a depth of about 28 km in the south (higher

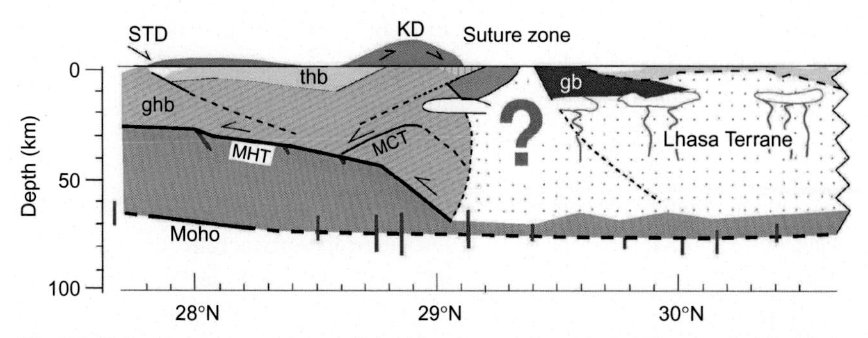

Fig. 7.3 Schematic cross-section of the Himalaya and southern Tibetan plateau along INDEPTH-1 profile. *MHT*, main Himalayan thrust; *STD*, south Tibetan detachment; *KD*, Kangmar dome; *ghb*, great Himalayan belt; *thb*, Tethyan belt; *gb*, Gangdese batholith. *(Modified from Hauck, M.L., Nelson, K.D., Brown, L.D., Zhao, W., Ross, A.R., 1998. Crustal structure of the Himalayan Orogen from project INDEPTH deep seismic reflection profiles at ~90°E. Tectonics 17, 481–500.)*

Himalaya) to about 50-km depth in the north (Kangmar dome in Tibet) and represents a thrust fault (main Himalayan thrust, or MHT) along which the Indian plate is presently underthrusting the southern Tibet (Fig. 7.3). This implies that the Indian continental crust dips gently beneath the southern Tibet and extends at least as far north as the middle of Tethyan Himalaya, some 200 km north of the Himalayan thrust front (the main boundary thrust). The crust above the main Himalayan thrust is characterized by a number of arched and north-dipping reflections, of generally lower amplitude than the reflections produced by the thrust. These reflections appear to flatten at or above the thrust. To the further north, a very prominent upper crustal reflection of high amplitude and coincident with negative polarity is seen, suggesting the presence of fluid in that part of the upper crust. Combined with the refraction and magnetotelluric studies, this zone is ascribed to a partially molten layer in large parts of southern Tibet. This molten layer is possibly a product of convergent crustal thickening.

At the deeper level two sets of reflections are seen in the southern part of the profile—an upper reflection band between 23 and 24 s TWT (70–75 km depth) and a lower band at about 25 s TWT (75–80 km depth). The first of these originates from the Moho at the base of the Indian continental crust underthrusting southern Tibet. These results are in agreement with the results of other seismic studies in the region and, viewed in combination with existing geological and seismological constraint, suggest that the thickness of the Indian crust between the main Himalayan thrust and the Moho is about 43 km. It also implies that the crust below the main Himalayan thrust is essentially an intact Indian continental crust that was thrust beneath the strata above it (main Himalayan thrust) as a consequence of continuing convergence between the Indian and Eurasian plates.

Postcritical reflections along the INDEPTH-II profile in the southern Lhasa block are very weak and do not provide a good measure of the velocities (Makowski and Klemperer, 1999). Even the near-vertical reflections from the Moho could not be seen on this profile probably due to its large (>65 km) depth and a steep velocity gradient zone (Kola-Ojo and Meissner, 2001). Here the cold subcrustal Indian lithosphere possibly carries with it a certain percentage of the lower crust, leading to lower mantle temperature.

On the INDEPTH-III profile a very low-velocity area is seen, in the postcritical reflections, to the north of the Lhasa block (Zhao et al., 2001; Haines et al., 2003). Combined with the seismological data this has been interpreted as a fault zone, extending down to at least 35 km, almost in the middle of the profile (Meissner et al., 2004). While to the south of this area the unusually thick (>70 km) crust of Tibet is formed by doubling of the rigid upper crust and also of the ductile lower crust, to its north only the doubling of the upper crust or a general mixing of hot crustal material seems to have taken place. The intruding subcrustal Indian lithosphere has probably reached the area of this suture in the middle of Tibet where it might delaminate vertically or even partly laterally.

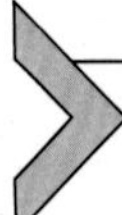

7.5 CRUSTAL SEISMIC STUDIES IN THE PAMIR-KARAKORAM REGION

Most of the earth models in the northwest Himalayan region are based solely on the seismological data, as very few studies using other geophysical methods have been carried out here. Analysis of the seismological data has revealed that the crust here is 65–70 km thick (Gupta and Naraian, 1967). In the northeast Himalayan region, the crust is estimated to be about 45-km thick (Cermak and Rybach, 1991) while in Tibet the estimate is 65–80 km (Chen and Molnar, 1981).

Seismic studies in this region were undertaken in the late 1970s as part of the international Pamir-Himalayan project along a profile (Tien-Shan-Srinagar) that started north of the Ferghana basin, crossed the Pamir and Karakoram and ended in the Jammu and Kashmir region (Fig. 7.4). Scanty refraction and postcritical reflection data to determine the models of the

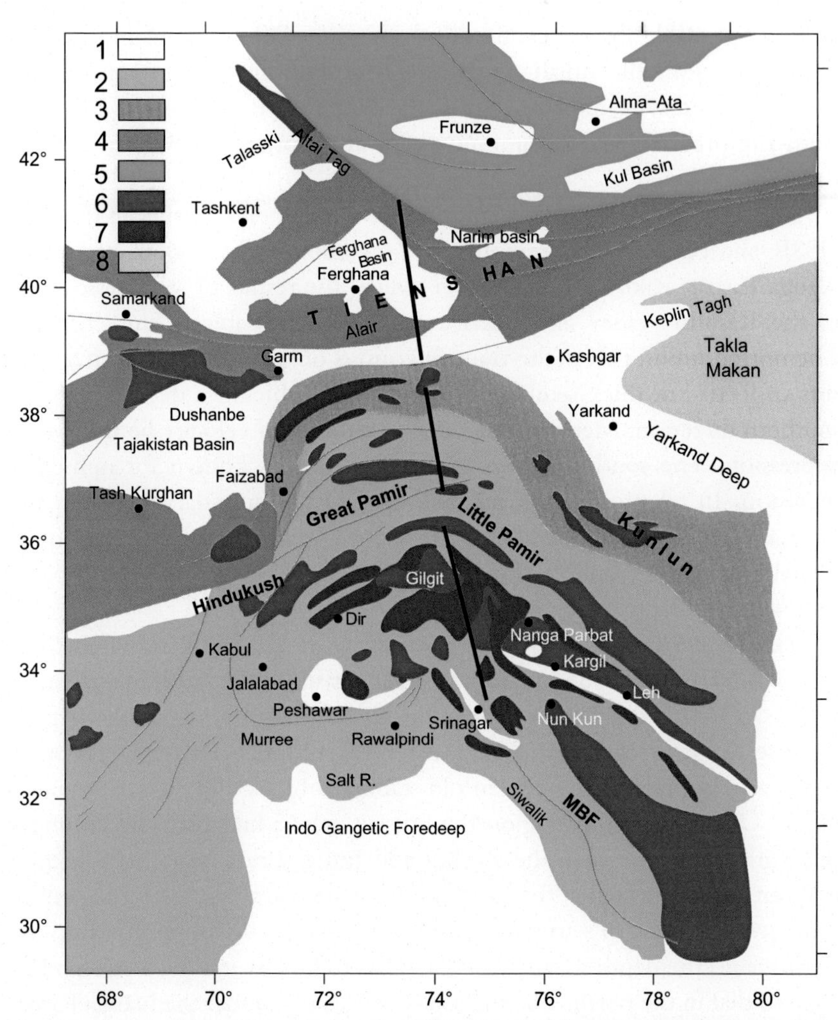

Fig. 7.4 The Tien-Shan-Srinagar seismic profile (N-S thick line) located on the geologic tectonic map of the region. *1,* Unconsolidated sediments and low density sedimentary rocks; *2,* Alpine folding; *3,* Hercynian-Alpine folding; *4,* Hercynian folding; *5,* Caledonian folding; *6,* Acidic igneous rocks; *7,* Basic igneous rocks; *8,* Platforms. Thick bold line—Profile line; Thin curve lines—Faults. *(Modified from Beloussov, V.V., Belyaevsky, N.A., Borazon, A.A., Volvovsky, B.S., Volvovsky, L.S., Resvoy, D.P., Tal-Viscid, B.B., Khambrabaev, I.K., Kaila, K.L., Narain, H., Marussi A., Finetti J., 1980. Structure of the lithosphere along the deep seismic sounding profile: Tien Shan-Pamir-Karakoram-Himalaya, Tectonophysics 70, 193–221.)*

crust are available for the profiles in this region. While the main profile is more or less in the north-south direction, another profile runs in the northeast-southwest direction across the Hazara Syntaxis toward Pakistan. Both the profiles have a common point at Astor, to the west of Nanga Parbat.

Seismic waves from large explosive shots, exploded in lakes of the then USSR and Nanga Parbat region, were recorded in the regions of USSR, Nanga Parbat, Pakistan and Kashmir valley (the results of reprocessed data in the Kashmir valley and Pakistan are described later in this chapter). The northernmost part of this transect consists of Fergana basin, which consists of 9–10-km thick sedimentary rocks. This thickness decreases in the southern direction. Metamorphic sequences of high velocity lie under this depression. The south Fergana deep-seated fault is known for ophiolitic intrusion. In all probability this fault serves as a boundary between the old stabilized block and early Hercynides of the southern Tien Shan. Individual blocks of crystalline rocks of supposedly Precambrian age probably represent the fragments of this basement.

The results of seismic studies (Beloussov et al., 1980) show that the crustal thickness is 50–55 km in the Fergana basin to the north but increases to 70–75 km in the middle of the profile (Karakul). The south-Fergana deep-seated fault corresponds to a rise of the Moho. This fault probably serves as a boundary between an old stabilized block and early hercynides block. Under the Pamir region the crust is 60–65 km thick, while under the Karakoram (between the Zorkul and Indus suture zone) its thickness is about 70 km. A fault at the Indus suture zone sharply intersects the entire thickness of the Earth's crust and disturbs the Moho discontinuity. Because of this fault, the depth to the Moho appears to be lowered by 5–10 km relative to that in the north. The thickness of the crust under the Nanga Parbat is estimated as 62 km (Fig. 7.5). The uppermost part of the crust shows a layer at depth of 20–25 km over the entire region. Maximum depth to the Phanerozoic sequence is 10 km. Its base is mostly formed of Precambrian crystalline complex that is exposed at Pamir, Karakoram, and in the inner and outer Himalaya. Local decrease in the seismic velocities is observed at 10–15 km depth. A distinct interface at about 40 km depth most probably marks the upper and lower crustal boundary. High velocity of 7.4 km s^{-1} is seen at the base of the crust in the Fergana basin. The Indus suture zone also shows indications of a higher velocity (7.0 km s^{-1}) above the Moho. In spite of the large depth to the Moho, the velocities of seismic waves in the upper mantle are almost equal to the normal values of 8.1–8.4 km s^{-1}. (The crustal model to the south of Nanga Parbat has since been revised and is

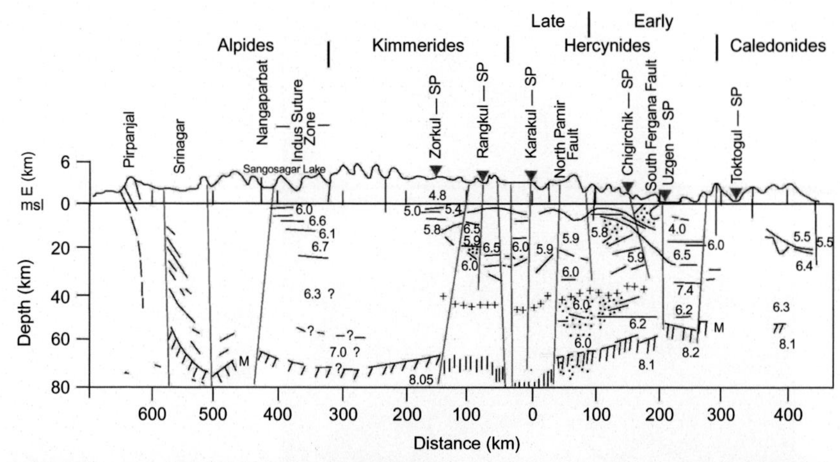

Fig. 7.5 Section of the Earth's crust along the Tien-Shan-Srinagar profile. The short horizontal line segments are the recorded velocity isolines. The seismic interface from refracted and reflected waves are shown. Interfaces from converted waves from earthquakes are also shown. The short lines with hatches are Moho. The short vertical lines in the elevation panel above the surface are the geologically interpreted faults, while the long vertical lines within the section are seismically interpreted faults. *(Modified from Beloussov, V.V., Belyaevsky, N.A., Borazon, A.A., Volvovsky, B.S., Volvovsky, L.S., Resvoy, D.P., Tal-Viscid, B.B., Khambrabaev, I.K., Kaila, K.L., Narain, H., Marussi A., Finetti J., 1980. Structure of the lithosphere along the deep seismic sounding profile: Tien Shan-Pamir-Karakoram-Himalaya, Tectonophysics 70, 193–221.)*

described separately in detail later in this chapter.) The gravity anomaly data suggests that the anomalies are caused by a combination of the lithospheric thickness as well as the asthenospheric layer.

The asthenosphere is at depths varying between 120 and 130 km in the central portion (Pamir) of the region and its depth decreases toward the south (the northern margins of the Indian shield) and north (Fergana basin). The increase in the thickness of the asthenosphere under the Pamir probably is accompanied by an appreciable decrease of the seismic wave velocity in its central portion.

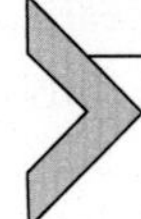

7.6 CRUSTAL SEISMIC STUDIES IN HINDUKUSH AND JAMMU AND KASHMIR REGIONS

The seismic studies in the northwest Himalaya included: (i) the 300-km long Lawrencepur-Astor (Nanga Parbat) profile, across the Hazara Syntaxis in the Hindukush region, in a southwest-northeast direction, and (ii) 250-km long Naoshera-Astor profile in the north-south direction (part

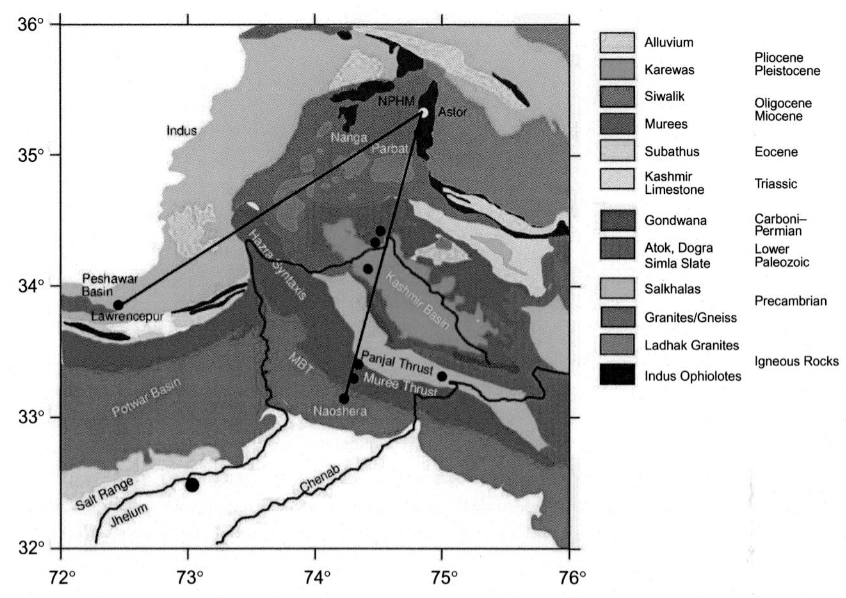

Fig. 7.6 Map of the northwest Himalaya showing the two seismic profiles Lawrencepur-Astor and Naoshera-Astor. *NHPM*, Nanga Parbat-Harmosh massif.

of the profile described earlier). The Lawrencepur-Astor profile (Profile I) starts in the southern Peshawar basin, crossing the Hazara Syntaxis. It ends at Astor near the Nanga Parbat-Harmosh massif (Fig. 7.6). This massif is composed of the high-grade Precambrian basement gneiss that has been overprinted by the Himalayan metamorphism (Shams and Ahmad, 1979).

The Lawrencepur-Astor profile was recorded at 28 locations from two shot points: Lawrencepur located on the Peshawar basin in Pakistan, and Astor situated to the east of the Nanga Parbat. However, published data are available only from 21 locations, the other data being rejected because of the noise problem. The first seismic velocity-depth model of this profile showed a number of low-velocity zones in the crust (Guerra et al., 1983). These data have since been reinterpreted and provide a revised model (Bhukta Subrata et al., 2006) showing that the near-surface velocities toward Lawrencepur are less than those toward Astor. In the top part of this velocity-depth model (Fig. 7.7) the layers have velocities of 3.5, 5.1, 5.4, and 5.8 km s^{-1} toward Lawrencepur and 5.4 and 6.0 km s^{-1} toward Astor. Correlation of these velocities with geological exposures indicates that the high Himalayan crystallines exposed at about 55-km distance from Lawrencepur and having a velocity of 5.4 km s^{-1} are at a depth of about

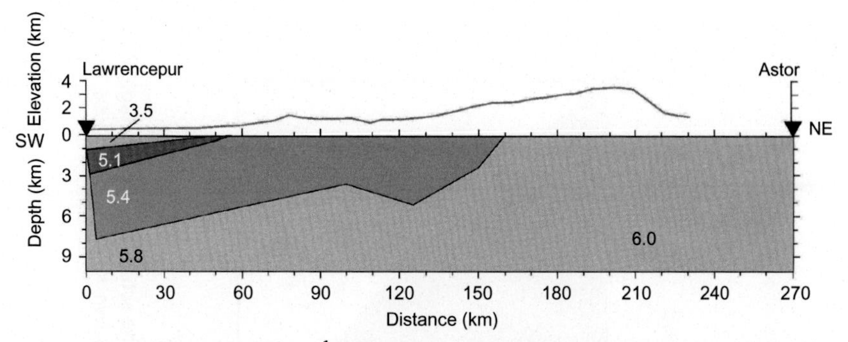

Fig. 7.7 Shallow velocity (km s^{-1})-depth model along Lawrencepur-Astor profile. The *dark lines* show the actual seismic coverage. *(Modified from Bhukta Subrata, K., Sain, K., Tewari, H.C., 2006. Crustal structure along the Lawrencepur-Astor profile across Nanga Parbat. Pure Appl. Geophys. 163, 1257–1277.)*

3 km at Lawrencepur. The velocity of 5.8–6.0 km s^{-1} belongs to the layer that forms the base of these crystallines.

Below the velocity of 5.8–6.0 km s^{-1} the velocities in various layers in the crust are 6.2, 6.4, and 6.8 km s^{-1} at depths of about 15, 25–30, and 36–46 km, respectively (Fig. 7.8). A velocity inversion is seen to the northeast of the Hazara Syntaxis where a low-velocity layer, with a velocity of 5.7 km s^{-1}, is seen below the layer of velocity 6.2 km s^{-1}. Maximum thickness of the low-velocity layer is about 5 km. The Moho discontinuity is at 50–62 km depth over an upper mantle of velocity 8.2 km s^{-1}. While the layers with velocities less than and including 6.2 km s^{-1} represent the upper crust, the layers with velocities 6.4 and 6.8 km s^{-1} represent the lower crust. All the crustal layers dip from the southwest to the northeast. One-dimensional velocity depth models at 50-, 100-, 150-, 200-, and 250-km distance, respectively, from Lawrencepur shot point show a change in depth with layer velocity along the profile.

Due to a limited amount of seismic data, the velocity-depth model of the crust has been configured in a very small part of the profile. The model in between has been computed with the help of the gravity data, despite the fact that the gravity contours for the region are not very reliable because of the difficult topography of the Himalayan region. A comparison of the observed Bouguer gravity values (Balakrishnan, 2003) is made with those computed on the basis of the seismic model. This comparison shows that, despite limited data, the seismic model is able to satisfy the gravity observations reasonably well and can be extended to a large part of Profile I (Fig. 7.9).

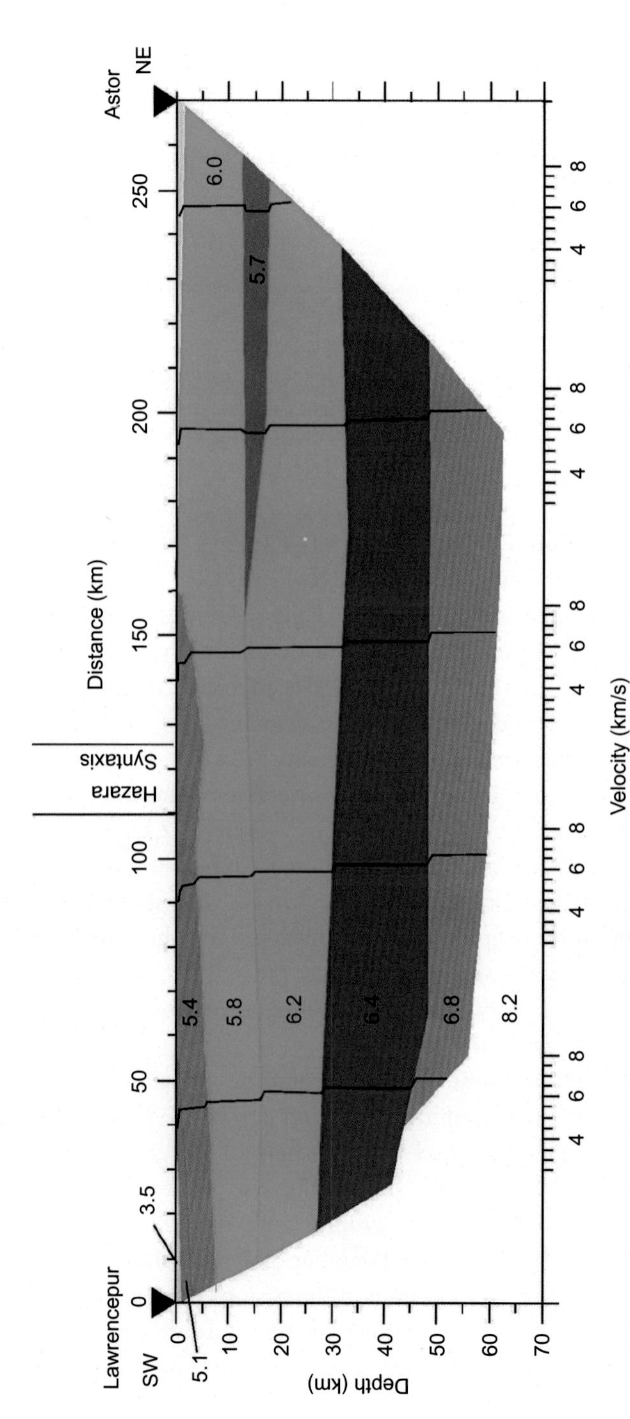

Fig. 7.8 Velocity (km s^{-1})-depth model of the crust along Lawrencepur-Astor profile. The *dark lines* show the actual seismic coverage. *White lines* are the velocity models computed at various places along the profile. *(Modified from Bhukta Subrata, K., Sain, K., Tewari, H.C., 2006. Crustal structure along the Lawrencepur-Astor profile across Nanga Parbat. Pure Appl. Geophys. 163, 1257–1277.)*

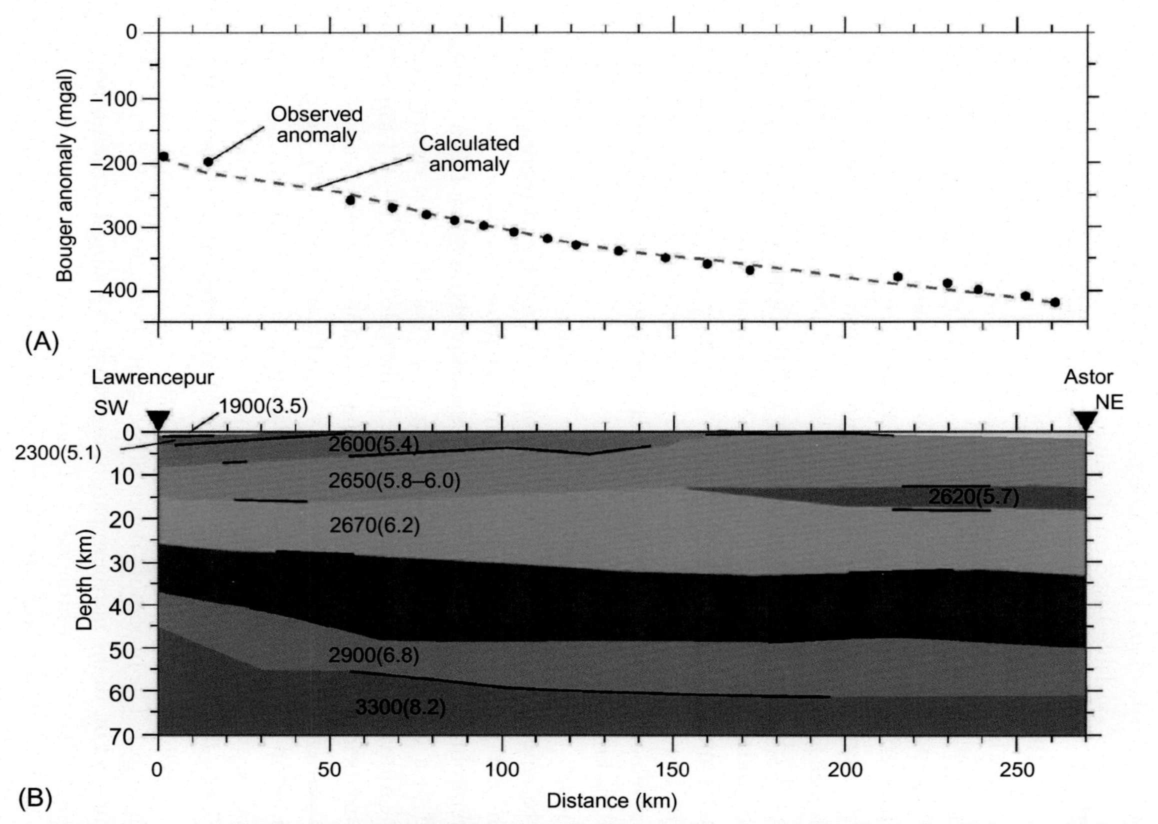

Fig. 7.9 Seismic-based gravity model along Lawrencepur-Astor profile showing (A) observed and calculated Bouger anomaly for (B) the depth model. The density and velocity (within bracket) values are in kg m^{-3} and km s^{-1}. The *dark lines* show the actual seismic coverage. *(Modified from Bhukta Subrata, K., Sain, K., Tewari, H.C., 2006. Crustal structure along the Lawrencepur-Astor profile across Nanga Parbat. Pure Appl. Geophys. 163, 1257–1277.)*

Seismic structure along the Naoshera-Astor profile (Profile II) was first determined based on the analysis of analog data (Kaila et al., 1984). This crustal model is also incorporated in the structural model of the Pamir-Karakoram region (Beloussov et al., 1980). A revised model along the Naoshera-Astor profile is now available based on the digitized data. This model determines the velocity-depth configuration of the sedimentary basin and the crust separately (Bhukta Subrata and Tewari, 2007). The seismic model for the sedimentary basin in the Jammu region (Naoshera-Thanamandi) shows three velocities of 1.9, 4.1, and 4.4–5.3 km s^{-1}, respectively, that are associated with the exposed upper Siwalik, middle/lower Siwalik and the Murree formations overlying the velocity of 5.9–6.1 km s^{-1} (Fig. 7.10).

The velocities of 4.1 and 5.3 km s^{-1} belong to different sequences within the Murree formation. The layer with velocity of 5.9–6.1 km s^{-1}, at depth of about 4.7 km, at Naoshera shows a steep updip to a depth of 3.0 km at the main boundary thrust, exposed at about 12-km distance from Naoshera. At this thrust this layer is uplifted by almost 1.5-km depth from a depth of about 3.0 km. The sedimentary column beyond the thrust mainly consists of the layer of velocity 5.3 km s^{-1} resting on the layer of velocity 5.9–6.1 km s^{-1} at about 1.5-km depth.

Though the velocity of 5.9–6.1 km s^{-1} is close to that of the Precambrian basement in the Indian shield region, it is unlikely to represent the same here. Hydrocarbon surveys in the Jammu region have shown that here the Precambrian basement is at a depth in excess of about 8 km. Deep wells, drilled to a depth of more than 6500 m, have not even penetrated the Eocene formations in this region. In Fig. 7.11 the velocity of 5.9–6.1 km s^{-1} probably belongs to an Eocene limestone formation. Similar formations have been encountered at more than 2 s in reflection two-way time, in hydrocarbon surveys in nearby areas of Pakistan.

The sedimentary velocity-depth model for the Kashmir region (Fig. 7.11) shows several faults and three layers that have velocities of 1.7–2.1 (Quaternary sediments), 4.0 (Triassic limestone), 5.5–6.0 km s^{-1} (Panjal Traps). The maximum depth of the layer of velocity 5.5–6.0 km s^{-1} is about 2.5 km. In the northern part of the profile it is almost at the surface. In a deep well in the Kashmir valley, Triassic limestone was encountered at a depth of about 500 m under the Quaternary (Karewa) formation. The bottom of this limestone could not be touched in that well, drilled to a depth of 1600 m. Panjal Traps (velocity 5.8 km s^{-1}) are exposed on both sides of the Kashmir basin and encountered in another well directly under the Karewa, at 1250 m depth (Bose and Arora, 1969).

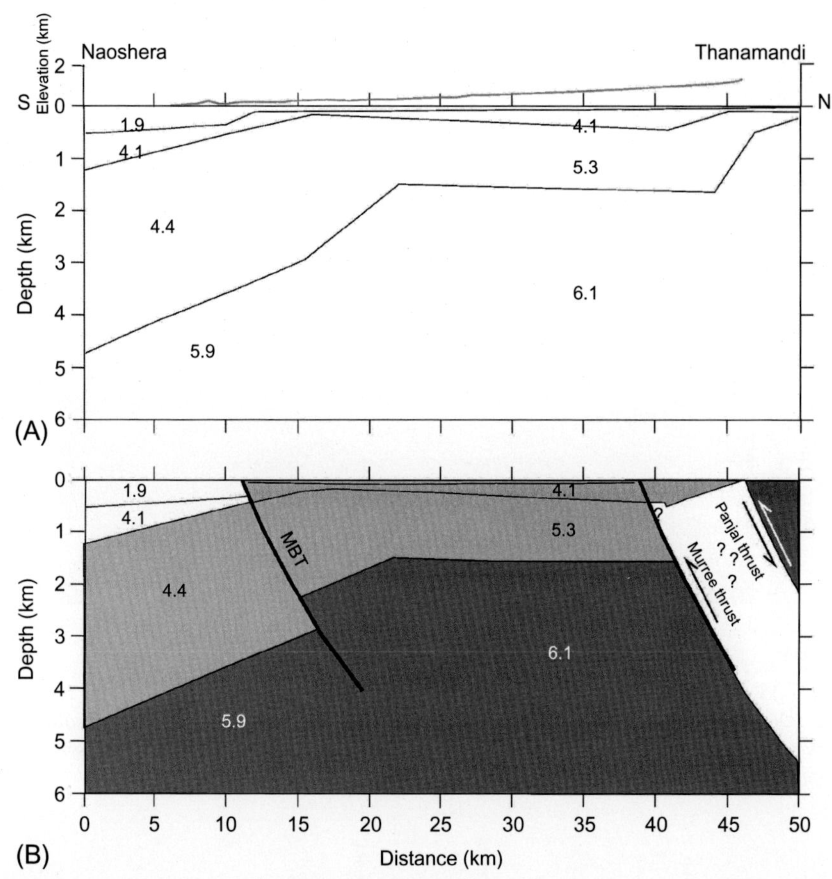

Fig. 7.10 Velocity (km s^{-1})-depth model of the sedimentary basin in Jammu region: (A) seismic model; (B) geological interpretation. The faults in the geological model are based on the steep dip seen in the refraction depth section and supported by the local geology. *(Modified from Bhukta Subrata, K., Tewari, H.C., 2007. Crustal seismic structure in Jammu and Kashmir region. J. Geol. Soc. India 69, 755–764.)*

Due to nonavailability of enough data at longer distances the crustal model has been inferred only sporadically. This model along the Naoshera-Astor profile shows a horizon (velocity 6.0–6.2 km s^{-1}) that is at a depth of about 10 km at 0–30 km distance from Naoshera, at about 6 km depth at 30–40 km distance and 8–10 km depth at 120–150 km distance (Fig. 7.12). Toward the Astor shot point the layer of velocity 6.0 km s^{-1} has come close to the surface. A low-velocity layer of velocity 5.7 km s^{-1} that lies in between the velocities of 6.0 and 6.2 km s^{-1}, similar to the one along the Lawrencepur-Astor profile, is seen toward Astor in this

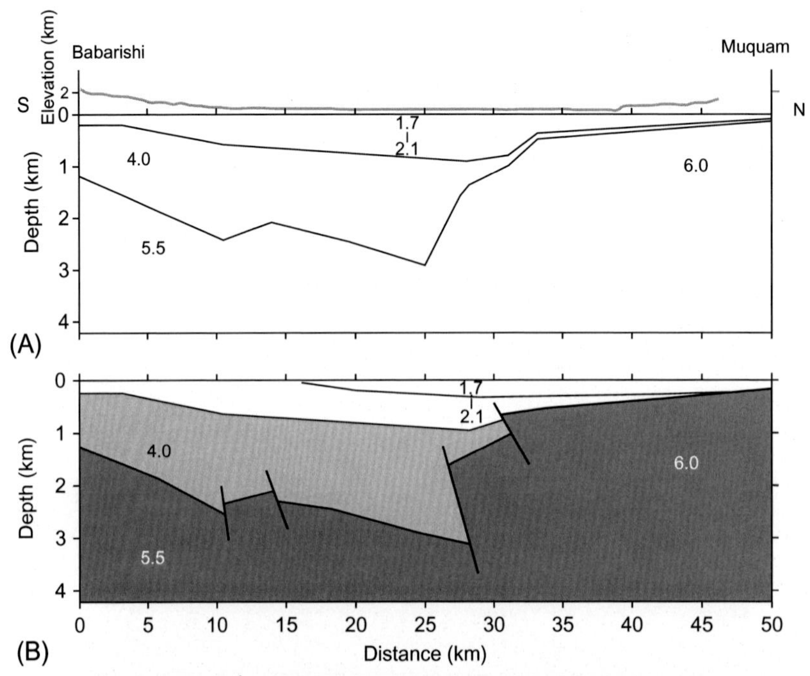

Fig. 7.11 Velocity (km s^{-1})-depth model of the sedimentary basin in Kashmir region: (A) seismic model; (B) geological interpretation. *(Modified from Bhukta Subrata, K., Tewari, H.C., 2007. Crustal seismic structure in Jammu and Kashmir region. J. Geol. Soc. India 69, 755–764.)*

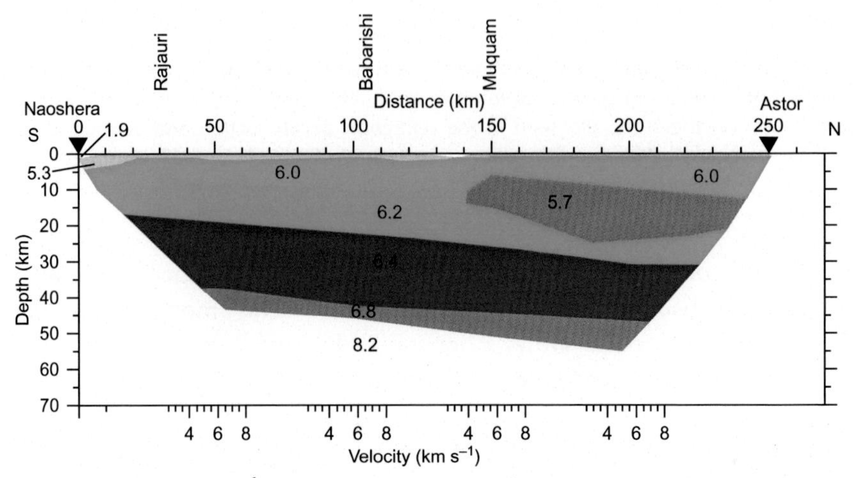

Fig. 7.12 Velocity (km s^{-1})-depth model of the crust along Naoshera-Astor profile. The *dark lines* show the actual seismic coverage. *(Modified from Bhukta Subrata, K., Sain, K., Tewari, H.C., 2006. Crustal structure along the Lawrencepur-Astor profile across Nanga Parbat. Pure Appl. Geophys. 163, 1257–1277.)*

profile as well. A horizon with velocity 6.4 km s^{-1}, at a depth varying between 18 and 32 km, indicates a change from the upper to the lower crust. The horizon with velocity 6.8 km s^{-1} at a depth of about 36 km at a distance of 55–65 km from Naoshera is part of the lower crust. Another horizon, at a depth of 40–43 km at 45–65 km distance and about 55 km depth at 170–200 km distance, represents the Moho above an upper mantle of velocity 8.2 km s^{-1}. The dip of this horizon is to the north, in agreement with the general northerly dip of Moho toward the higher Himalaya.

Similar to the model along Profile I, this model is also constrained by comparing the observed gravity values (Banerjee and Satyaprakash, 2003) along the profile with those computed based on the seismic model. Though the match of calculated anomaly is not very good with the observed data in the middle portion of the profile (Fig. 7.13), the structural trend of the seismic model is more or less undisturbed.

7.7 TECTONIC SIGNIFICANCE OF SEISMIC RESULTS

The seismic results, though very limited, are important in understanding the tectonics of this region. The Hazara Syntaxis is one of the most important structural elements in the northwest Himalaya. The seismic results show that the upper crust to the northeast of this syntaxis has a different character as compared to its southwest. While the top of the upper crustal layer of velocity 6.2 km s^{-1} represents a detachment zone (Seeber et al., 1981) to the southwest of the syntexis, the upper crustal low-velocity zone to its northeast conforms to the model that shows a two-phase overthrusting responsible for the development of the syntexis (Bossart et al., 1988). The low-velocity zone may represent the leading edge of a slab detached from the basement of the Indian plate. This zone, at a depth between 7 and 13 km on Profile I, extends up to about a 20-km depth in the northern part of Profile II. Presence of the low-velocity zone compares well with high attenuation of the seismic waves, the midcrustal conductor (Arora et al., 2005) and heat-flow values in excess of 200 mW m^{-2} (Ravishanker, 1988), and suggests a widespread partial melt in the deep crust beneath the Indus-Tsangpo suture zone (Gokaran et al., 2002). A number of geophysical studies in the Nanga Parbat region also suggest lower velocities and presence of fluids at about a 20-km depth below it (Chamberlain et al., 2002).

Depth of the Moho appears to be close to that obtained by other workers in the northwest Himalayan region (Ni et al., 1991). Comparison of the Moho depths of the Lawrencepur-Astor and Naoshera-Astor profiles shows

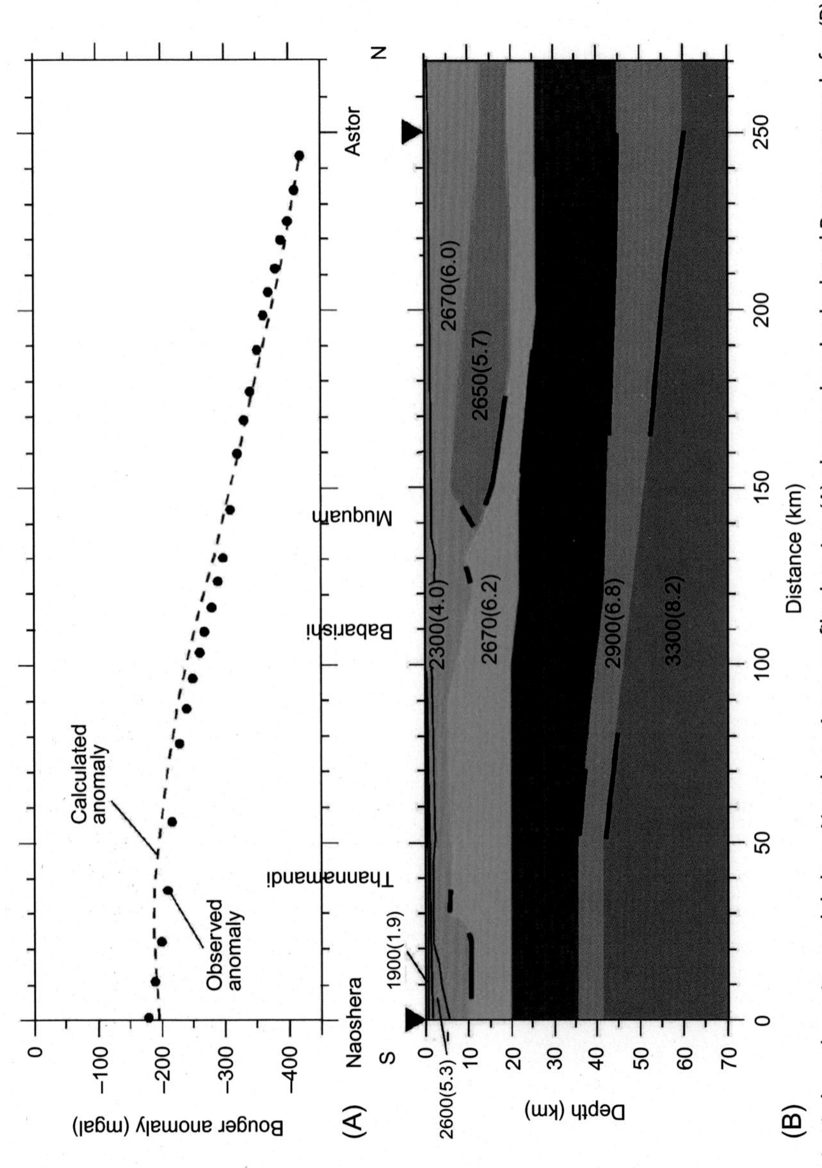

Fig. 7.13 Seismic-based gravity model along Naoshera-Astor profile showing (A) observed and calculated Bouger anomaly for (B) the depth model. The density and velocity (within bracket) values are in kg m^{-3} and km s^{-1}. The *dark lines* show the actual seismic coverage. *(Modified from Bhukta Subrata, K., Tewari, H.C., 2007. Crustal seismic structure in Jammu and Kashmir region. J. Geol. Soc. India 69, 755–764.)*

that the thickness of the crust increases sharply to the north and also to the northwest. This indicates that collision of the Indian plate with the Eurasian plate resulted in underthrusting of the Indian plate beneath Tibet as well as Hindukush, conforming to the fold orientation in the northwest Himalaya. The tomography study shows well-pronounced traces of subducted Indian slab in the region of Hindukush (Koulakov et al., 2002).

7.8 DEEP REFLECTION STUDIES IN HIMACHAL PRADESH

In the Sub-Himalaya and lesser Himalaya of NW India, the only crustal-scale seismic studies have been pre-1980 deep seismic soundings and modern teleseismic receiver-function measurements (Rai et al., 2006), both of which have a lower spatial resolution. Shallow seismic reflection studies carried out by ONGC in the Sub-Himalayan region for exploration of hydrocarbons (Raiverman, 2002) were focused on the structures in the Tertiary sedimentary formations. A deep seismic multifold reflection study, under the HIMPROBE program in India (Fig. 7.14), was recorded in the Sub-Himalayan region of Himachal Pradesh (Kangra recess) to study the nature of the crust (Rajendra Prasad et al., 2007). This is the first such study located anywhere in the 2500-km long Sub-Himalayan region and targeted structures within the basement from the MHT down to the Moho. This region, located in the foothills of the uplifted Himalayan foreland basin, consists of predominantly fluvial Late Tertiary sediments derived from rising Himalaya during the Cenozoic. In the south the Sub-Himalayan fold belt rises abruptly (500–2000 m) above the foredeep of the Holocene Indo-Gangetic alluvial plains along the main frontal thrust. In the north it is overridden by the Proterozoic lesser Himalayan sequence along the main boundary thrust.

Hydrocarbon-related seismic studies and lithostratigraphy of the wells in this region reveal that maximum thickness of the Tertiary sediments in the foreland basin is around 4 s TWT (~7000 m). The thickness of pre-Tertiary sediments extending deeper than 4 s TWT is not well known. Deep wells drilled in the anticline regions show several sedimentary formations, which are 2500–3500 m thick. Among the five oil wells drilled between the main boundary thrust and main frontal thrust, only one well reached the basement at a depth of 5027 m. In one of these wells, the basement was not reached even at a depth of about 6700 m. Since all the wells are drilled over anticlines and thrust zones, the average thickness of the sedimentary column may be much greater than the value indicated by those studies.

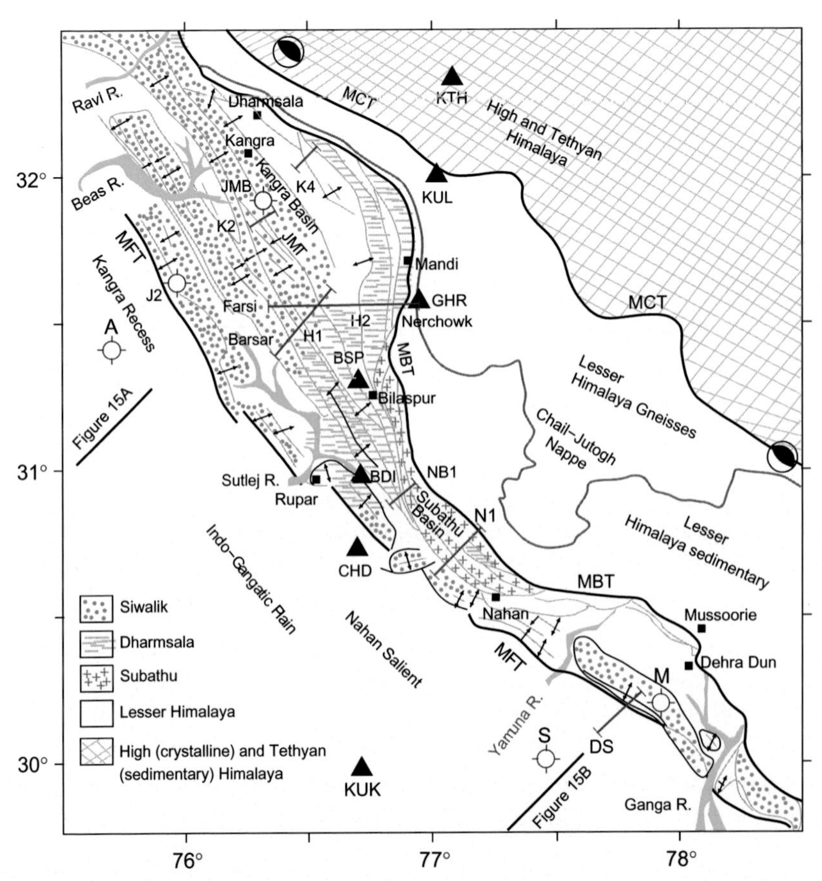

Fig. 7.14 Simplified geology and tectonic map of the study region, in Himachal Pradesh. Faults: *MFT*, main frontal thrust; *BT*, Barsar Thrust; *JMT*, Jwalamukhi Thrust; *MBT*, main boundary thrust; *MCT*, main central thrust. Seismic profiles shown as continuous *gray lines*: *H1*, *H2*, Himprobe-1 and -2; *K2*, *K4*, Kangra 2 and 4; *DS*, Doon-South, of the oil industry: *N1*, Nahan-1. Exploration wells: *A*, Adampur; *J2*, Janauri-2; *JMB*, Jwalamukhi-Baggi; *M*, Mohand; *S*, Saharanpur. *Black triangles* with three-letter station codes are broadband seismic stations. Example thrust focal mechanisms are (in the SE) the 1991 Uttarkashi mb-6.5 earthquake displaced 40 km along strike to appear in the map area and (in the NW) the Kangra earthquake. Interpreted sections across Kangra recess and Doon South are shown in Fig. 7.16A and B. *(Modified after Rajendra Prasad, B., Klemperer, S.L., Vijaya Rao, V., Tewari, H.C., Khare, P., 2011. Crustal structure beneath the Sub-Himalayan fold-thrust belt, Kangra recess, northwest India from seismic reflection profiling: implications for Late Palaeoproterozoic orogenesis and modern earthquake hazard. Earth Planet Sci. Lett. 308, 218–228.)*

The 100-km long, multifold deep seismic reflection data were acquired in this region along two profiles (Fig. 7.14): the SW-NE (H1) 35-km long profile crossing the Barsar and Jwalamukhi Thrusts, and the E-W (H2) 65-km long across the Barsar Thrust to the main boundary thrust. Both profiles cross logistically very difficult and rugged hilly terrain of the Siwalik ranges within the Kangra recess. The final seismic stack sections (Fig. 7.15A), and schematic line drawings (Fig. 7.15B) highlight the prominent reflections along these two profiles (Rajendra Prasad et al., 2011).

The reflection data show no reflections from the outcropping Tertiary sequence, above a bright reflection band, from the main Himalayan thrust (MHT) and immediately subjacent sedimentary rocks, at 3.0–3.5 s TWT. Reflections dipping west and southwest appear throughout the basement continuously to 12 s TWT, below which subhorizontal reflections, locally bright, continue diffusely to 16 s TWT. These are most probably from the reflection Moho. The reflection times of 3.0–3.5, 12, and 16 s are interpreted to be equivalent to depths of 6–8, 37, and 51 km, respectively. The nearby hydrocarbon-exploration reflection profiles of the oil industry also show complex folding and thrusting above the MHT and were interpreted as showing the Sub-Himalaya as formed above high-angle reverse and wrench faults (Raiverman, 2002), but with the information from deep reflection studies these are now recognized as MHT (Rajendra Prasad et al., 2011). Because the Sub-Himalayan décollement is at a depth of only 6–8 km and has only a modest displacement (a few tens of kilometers), the thrust should be a brittle fault and is likely to be of very limited thickness. The bright reflections span up to 1 km vertically and represent the Vindhyan (Neoproterozoic) sediments immediately beneath the décollement, the planar top to the bright reflections representing the MHT. Recent earthquakes on the MHT (1991-Uttarkashi EQ mb = 6.5) to the southeast places the MHT at a 10–12-km depth beneath the surface trace of the main central thrust (Kayal, 1996). Based on a downwards decrease in microseismicity on the north margin, the MHT is estimated to be at 15 ± 2 km depth beneath the surface trace of the main central thrust (Kumar et al., 2009). However, Rajendra Prasad et al. (2011) feel that microseismicity is not a reliable way to infer the depth to the MHT, and that the MHT is several kilometers shallower than that inferred by Kumar et al. (2009).

The character of the reflectivity changes below the bright reflection band of 3.5 s TWT. It is more prominent than above 3 s TWT and has apparent dips of 15 degrees down to 12 s TWT on the west-east profile, and (less clearly) as much as 30 degrees on the SW-NE profile to at least 8 s TWT

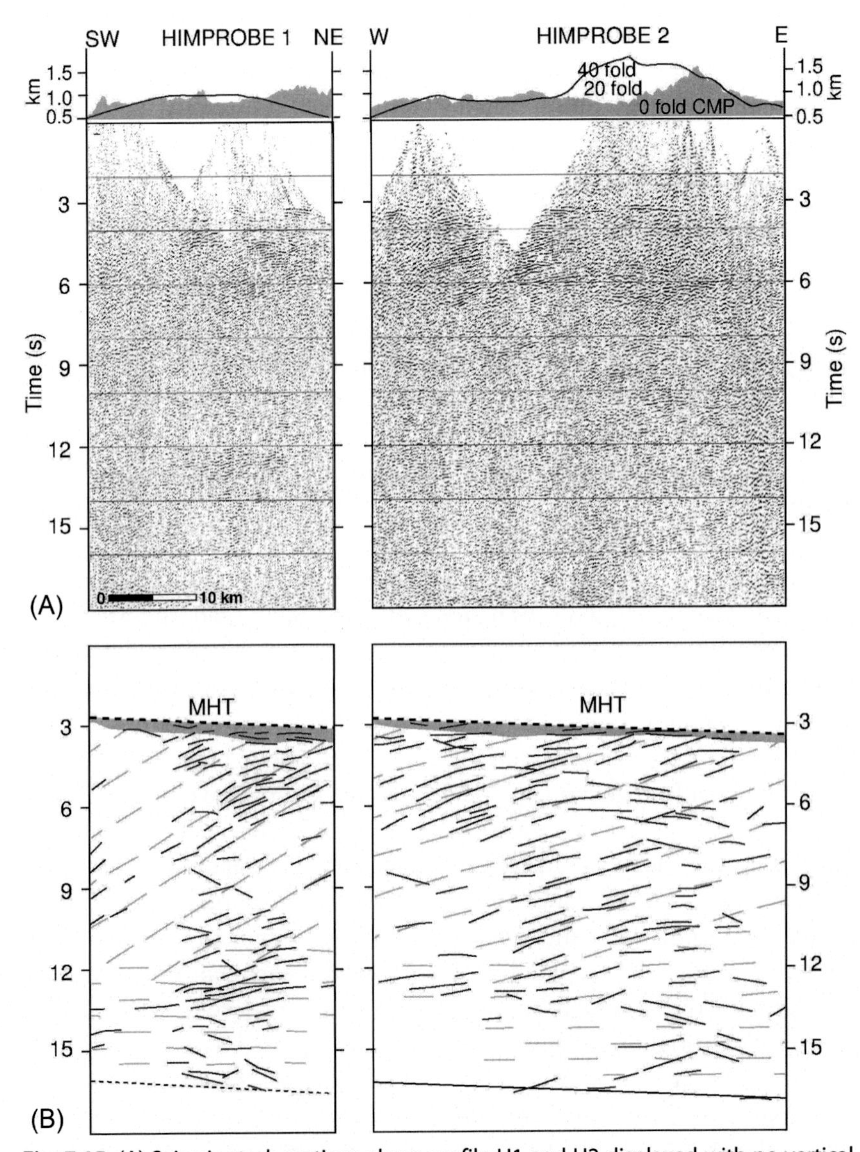

Fig. 7.15 (A) Seismic stack sections along profile H1 and H2 displayed with no vertical exaggeration for a velocity of 6 km s^{-1}. Gray shading above seismic sections is topography along profiles; *dashed line* is common midpoint stacking fold. (B) Line drawings and interpretive cartoons of profiles. *Thick black lines* were drawn directly on reflections visible on stack sections. *Thin vertical line*: profile intersection. *Dot-dash lines*: reference lines dipping 2.5 degrees (dip of MHT). *Gray rectangles* at Moho depth: range of Moho travel times from surface wave-receiver function inversions (Rai et al., 2006). *Gray shading*: region of subhorizontal reflections inferred to be Proterozoic sedimentary rocks. *Dipping gray lines*, dashed: dipping 30 degrees on SE-NW and 15 degrees on E-W, in region dominated by west-southwest dipping reflectivity. *(After Rajendra Prasad, B., Klemperer, S.L., Vijaya Rao, V., Tewari, H.C., Khare, P., 2011. Crustal structure beneath the Sub-Himalayan fold-thrust belt, Kangra recess, northwest India from seismic reflection profiling: implications for Late Palaeoproterozoic orogenesis and modern earthquake hazard. Earth Planet Sci. Lett. 308, 218–228.)*

(Fig. 7.15). Since the dipping reflectivity lies below the MHT, it almost certainly represents pre-Himalayan structure. Combined with the oil exploration data (Raiverman, 2002), it is suggested that the SW-dipping reflections exist for at least 150 km northwest-southeast along the strike, as well as for the 50 km across the strike observed on the HIMPROBE profiles, and that the dipping reflections are a single tectonic element down to 12 s TWT, thereby forming the major part (25–30 km) of the Indian basement thickness. The prominent dipping reflectivity is underlain by more horizontal reflections. There is no clear base to the reflectivity that can be interpreted as a sharp tectonic or compositional Moho, but the reflectivity disappears around 16 s TWT, which allows recognition of a diffuse reflection Moho (Rajendra Prasad et al., 2011). Joint inversion of receiver-functions with surface wave data also indicates a Moho depth of 50–55 km (Oreshin et al., 2008; Priestley et al., 2007; Rai et al., 2006).

The velocity inversions suggest a gradual increase in seismic velocity from 6.5 to 7.6 km s^{-1} in the lower 15 km of the crust (Rai et al., 2006), consistent with a diffuse reflection Moho as due to a gradually increasing density of mafic sills intruded into felsic-intermediate crust in a broad crust-mantle transition zone (Nelson, 1991) and with global compilations showing that Proterozoic crust typically has a high-velocity basal layer (Durrheim and Mooney, 1994). The 50–55-km crustal thickness inferred from the deep reflection profiles and the teleseismic recordings represents 6–8 km of Himalayan thrust sheets above a 42–49-km cratonic basement.

Given the long history of the Indian craton, many possible choices exist for the observed WSW-dipping reflections: (1) an original Archean fabric of the north India block; (2) a Proterozoic orogenic event on the margin of the north India block; (3) a Proterozoic (meta-) sedimentary basin; (4) a "Pan-African" Cambro-Ordovician event; or (5) a "Himalayan" Cenozoic tectonic fabric. Rajendra Prasad et al. (2011), however, prefer the second explanation that the dipping reflection fabric represents accretion of a Paleoproterozoic arc, as it is consistent with Neoproterozoic truncation above and possibly younger overprinting below the dipping fabric due to the moderate (35 degrees) dip of the reflectivity and the increased crustal thickness, as depicted in the conceptual models across the Kangra recess and the Doon recess (Fig. 7.16). The existence of a geologic event of this age is implied by the widespread occurrence of 1.8–1.9 Ga granitoids in the lesser Himalayan series including the Wangtu orthogneiss just 100 km east of the reflection profiles and Ulleri orthogneisses throughout Nepal. The deepest horizontal reflectivity in the profiles and the possible velocity increase above the Moho could mark mafic sills related to a rifting episode.

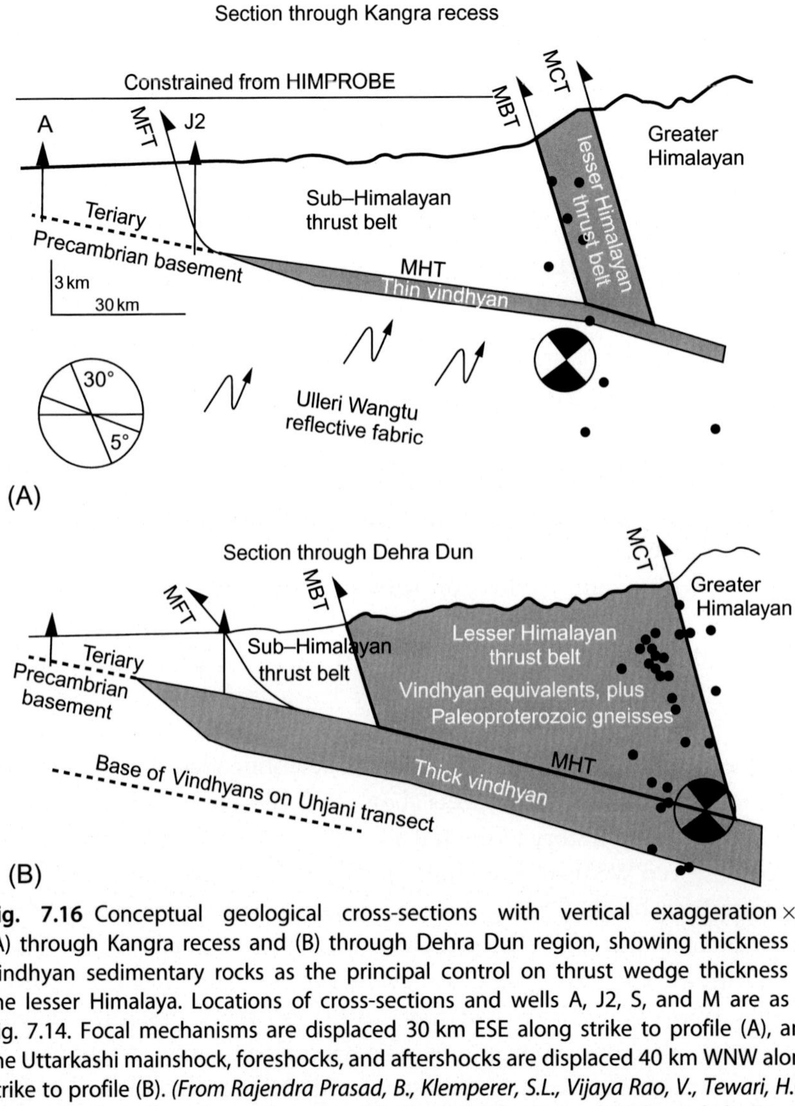

Fig. 7.16 Conceptual geological cross-sections with vertical exaggeration × 4: (A) through Kangra recess and (B) through Dehra Dun region, showing thickness of Vindhyan sedimentary rocks as the principal control on thrust wedge thickness in the lesser Himalaya. Locations of cross-sections and wells A, J2, S, and M are as in Fig. 7.14. Focal mechanisms are displaced 30 km ESE along strike to profile (A), and the Uttarkashi mainshock, foreshocks, and aftershocks are displaced 40 km WNW along strike to profile (B). *(From Rajendra Prasad, B., Klemperer, S.L., Vijaya Rao, V., Tewari, H.C., Khare, P., 2011. Crustal structure beneath the Sub-Himalayan fold-thrust belt, Kangra recess, northwest India from seismic reflection profiling: implications for Late Palaeoproterozoic orogenesis and modern earthquake hazard. Earth Planet. Sci. Lett. 308, 218–228.)*

CHAPTER EIGHT

Velocity Structure of the Indian Crust

The velocity structure of the continental crust is one of the most important parameters in identification of different regions within the crust. Keeping this in mind, several attempts have been made to get a generalized crustal velocity configuration of the Indian continental crust. Studies through the body and surface waves, generated by the earthquakes, have provided the bulk velocities in the crust of the Indian peninsula. Aftershocks of crustal phases from the Koyna earthquake (Dubey et al., 1973) have shown a two-layered crust, where the P-wave velocities in the upper and lower crust are 5.78 and 6.58 km s^{-1}, respectively. Respective S-wave velocities are 3.42 and 3.92 km s^{-1}. In the upper mantle the P- and S-wave velocities are given as 8.19 and 4.62 km s^{-1}, respectively. Another study based on the group velocity dispersion (Bhattacharya, 1981) gave more or less similar velocities for the crust of the central Indian region. In addition, the S-wave velocities of the mantle were found to be between 4.6 and 4.9 km s^{-1}. Another model based on the study of the Love and Rayleigh waves generated by the Jabalpur earthquake in central India (Singh et al., 1999) gives the S-wave velocity for the upper crust as 3.55 km s^{-1}, lower crust 3.85 km s^{-1} and upper mantle 4.65 km s^{-1}.

It is well known by now that the Indian subcontinent is made up of different crustal blocks that are geologically unrelated and have been juxtaposed/sutured together during different periods of the Earth's history. Based on what has been discussed in the earlier chapters it is, therefore, obvious that the crustal velocity models of these zones would be different from each other and a generalized model for the whole of the Indian crust is not justified.

The first serious attempt to identify different velocity regions of the Indian subcontinent describes all the models that were known at that time and compares the P- and S-wave velocities of several regions (Mahadevan, 1994). High-velocity values for the head waves from the Moho discontinuity (Pn) to the west of the Cuddapah basin (>8.4 km s^{-1}, Chapter 2) are identified with a cratonic crust and possible presence of an eclogite layer below that region.

However, more serious attempts to study the P-wave velocity structure on the basis of controlled source seismic studies, particularly of the Indian

Structure and Tectonics of the Indian Continental Crust and Its Adjoining Region
https://doi.org/10.1016/B978-0-12-813685-0.00008-X

peninsular shield where a large number of seismic profiles are recorded, were attempted later. The first such attempt (Kaila and Sain, 1997), on the basis of one-dimensional velocity models of various regions, divides the Indian subcontinent into six zones. Zone 1a consists of the Dharwar craton, Cuddapah basin, Godavari graben and the Koyna region; and zone 1b consists of the western margin of the Bengal basin and Singhbhum craton. Zone 2 consists of the West Bengal and Mahanadi sedimentary basins. Zone 3 encompasses the Bundelkhand craton and the Vindhyan basin. Zone 4 consists of the Cambay basin to the north of the Narmada-Son lineament, Saurashtra peninsula, Kutch region and the Aravalli-Delhi fold belt. Zone 5 comprises the region to the south of the Narmada-Son lineament that covers the Tapti Graben, and zone 6 the northwestern Himalaya. Zones 1a, 1b, and 3 have the velocity models typical for the shield regions in which the upper crust, of velocity up to 6.4 km s^{-1}, is 20–24 km thick and the lower crust, with velocity of 6.8–6.9 km s^{-1}, is about 15–20 km thick. Depth to the Moho varies between 35 and 42 km. Zone 2 has up to 10-km thick sediments and a thinner (about 10 km) lower crust of somewhat higher velocity of 6.8–7.0 km s^{-1}, the Moho being at 32–35 km depth. Zone 4 shows rift-type structure where the sediments are thick and the upper crust is thinner as compared to the shield regions, while the lower crust has higher velocity (6.9–7.4 km s^{-1}) and is much thicker. Depth to the Moho discontinuity is similar to that in zone 2. Velocities and depths in zone 5 are more or less similar to those in zone 4. In zone 6 the upper crustal velocities are smaller than the shield region and the depth to the Moho varies between 45 and 65 km.

In another attempt (Fig. 8.1) the regional velocity-depth relationship is limited to the peninsular shield regions and the Cambay, Narmada and the east coast basins, as velocity data in the crust in other parts of the Indian subcontinent is negligible (Kumar et al., 1999). Iso-velocity interfaces for average velocities (not the P-wave velocities of the individual layers) of 4.5, 5.0, 5.5, and 6.0 km s^{-1} indicate the depths to these velocities over these previous regions.

With the help of these interfaces, the average velocity distribution with depth can be visualized at a glance (average velocity is the average of all the velocities encountered in a particular column). To avoid any confusion between the velocities and the geological interfaces, some explanation is needed here. Lower average velocity at a particular depth means larger thickness of the low-velocity material, e.g., sediments, while a higher average velocity means a smaller thickness. For example, a larger depth, between 5 and 15 km, to the average velocity interface of 4.5 km s^{-1} in the regions of the Quaternary and Tertiary sedimentary basins of Cambay and the east coast means a larger

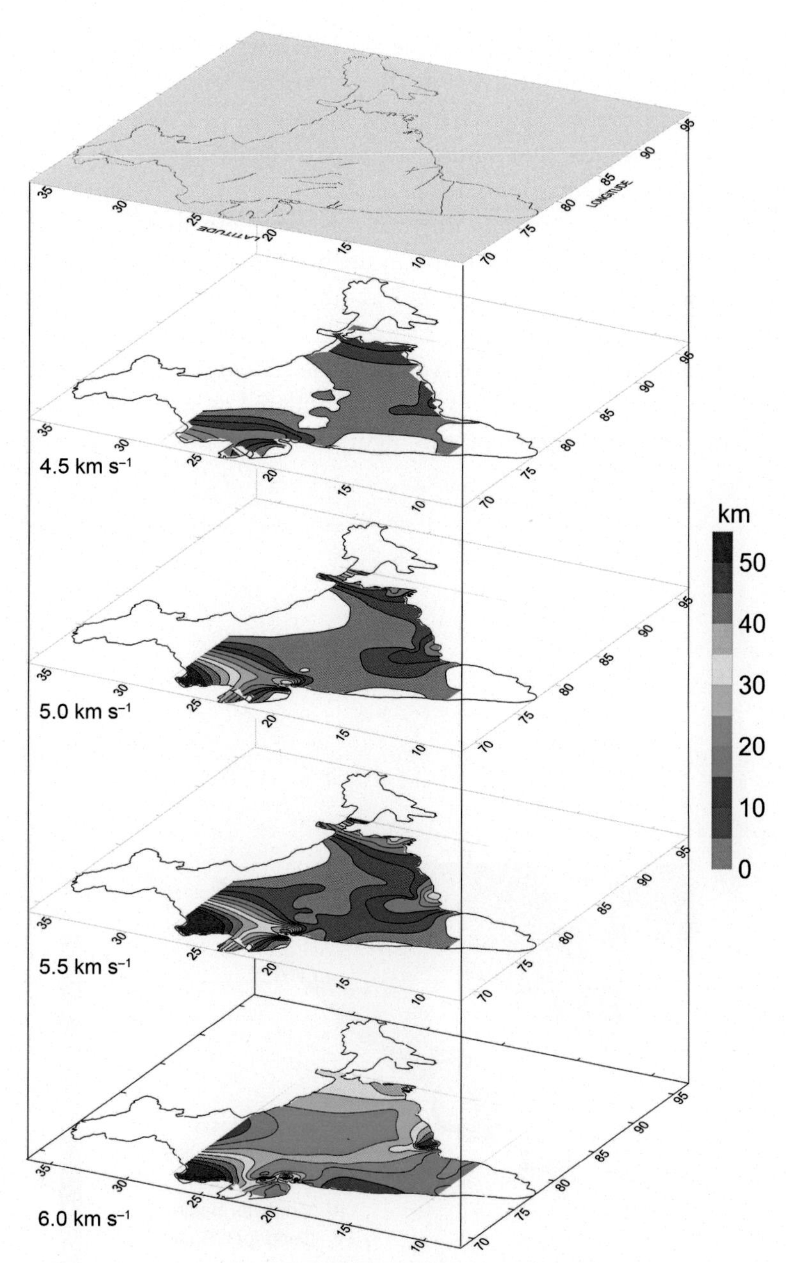

Fig. 8.1 Iso-velocity interfaces in the Indian shield for average velocities of 4.5, 5.0, 5.5, and 6.0 km s^{-1}. *(After Kumar, P., Tewari, H.C., Sain, K., 1999. Velocity-depth relationship in selected parts of Indian crust. J. Geol. Soc. India 54, 129–136).*

thickness of the lower-velocity sediments. This does not indicate the actual thickness of sediments in these regions. Regions of predominant higher velocity exposures are narrowed down by these interfaces.

Specific depth interfaces for the basement (velocity varying between 5.8 and 6.0 km s^{-1}) and the Moho, based on the P-wave velocity (not the same as average velocity), represent variations in the depth to these particular interfaces. The contour for variation in the basement depth (Fig. 8.2) indicates the thickness of sediments as well as other rocks with P-wave velocity of less than 5.5 km s^{-1}.

Rocks like the Deccan Traps (velocity 4.4–5.3 km s^{-1}) and Proterozoic exposures (velocity 5.0–5.5 km s^{-1}) are considered as sediments (and not basement) for the purpose of this figure. Due to this, the regions of

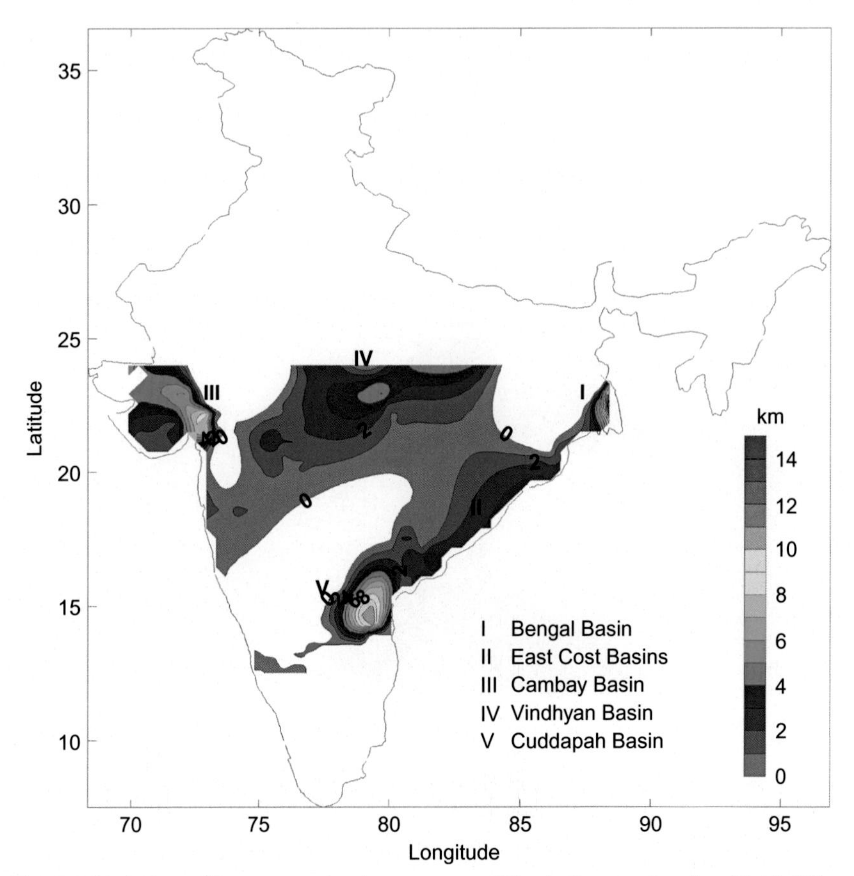

Fig. 8.2 Variation of basement depth over part of the Indian peninsular shield. *(After Kumar, P., Tewari, H.C., Sain, K., 1999. Velocity-depth relationship in selected parts of Indian crust. J. Geol. Soc. India 54, 129–136).*

Proterozoic basins, such as the Cuddapah and Vindhyan, can be identified in this figure and show maximum depths of the Proterozoic sediments as 10 and 5 km, respectively. In this figure, areas of the Archean exposures, having a velocity larger than 5.5 km s^{-1} are clearly differentiated from the Quaternary/Tertiary/Proterozoic basins. In most parts of central India and the Dharwar craton, the basement is exposed and shows zero depth. In the region of the exposed Deccan Traps, including the Narmada Graben, the basement depth values are 1–2 km. The eastern part of the West Bengal basin shows a maximum depth of 13 km. At the onland Godavari and Mahanadi basins, the maximum basement depths are about 4 and 3 km, respectively. The Cambay basin shows a maximum depth of 8 km.

The Moho depth contours (Fig. 8.3) show that in the West Bengal basin the depth to this interface increases from about 24 km in the east to 34 km in the west, while in the Mahanadi basin it is 32 km. In the Cuddapah basin, the increase is from about 35 km in the middle to 41 km at the east (Archean

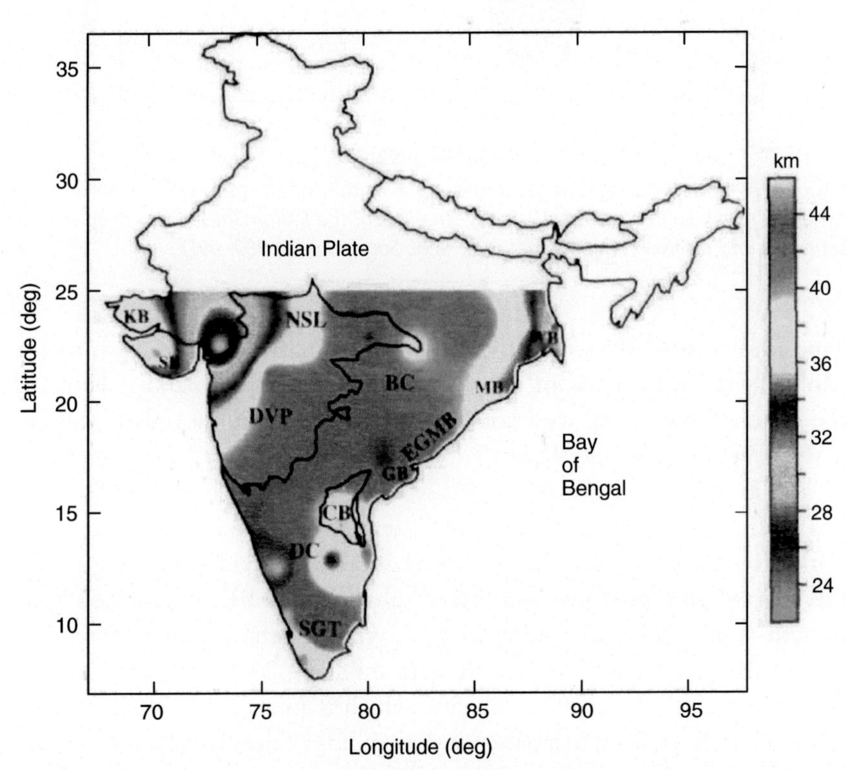

Fig. 8.3 Moho depth contour map of the Indian peninsular shield. *(After Behera, L., Sain, K., 2006. Crustal velocity structure of the Indian Shield from deep seismic sounding and receiver function studies. J. Geol. Soc. India 68, 989–992).*

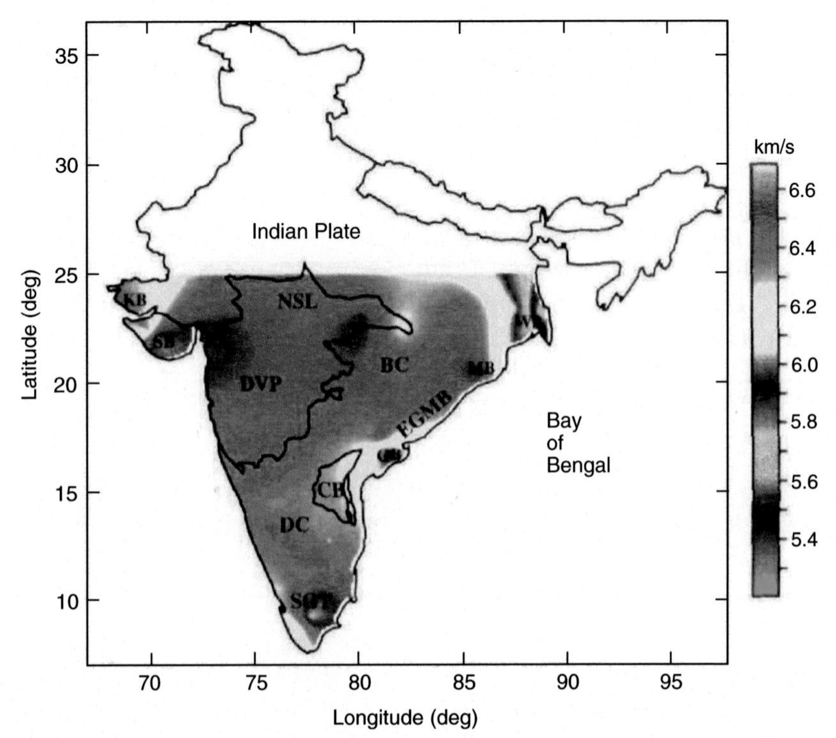

Fig. 8.4 Average velocity image of the crust in the Indian peninsular shield. *(After Behera, L., Sain, K., 2006. Crustal velocity structure of the Indian Shield from deep seismic sounding and receiver function studies. J. Geol. Soc. India 68, 989–992).*

exposures) and 42–43 km in the Godavari basin. Maximum depth to the Moho in the large parts of the Cambay basin is about 32 km, while in the regions close to the west coast of India and the central Indian regions, it varies between 38 and 42 km.

Addition of data from the recent seismic studies and broadband seismic observations in the Southern Granulites terrane and also in the Deccan volcanic province shows that the northwest-southeast corridor, connecting the Deccan volcanic province and eastern ghat Mobile belt, has large crustal growth with a thicker (~43 km) crust, while crustal thinning is seen in the Southern Granulites terrane (Behera and Sain, 2006).

Estimation of the average velocity, which is a measure of bulk composition of the crust, leads to a better understanding of the crustal composition. Low average velocity values (6.0–6.5 km s^{-1}) indicate a dominantly felsic composition while the higher values (6.5–6.8 km s^{-1}) imply a more mafic crust (Christensen and Mooney, 1995). The average crustal velocities in the Indian shield are less than or equal to the global average of 6.44 km s^{-1} (Behera and Sain, 2006), except in the western and eastern coastal regions (Fig. 8.4). This means that the Indian crust is dominantly felsic to intermediate in tectonic deformations (Fleitout and Froidevaux, 1982). The west coast of India and the periphery of the Cambay basin show considerable relief at the Moho level, indicating that these regions may be under greater stress than the Indian shield and central Indian regions (Kumar et al., 1999).

The crustal structure and velocity pattern of the Indian shield show that its units differ from each other in several details. Continued tectonic movements during the Proterozoic, passage of the subcontinent over several plumes at different geological times, northward movement of the Indian plate after its separation from Gondwanaland, and its collision with the Eurasian plate have probably given rise to most of these complexities.

CHAPTER NINE

Global and Indian Scenario of Crustal Thickness

9.1 GLOBAL CRUSTAL THICKNESS AND NATURE OF THE MOHO

Global observations of the crustal thickness show that the nature of the Moho is not quite consistent through the tectonic regions. At a few places it is a sharp boundary, whereas at most of the other places it is diffused. Seismically, refraction phases (Pn and Sn) constrain the velocity contrast across the Moho (Fig. 9.1); however, their observation is not always easy. In most cases, identification of Pn (or Sn) is a challenging job due to the weak amplitude. In most regions of the world, active seismic experiments show that the Moho is not a single interface but is a composite of many layers or a band of layers that acts as a transition.

The seismic characteristics of Archean, Proterozoic, and Phanerozoic crust (Fig. 9.2) are different over geologic time as their formation processes are different, thus favoring secular change in the seismic character of the crust (Meissner, 1986).

A number of models do exist to describe the global crust in diverse geological terrains. Fig. 9.3 shows CRUST1.0, the global distribution of Moho depth with 1 degree × 1 degree (http://igppweb.ucsd.edu/~gabi/crust1.html). Fig. 9.4 shows the crustal structure with seismic velocity of different geological shields, provinces, and cratonic regions. The figure is based on crustal models, mostly computed using the active seismic data available up until 1995 (Mooney et al., 1998). These unified models suggest that the average Archean crust is ~35 km thick, whereas Proterozoic crust is significantly thicker (~45 km). However, the cratons have highly variable crustal thickness. Archean and Paleoproterozoic crusts (more than ~1.7 Ga) have crustal thickness of 42–44 km and crust younger than 1.7 Ga is 35–38 km thick. In the case of the crystalline crust of the Archean to the Paleoproterozoic, the crustal thickness varies between 41 to 33 km.

Structure and Tectonics of the Indian Continental Crust and Its Adjoining Region
https://doi.org/10.1016/B978-0-12-813685-0.00009-1

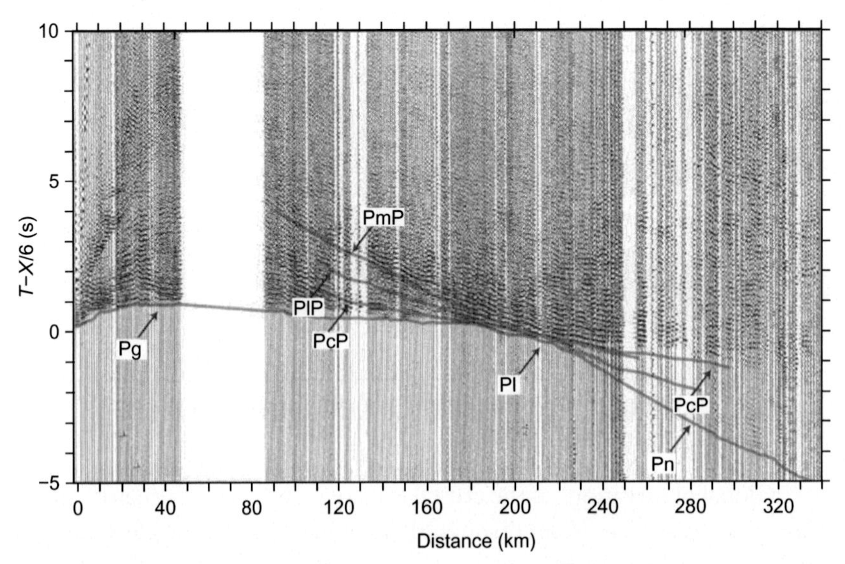

Fig. 9.1 An example of a seismic section where different phases are marked. The refracted wave from the Moho (Pn) can be identified at large offset (Snelsonl et al., 2005).

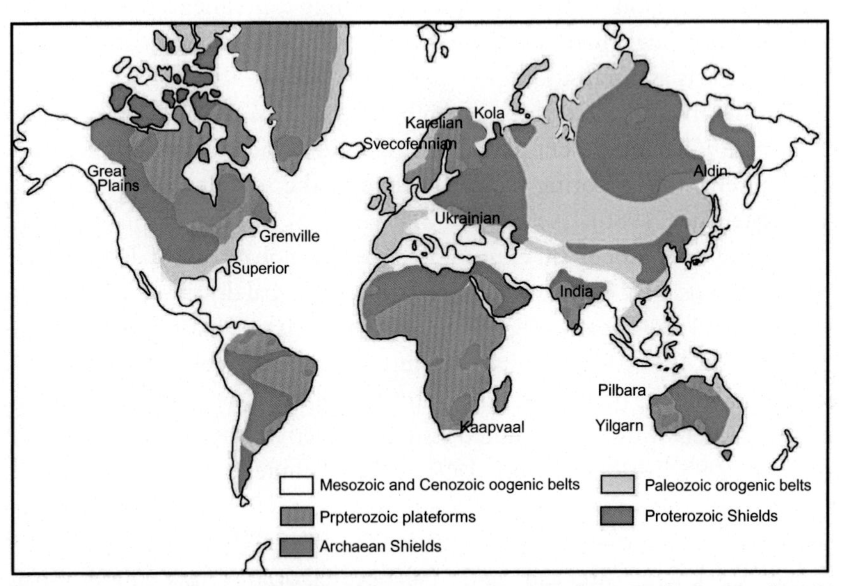

Fig. 9.2 Ages of different geological provinces. *(After Miyashiro, A., Aki, K., Şengör, A.M. (Eds.), 1982. Orogeny. John Wiley, New York; Durheim, R.J., Mooney, W.D., 1994. Evolution of the Precambrian Lithosphere: geological and geochemical constraints. J. Geophys. Res. 99, 15359–15374).*

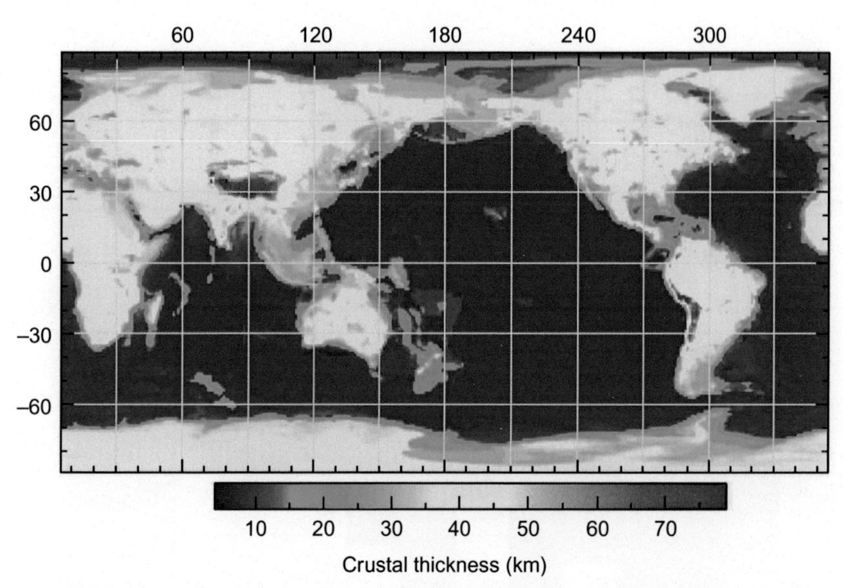

Fig. 9.3 Global crustal thickness (Crust1.0) (http://igppweb.ucsd.edu/~gabi/crust1.html).

Meso- and Neoproterozoic terranes are 33–35 km thick, while the Meso-Cenozoic crust is ~30 km thick (Figs. 9.4 and 9.5).

In the following paragraphs we describe, in brief, the crustal thickness scenario across the globe (Fig. 9.4).

In the southeastern United States, the Moho is relatively horizontal beneath the late Paleozoic orogenic belt (Southern Appalachians) and is discontinuous across the orogen (Pratt et al., 1988; Nelson et al., 1985). Its depth varies between 32 and 40 km at Archean tectonic terranes of the Abitibi within the Superior Province. Beneath the Grenville Province and the Superior Craton, its depth varies between 35 and 45 km (White et al., 2000). The upper to middle Archean Kapuskasing Structural Zone of North America has a crustal thickness of ~53 km (Boland and Ellis, 1989; Percival and Card, 1985; Fountain et al., 1990), while in the Wyoming Province it is in the range of 40–50 km (Mooney and Braile, 1989). The Northern Sierra Nevada has a crustal thickness in the range of 41–46 km (Eaton, 1966; Prodehl, 1979; Leaver et al., 1984). It is ~40 km in the southern part of the Oregon Cascade (Leaver et al., 1984) and ~45 km in Colombian Andes (Hurtado and Leuro, 2000; Poveda et al., 2015). The thickness of the crust is high (~49 km) with high average crustal velocity at the Minnesota River Gneiss Terrane, situated on the southern margin of the Superior Province (Boyd et al., 1992).

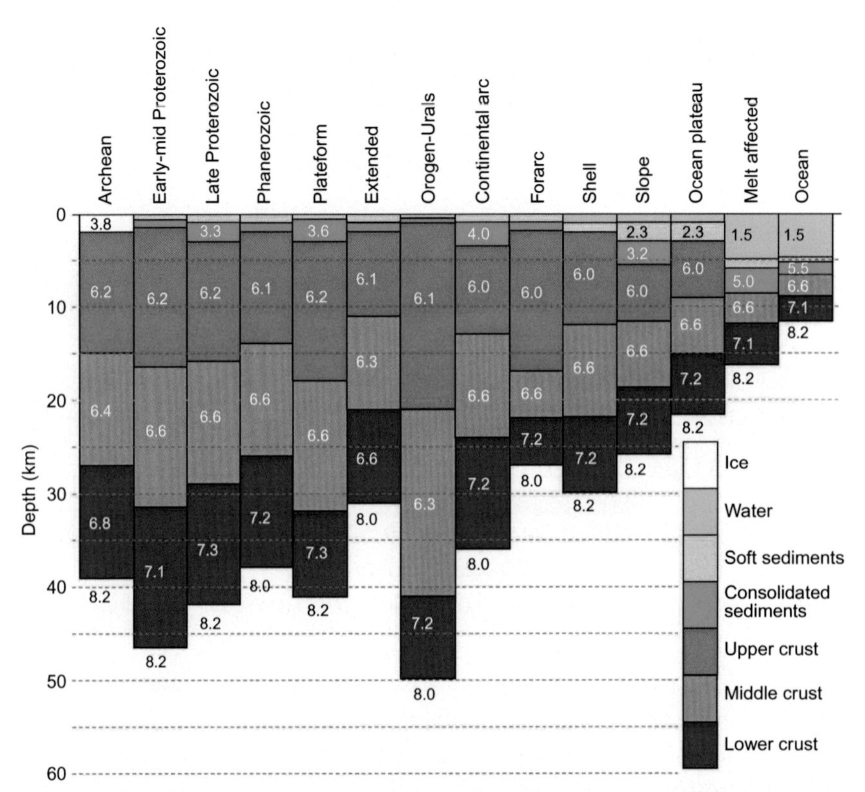

Fig. 9.4 Representative seismic-velocity-depth functions for 14 diverse provinces. *(From Mooney, W.D., Laske, G., Masters, G., 1998. CRUST 5.1: a global crustal model at 5 degrees × 5 degrees. J. Geophys. Res. 103, 727–747).*

In ice-covered Greenland it is 39–49 km thick (Kumar et al., 2007, 2014a,b; Dahl-Jensen et al., 2003). Offshore seismic refraction profiles estimate the crustal thicknesses of 40–45 km both in southern West Greenland (Chian and Louden, 1992) and southern East Greenland (Dahl-Jensen et al., 2003).

In South America the crustal thickness varies between 40 and 50 km from the southern Andean system of Colombia to the northern central/eastern Cordillera (Ojeda and Havskov, 2001; Hernandez-Pardo et al., 2007; Poveda et al., 2015). The central Andes of Peru is underlain by a remarkably thick crust of ~60 km (Mooney et al., 1998). The subducting oceanic Nazca plate under the South American continent between the Peru-Chile trench and the Central Andes in northern Chile shows a portion of the Moho dipping eastward from 30–50 km near the coast to 55–64 km up to about 240 km

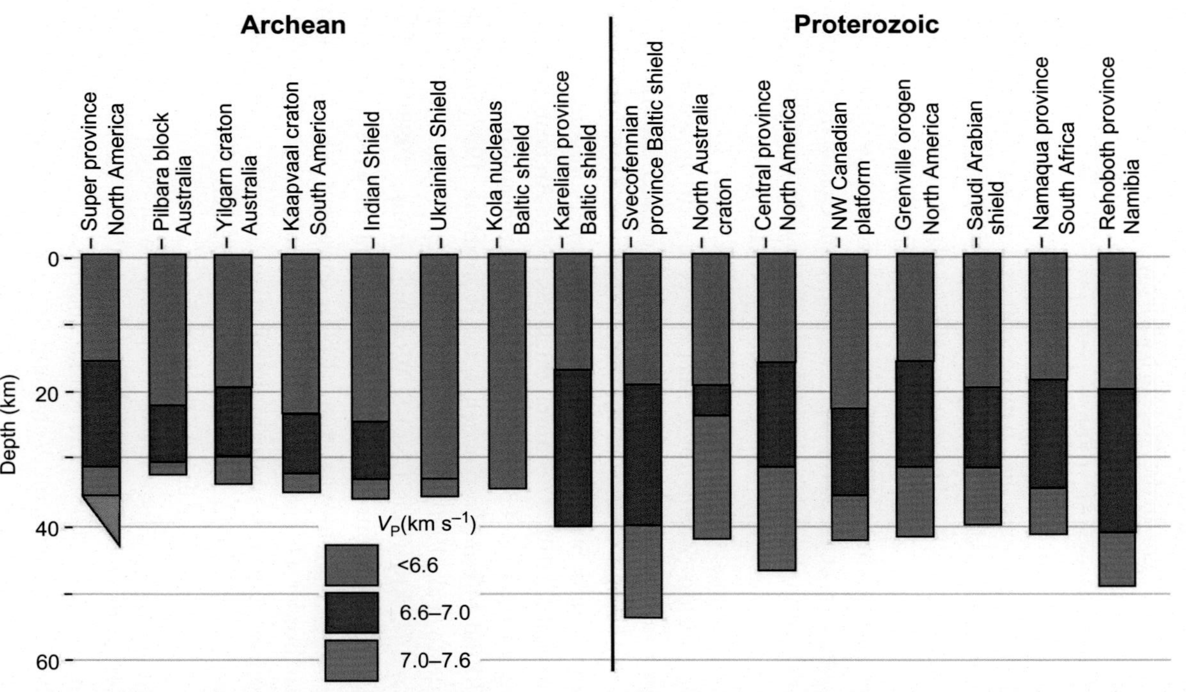

Fig. 9.5 Representative seismic-velocity-depth functions for Archean and Proterozoic provinces. The crust in Archean provinces is relatively thin and lacks a substantial high-velocity basal layer when compared with Proterozoic provinces. Thickness of the Archean crust is ~35 km, whereas that of Proterozoic crust is ~45 km. The basal high-velocity layer, probably indicating rocks with mafic average composition, is substantially thicker in Proterozoic provinces. *(After Durrheim, R.J., Mooney, W.D., 1991. Archcan and Proterozoic crustal evolution: evidence from crustal seismology. Geology 19, 606–609).*

inland. Crustal thickness beneath the western Cordillera and the Altiplano is at least 70 km. In the Precordillera a pronounced Moho discontinuity is detected at a depth of 50 km. Along the coast, the oceanic Moho boundary can be identified at a depth of 40–45 km. In the Andean backarc region, tectonic compression has produced pronounced crustal thickening and, locally, a doubling of the Moho discontinuity (Giese et al., 1999).

Across the Trans-European Suture Zone (TESZ) of Phanerozoic Europe, the crustal thickness lies between 30 and 45 km (Grad et al., 2006). Southcentral Finland has the deepest Moho (60–65 km) in continental Europe (Tiira et al., 2006). The margins of Greenland, western Norway and the British Isles have a more or less uniform crustal thickness between 32 and 36 km and in the Arctic shelf it varies between 20 and 24 km (Zorin, 1999; Makarov et al., 2010). The East European Craton has a crustal thickness of ~60 km in the Baltic Shield. In the Arabian-Nubian shield, the average crustal thickness is ~38 km (Mooney et al., 1985; Stern and Johnson, 2010). Crustal thicknesses in the Central Asian Orogen vary between 40 and 45 km (Zorin, 1999; Makarov et al., 2010), whereas for the Karelian province of the Baltic Shield and the Volga-Uralia subcraton it varies between 50 and 60 km. In the Alpine belt, the crustal thickness varies between 20 and 50 km from older to younger orogens (Zorin, 1999; Makarov et al., 2010). The crustal thickness in the Archean Karelian province varies in the range of 37–45 km and for the Archean Kola province it is between 30 and 44 km (Kostyuchenko and Romanyuk, 1997; Sollogub et al., 1973; Luosto et al., 1990). In the Sveconorwegian province it changes between 32 and 42 km (EUGENO-S Working Group, 1988; Thybo, 2001). The Gothian province and its accreted terranes have a crust of 40–45 km thickness (Abramovitz et al., 1997), consistent with the crust of the western border of Latvia (40–60 km) (Ankudinov et al., 1991). The East European craton is similar to the Siberian craton but has thicker (~45 km) crust compared to other Precambrian cratons (Cherepanova et al., 2013). Riphean and Palaeozoic rifting have a very thick crust of ~60 km (Trofimov, 2006). Beneath the Karpinsky Swell (southern margin of the East European craton) the crustal thickness varies in the range of 42–48 km (Saintot et al., 2006). The crustal thicknesses of two Ukrainian shields, i.e., the Paleoproterozoic Volyn and the Archean Podolian blocks (Thybo et al., 2003), vary between 45 and 50 km, whereas the Mezen rift province has a crustal thickness of 30–37 km.

Artemieva and Thybo (2013) present new tectonic maps of Europe by analyzing regional trends in crustal structure and conclude that: each tectonic setting has a different crustal structure and depth to Moho; global averages of crustal parameters are outside of observed ranges for any tectonic setting in Europe; variation of V_p with depth in the sedimentary cover does not follow commonly accepted trends; the thickness ratio between upper-middle and lower crystalline crust is indicative of either oceanic, transitional, platform, or extended crustal origin; continental rifting generally thins the upper-middle crust significantly without changing V_p. The lower crust experiences less thinning, also without changing V_p, suggesting a complex interplay of magmatic underplating, gabbro-eclogite phase transition and delamination; the crustal structure of the Barents Sea shelf differs from rifted continental crust; and most of the North Atlantic Ocean north of 55°N has anomalously shallow bathymetry and anomalously thick oceanic crust.

Salmon et al. (2013) summarized the crustal pattern in Australia and New Zealand. The Australian continent is an assemblage of the Yilgarn craton of Archean age, Pilbara craton of Archean to Proterozoic age, and Gawler craton and Willyama block of Archean to Proterozoic age. The oldest portions of the West Australian craton, north Yilgarn and Pilbara cratons have crustal thickness in the range of 30–35 km. The crust thickens slightly beneath the Neoarchean Hamersley block at the southern edge of the Pilbara. The Archean terranes have a crust of thickness ranging between 30 and 32 km. In the Capricorn Orogen the Moho depth exceeds 40 km. The thickest crust with a rather indistinct base is found in the Glenburgh terrane at the western edge of the Capricorn Orogen. A Moho depth greater than 40 km extends across the entire Kimberley block. In the north of the Central Australian zone, the crust appears to thin in the Pine Creek Inlier to less than 35 km thick. The Tennant Creek block also has a crustal thickness up to 50 km. Much of the Arunta block also has a rather thick crust (more than 50 km). The Gawler craton in the south has a more than 40 km thick crust. The Mt Isa block also has a thick crust (greater than 40 km). At the Georgetown Inlier, the crust thins rapidly to around 35 km.

Offshore continental plateaus of New Zealand have a crustal thickness of ~25 km. The northwestern North Island also has a crustal thickness of 25 km. There are three areas of thickened crust (~42 km): one between the North and South Islands (Wanganui Basin), one along the spine of the South Island (Southern Alps), and in the southwest corner of the South Island inland. The Wanganui Basin is a region of thickened crust.

The structural variability in platform, orogens and depression structures of the Antarctic is reflected both in thickness and physical properties of the crust (Baranov and Morelli, 2013). The oldest Archean and Proterozoic crust of East Antarctica has a thickness of 36–56 km (average ~41 km). The continental crust of the Transantarctic Mountains, the Antarctic Peninsula and Wilkes Basin has a thickness of 30–40 km (average ~30 km). The youngest rifted continental crust of the West Antarctic Rift System has a thickness of 16–28 km (average ~26 km).

The northern Honshu province of Japan has a crustal thickness of ~32 km. Most of the crust in Indonesia is of continental or borderland type of intermediate (30 km) thickness. In the Gulf of Papua, the Moho is 27–29 km deep along the southwestern coast of the peninsula and shallows beneath the Moresby Trough to 19 km. This depth is greater than normal for the Moho beneath oceanic crust, but it is not unreasonable when compared to other measurements in the area. Curray et al. (1977) found the Moho to be at 18 km depth south of the Island of Bali and 21–25 km south of the Island of Java. Beneath the Eastern Plateau, the Moho again deepens to 25 km (Drummond, 1979). With a lower crustal layer of 9 km in thickness, the crust beneath the Papuan Peninsula is continental. Offshore, the crust in the Moresby Trough is about 19 km thick, with sediments comprising approximately the top 10 km. Therefore, the crust below the sediments has a thickness typical of oceanic crust. The northern extent of the thin crust beneath the Moresby Trough is uncertain.

Crustal thickness beneath the Chinese mainland (Teng et al., 2013) varies in the range of 10–85 km. The deepest Moho discontinuity—at approximately 70–85 km beneath the Tibetan Plateau—was formed by ongoing continent-continent collision. The Moho beneath the eastern North China craton, at a relatively constant 30–35 km, has endured mantle lithosphere destruction. In the northwestern South China the Moho is deep (up to 60 km), while it is at smallest depth in the southeast. All of western China is underlain by 45–70 km thick crust that is characteristic of young orogenic belts worldwide and is separated from the 30–45 km thick crust of Eastern China by a seismic belt that trends roughly north-south. A low-velocity, perhaps fluid-rich or partially molten, zone in the middle crust (Zhao et al., 1993; Nelson et al., 1996; Makovsky et al., 1996a,b) is a unique crustal feature of the Tibetan Plateau.

The crustal picture from the Arabian plate has gaps due to lack of data coverage, especially in the interior of the Arabian plate (Mechie and Kind, 2013). Beneath the interior of the Arabian plate the Moho lies between 32 and 45 km depth. Across the northern margin with the Eurasian plate, the Moho depths increase to over 50 km beneath the Zagros mountains. Across the western margin, the Dead Sea Transform, Moho depths decrease from almost 40 km beneath the highlands east of the transform to about 21–23 km under the southeastern Mediterranean Sea. This decrease seems to be modulated by a slight depression in the Moho beneath the southern transform. The southwestern and southeastern margins of the Arabian plate also show the Moho shallowing from the plate interior towards the plate boundaries.

Large portions of Africa have not been investigated for crustal studies. The axial trough of the southern Red Sea rift has an oceanic crust floor and thinned continental crust underlies the margins of the Red Sea, Afar depression, and northern Kenya rift. In contrast, 30–35-km thick continental crust is found under the Jordan-Dead Sea rift. The transition from thinned continental crust to 5–6-km thick oceanic crust in the center of the Red Sea appears to be gradual. In contrast, the transition from the thin crust of the East African rift to undisturbed continental crust of ~40 thickness is rather abrupt. Beauchamp et al. (1999) report thin crust (18–20 km) beneath the Moroccan shelf and a thickness of 30 km south of the High Atlas Mountains. The crust thins to 24 km toward the Atlantic margin, and attains an average crustal thickness of 35 km to the north of the mountains. The crust is 39 km thick beneath the mountains.

The results from the Kaapvaal and Zimbabwe cratons of South Africa show a relatively thin crust of 35–40 km with a flat crust/mantle boundary (Nair et al., 2006; Nguuri et al., 2001; Niu and James, 2002; Yang et al., 2008). In the Limpopo Belts, the crustal thickness is 30–40 km. The crust is about ~42 km beneath the Namaqualand Metamorphic Complex (Green and Durrheim, 1990). Gradational crust is observed at the central zone of the Damara Orogen Namibia, with crustal thickness of ~47 km (Green, 1983).

9.2 OCEANIC CRUSTAL STRUCTURE

To understand the mechanism prevailing at the time of origin of the continental crust, it is also necessary to understand the structure and composition of oceanic crust, which is relatively much younger. To understand this, the seismic structure and composition of oceanic crust from five sites located near to the midoceanic ridges is presented here.

Earlier, it was believed that the oceanic crust is formed at a slow spreading midoceanic ridge where the spreading rate is <30 mm/year (Lagabrielle et al., 1998; Osler and Louden, 1995; Grevemeyer et al., 1997). However, recent studies suggest that the slow spreading segments are unable to generate a continuous crust, where a heterogeneous mixture of mantle (peridotite) and magmatic rocks exist. Moho is observed as a consistent feature for all spreading rates, but appears quite variable, being very distinct in some areas, complex in form in others, and absent in many regions (Mutter and Hélène, 2013). The convective cooling of the Earth forms new seafloor at the midoceanic ridges and hot mantle ascending beneath them melts partially to create a basaltic oceanic crust. This crust and the residual mantle after melting (depleted lithospheric mantle) are compositionally less dense than the mantle before melting (the asthenosphere) (Oxburg and Parmentier, 1977). The oceanic crust, extending 5–10 km beneath the ocean floor, is mostly composed of different types of basalts ("sima," which stands for silicate and magnesium, the most abundant minerals in oceanic crust). It is constantly formed at midocean ridges, where tectonic plates are tearing apart from each other. As the magma wells up from these rifts in the Earth's surface and then cools, it becomes young oceanic crust. The crust that is generated later is pushed away from the midocean ridges.

Two models are generally considered for formation and evolution of the oceanic crust. One is called the *gabbro glacial model*, in which the magma rises in the form of melted lenses at the midocean ridge and the oceanic crust is evolved. Due to this process magma cumulates at the bottom and, subsequently, the crystallized magma sinks to form the layer of gabbros (Morgan and Chen, 1993; Henstock et al., 1993). The second model suggests that gabbros are formed due to the in-situ crystallization of sills emplaced in the lower crust (Kelemen et al., 1997; Macleod and Yaouancq, 2000) (Fig. 9.6).

Most of our knowledge about the oceanic crust comes from active seismic observations using Ocean Bottom Seismometer data (Edwards et al., 2008; Shinohara et al., 2008). However, in recent times seismological studies such as surface wave dispersion and receiver functions analysis are also widely used to decipher the oceanic crust and lithospheric structures (e.g., Bagley et al., 2009; Shinohara et al., 2008; Kumar and Kawakatsu, 2011). The high-resolution 3D seismic tomography studies using active source data (Evangelidis et al., 2004) reveal fine layering within the crust in the volcanic edifices of Ascension Island. Seismic reflection/refraction studies have also been carried out near the ridges (Bjarnason et al., 1993; Evangelidis et al., 2004; Rodgers and Harben, 1999) in and around the Atlantic. All these studies reveal a highly variable crustal structure. Attempts to understand the structure of the oceanic plates have been made through a number of seismological studies at different locations beneath the Pacific Ocean (Gaherty et al., 1999; Tan

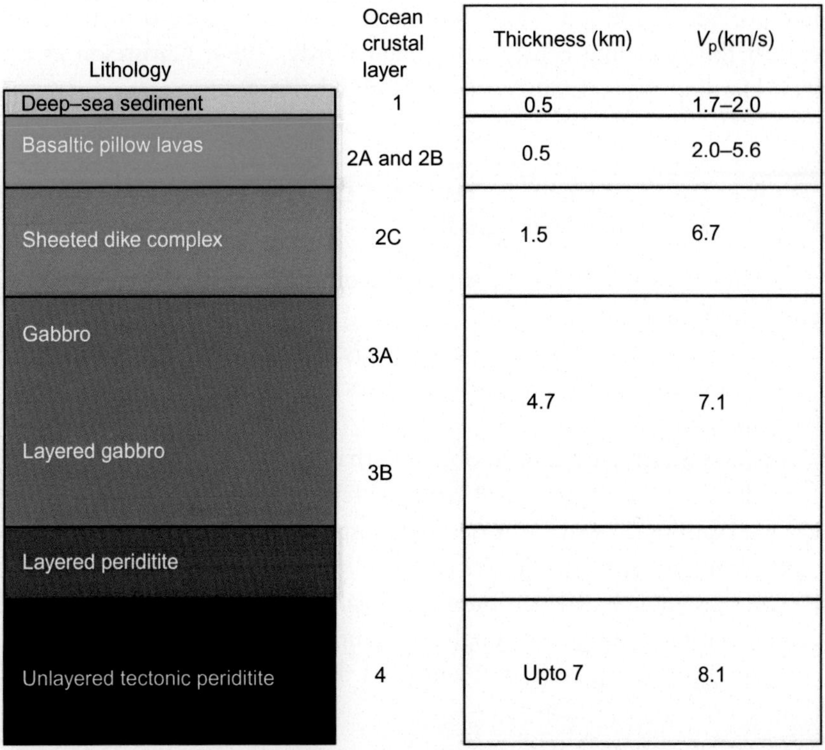

Lithology	Ocean crustal layer	Thickness (km)	V_p(km/s)
Deep–sea sediment	1	0.5	1.7–2.0
Basaltic pillow lavas	2A and 2B	0.5	2.0–5.6
Sheeted dike complex	2C	1.5	6.7
Gabbro	3A	4.7	7.1
Layered gabbro	3B		
Layered periditite			
Unlayered tectonic periditite	4	Upto 7	8.1

Fig. 9.6 Schematic diagram showing a typical oceanic crustal section. *(Modified after Brown, G.C., Mussett, A.E., 1993. The Inaccessible Earth, An Integrated View to Its Structure and Composition. ISBN: 978-0-412-48160-4 (Print) 978-94-011-1516-2).*

and Helmberger, 2007; Kawakatsu et al., 2009; Kumar and Kawakatsu, 2011; Kumar et al., 2012; Rychert and Shearer, 2011; Schmerr, 2012, etc.). Logistic problems and water reverberations that contaminate the vertical components of the seismograms of the ocean bottom seismometers impose constraints in acquisition of oceanic data (Kumar et al., 2012).

The high-resolution seismic and other studies have revealed that the oceanic crust is composed of ophiolites, which occurred in a sequence as pillow basalt, dikes and gabbros either in massive or layered form. Therefore, the oceanic crust is characterized by three layers (Fig. 9.5): Layer 1, ~1 km thick, consisting of pelagic sediments derived from weathering and erosion of the continents; Layer 2, composed of two layers, i.e., basalt pillow lavas and sheeted dike complex; and Layer 3, believed to be mostly gabbros, crystallized from magma generating from a shallow chamber that feeds the dikes and basalts. The layering sequence and internal structures of its major rock units provide significant constraints on the mode of genesis of the crust beneath the spreading centers (Bratt and Purdy, 1984; Christeson et al., 1994; Harding et al., 1989; Vera and Diebold, 1994).

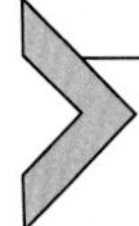

9.3 DIFFERENCES BETWEEN CONTINENTAL AND OCEANIC CRUST

The major differences between the continental and oceanic crust can be summarized as follows:

1. The oceanic crust is younger and mostly mafic in nature (basaltic composition) whereas the continental crust is much older and also has a felsic component.
2. Seismologically, it has been observed that the oceanic crust is 6–12 km thick, unlike the continental crust, which is 30–70 km thick.
3. Observations show that the oceanic crust is denser (2.8–3.0 g/cm^3) than the continental crust (2.6–2.7 g/cm^3).
4. The oceanic crust has less buoyancy than the continental crust.
5. The oceanic crust goes through a recycling process through subduction, while this phenomenon is absent from the continental crust.

9.4 INDIAN CONTINENTAL CRUST

The Indian shield is a mosaic of diverse terrains bearing the imprints of various tectonic episodes in geological history from Archean to the late Proterozoic eon. There are five major cratons demarcated by mobile belts, rifts or sutures. Most of the knowledge about the crustal structure of the Indian subcontinent comes from several seismic studies, as discussed earlier in this book. In addition to these, several other studies have helped in understanding the nature of the Indian continental crust. These include: (i) body wave studies of shallow earthquakes (Roy, 1939; Mukherjee, 1942; Tandon, 1954; Chakravorty and Ghosh, 1960; Chouhan and Chouhan, 1965; Kaila et al., 1968; Reddy, 1971), (ii) surface wave studies (Bhattacharya et al., 2009; Mohan et al., 1997; Rai et al., 2003; Mitra et al., 2006) and (iii) receiver function studies (Gaur and Priestley, 1997; Zhou et al., 2000; Saul et al., 2000; Kumar et al., 2001, 2004, 2013; Sarkar et al., 2003; Gupta et al., 2003a,b; Rai et al., 2003; Ramesh et al., 2005). There is a wide variation in the crustal thickness in different geological provinces of the Indian shield. It varies from a minimum of ~25 km (coastal plains of the peninsular India) to a maximum of ~80 km (mountain ranges of Himalaya) depth, which also leads to changes in average crustal velocity. The Himalaya-Tibet collision zone has low shear velocity ($V_s \sim 3.57$ km s^{-1}) compared to that of the Indian shield (~3.7–3.75 km s^{-1}). The Himalayan crustal thickness clearly follows a trend with elevation. The rift zones of the Godavari graben and Narmada-Son Lineament show larger thickness of the crust as compared to their surroundings. Indian cratonic fragments, i.e., Bundelkhand, Bastar and Singhbhum, show thick crusts in comparison to the Eastern Dharwar craton.

The Indian crust is heterogeneous (Julia et al., 2009) in nature. The average crustal thickness (~35 km) in the Eastern Dharwar craton is less (Gupta and Rai, 2005) compared to that in the Western Dharwar craton (~45 km) (Kumar et al., 2001). The western Dharwar craton has a complex geology and gradational transition from crust to Moho (Gupta et al., 2003a,b; Sarkar et al., 2003; Rai et al., 2003). The S-wave velocity in the Western Dharwar craton ($V_s \sim 3.73$ km s^{-1}) is higher than that in the Eastern Dharwar craton ($V_s \sim 3.71$ km s^{-1}) (Borah et al., 2014; Gupta et al., 2003a,b; Sarkar et al., 2003). Gupta et al. (2003a,b) suggest a very thick crust (~60 km) in the central portion of the Western Dharwar craton where Julia et al. (2009) reported Moho at about ~45 km depth. Active seismic studies (wide angle reflection and refraction) across the central part of the Dharwar craton suggest that the crust is broken into blocks, with Moho depth varying from ~34 km in the east to ~41 km in the west. The Poisson's ratio varies from 0.24 to 0.27 for some parts of the Dharwar craton, Southern Granulite terrain and basins such as Cuddapah (Singh et al., 2015). The Poisson's ratio is less for the Western Dharwar craton (Gupta et al., 2003a,b; Sarkar et al., 2003) as compared to the Eastern Dharwar craton, due to a more felsic upper crust where Poisson's ratio is ~0.25, while the lower portion of the crust is mafic in nature with a correspondingly higher Poisson's ratio (Sarkar et al., 2003; Kiselev et al., 2008). Though the crustal thickness values are different in the WDC and EDC, the V_p/V_s ratio values are mixed in nature. The Poisson's ratio of the Deccan Volcanic Province (DVP) is ~0.26, but the crustal composition is not well determined due to an uneven distribution of sampling of stations close to the Western Ghats, or at the easternmost boundary of the Deccan Volcanic Province. The crust of the DVP is classified as more felsic-to-intermediate in nature (Kumar et al., 2001). Kiselev et al. (2008) suggested that the crustal composition of the Deccan Volcanic region is mafic in nature. The

Narmada-Son Lineament and the Western Ghats have higher Poisson's ratios, i.e., ~0.28 (Jagadeesh and Rai, 2008; Rai et al., 2005) than the normal. The crust beneath the Narmada-Son Lineament is found to be more mafic in nature with high V_p/V_s ratios (1.84–1.91) than its surroundings and is suggestive of a high-density mafic mass at a depth that compensates the crustal root. This view is also supported by the existence of small topographic variation (~200 m) across the lineament (Rai et al., 2005). High variation in Poisson's ratio (0.26–0.32) is observed in the Himalaya-Tibet collision zone. The high Poisson's ratio beneath most of the Tibet plateau suggests the presence of partial melts/fluids, as interpreted in various studies (Li et al., 2006, 2009; Owens and Zandt, 1997).

9.5 GENERAL CHARACTERISTICS OF CONTINENTAL CRUSTAL STRUCTURE

According to Mooney et al. (1998), the continental lithosphere shows several important regularities, as listed here. (1) Crustal thickness and average velocities increase from continental margins toward the interior. (2) The crust of old, stable shields and platforms is 35–45 km thick, whereas young, nonorogenic crust is significantly thinner (25–35 km). (3) The crust is thicker under orogenic belts and thinner under basins. (4) An average petrological model may be presented for stable continental interiors in terms of three layers. The two upper layers have silicic-to-intermediate composition, but the middle crust is composed of rocks with a higher degree of metamorphism. The third layer is composed of mafic rocks of granulite-grade metamorphism. (5) The upper mantle, including the subcrustal lithosphere and possibly the asthenosphere, is characterized with fine-scale seismic and rheological stratification. Rheologically weak layers appear in the middle and lower crust and the crust/mantle boundary.

REFERENCES

Abbott, D.H., Mooney, W.D., VanTongeren, J.A., 2013. The character of the Moho and lower crust within Archean cratons and the tectonic implications. Tectonophysics 609, 690–705.

Abramovitz, T., Thybo, H., Berthelsen, A., 1997. Proterozoic sutures and terranes in the southeastern Baltic Shield interpreted from BABEL deep seismic data. Tectonophysics 270, 259–277.

Acharya, S.K., Kayal, J.R., Roy, A., Chaturvedi, R.K., 1998. Jabalpur earthquake of May 22, 1997: constraint from aftershock study. J. Geol. Soc. India 51, 295–304.

Agrawal, R.P., 1995. Hydrocarbon Prospects of the Pranhita-Godavari Graben, India, Petrotech-1995. vol. 1. B.R. Publishing Co, Delhi. pp. 115–122.

Agrawal, P.K., Pandey, O.P., 2004. Unusual lithospheric structure and evolutionary pattern of the cratonic segments of the south Indian shield. Earth Planets Space 56, 139–150.

Allegre, C.J., Brik, J.L., Campers, F., Courtlier, V., 1999. Age of the Deccan traps using 187Re-187Os systematics. Earth Planet. Sci. Lett. 170, 197–204.

Anderson, D.L., 1994. The sub-lithospheric mantle as the source of continental flood volcanics; the case against the continental lithosphere and plume head reservoirs. Earth Planet. Sci. Lett. 123, 269–280.

Ankudinov, S.A., Brio, H.S., Sadov, A.S., 1991. Deep structure of the crust in the Baltic republics based on DSS seismic studies. Seismol. Bull. Belorussia 1, 111–117.

Arndt, N., 2013. Formation and evolution of the continental crust. Geochem. Perspect. 2, 405–530.

Arora, B.R., Waghmare, S.Y., Mahashabde, M.V., 1995. Geomagnetic depth sounding along the Hirapur-Bhandara Profile, Central India. In: Sinha-Roy, S., Gupta, K.R. (Eds.), Continental Crust of NW and Central India. Geol. Soc. India, Mem. 31, 519–535.

Arora, B.R., Rawat, G., Unsworth, M.J., 2005. Deep structures of the NW Himalayas collision zone: constraints from long period magnetotelluric data. In: AOGS Singapore. Abstract Volume.

Artemieva, I.M., Thybo, H., 2013. EUNAseis: a seismic model for Moho and crustal structure in Europe, Greenland, and the North Atlantic region. Tectonophysics 609, 97–153.

Ashwal, L.D., 1989. Introduction to "growth of the continental crust". Tectonophysics 161, 143–145.

Auden, J.B., 1975. Seismicity associated with the Koyna reservoir, Maharashtra. UNESCO, Paris. Tech. Rep., RP/1975–76/2.222.3.

Awasthi, D.N., Ramakotaiah, G., Varadarajan, S., Rao, N.D.J., Behl, G.N., 1971. Study of the Deccan Traps of Cambay basin by geophysical methods. Bull. Volcanol. 35, 743–749.

BABEL Working Group, 1990. Evidence for early Proterozoic plate tectonics from seismic reflection profiles in the Baltic shield. Nature 348, 34–38.

Bagley, B., Courtier, A.M., Revenaugh, J., 2009. Melting in the deep upper mantle oceanward of the Honshu slab. Phys. Earth Planet. Inter. 175, 137–144. https://doi.org/10.1016/j.pepi.2009.03.007.

Baishya, N.C., Singh, S.N., 1986. DSS profiling in Mahanadi on shore areas: a case study. In: Kaila, K.L., Tewari, H.C. (Eds.), Deep Seismic Soundings and Crustal Tectonics. A.E.G., Hyderabad, pp. 1–9

Baksi, A.K., 1995. Petrogenesis and timing of volcanism in the Rajmahal Flood Basalt Province, North-eastern India. Chem. Geol. 121, 73–90.

Balakrishnan, T.S., 2003. Impact of gravity and other geophysical data on the geology of Indian subcontinent. J. Virtual Explor. 12, 83–92.

Banerjee, P., Satyaprakash, W., 2003. Crustal configuration across the northwestern Himalayas as inferred from gravity and GPS aided geoid undulation studies. J. Virtual Explor. 12, 93–106.

Baranov, A., Morelli, A., 2013. The Moho depth map of the Antarctica region. Tectonophysics 609, 299–313.

Bastow, I., Thompson, D., Wookey, J., Kendall, J., Helffrich, G., Snyder, D., Eaton, D., Darbyshire, F., 2011. Precambrian plate tectonics: seismic evidence from northern Hudson Bay, Canada. Geology 39 (1), 91–94. https://doi.org/10.1130/G31396.1.

Basu, A.R., Renne, P.R., Dasgupta, D.K., Teichman, F., Poreda, R.J., 1993. Early and late Alkali igneous pulses and a high-He plume origin for the Deccan flood volcanics. Science 261, 902–906.

Bean, C.J., Jacob, A.W.B., 1990. P-wave anisotropy in the lower lithosphere. Earth Planet. Sci. Lett. 99, 58–65.

Beauchamp, W., Allmendinger, R., Barazangi, M., Demnati, A., El Alji, M., Dahmani, M., 1999. Inversion tectonics and the evolution of the High Atlas Mountains, Morocco, based on a geological-geophysical transect. Tectonics 18, 163–184.

Behera, L., Sain, K., 2006. Crustal velocity structure of the Indian Shield from deep seismic sounding and receiver function studies. J. Geol. Soc. India 68, 989–992.

Behera, L., Sain, K., Reddy, P.R., Rao, I.B.P., Sarma, V.Y.N., 2002. Delineation of shallow and the Gondwana graben in the Mahanadi delta, India, using forward modelling of first arrival seismic data. J. Geodyn. 34, 127–139.

Behera, L., Sain, K., Reddy, P.R., 2004. Evidence of under plating from seismic and gravity studies in the Mahanadi delta and its tectonic significance. J. Geophys. Res. 109 (B12311), 1–25.

Beloussov, V.V., Pavlenkova, I., 1985. Types of the crust of the Earth. Geotectonics 19, 1–9.

Beloussov, V.V., Belyaevsky, N.A., Borazon, A.A., Volvovsky, B.S., Volvovsky, L.S., Resvoy, D.P., Tal-Viscid, B.B., Khambrabaev, I.K., Kaila, K.L., Narain, H., Marussi, A., Finetti, J., 1980. Structure of the lithosphere along the deep seismic sounding profile: Tien Shan-Pamir-Karakoram-Himalaya. Tectonophysics 70, 193–221.

Besse, J., Courtillot, V., 1988. Paleogeographic maps of the continents bordering the Indian Ocean since the early Jurassic. J. Geophys. Res. 93, 11791–11808.

Bhaskar Rao, Y.J., Janardhan, A.S., Vijay Kumar, T., Narayana, B.L., Dayal, A.M., Taylor, P.N., Chetty, T.R.K., 2003. Sm–Nd model ages and Rb–Sr isotopic systematics of charnockite and gneisses across the Cauvery Shear Zone, southern India: implications for the Archaean-Neoproterozoic terrane boundary in Southern Granulite Terrene. In: Ramakrishnan, M. (Ed.), Tectonics of Southern Granulite Terrene, Kuppam-Palani Geotransect. Geol. Soc. India, Mem. 50, 297–317.

Bhattacharya, S.N., 1981. Observation and inversion of surface wave group velocities across central India from surface wave dispersion. Bull. Seismol. Soc. Am. 71, 1489–1501.

Bhattacharya, S.N., Suresh, G., Mitra, Supriyo, 2009. Lithospheric S-wave velocity structure of the Bastar Craton, Indian Peninsula, from surface-wave phase-velocity measurements. Bull. Seismol. Soc. Am. 99, 2502–2508.

Bhukta Subrata, K., Tewari, H.C., 2007. Crustal seismic structure in Jammu and Kashmir region. J. Geol. Soc. India 69, 755–764.

Bhukta Subrata, K., Sain, K., Tewari, H.C., 2006. Crustal structure along the Lawrencepur-Astor profile across Nanga Parbat. Pure Appl. Geophys. 163, 1257–1277.

Bilham, R., Larson, K., Freymueller, J., Project Idyl Him members, 1997. GPS measurements of present-day convergence across the Nepal Himalaya. Nature 386, 61–64.

Biswas, S.K., 1987. Regional tectonic framework, structure and evolution of western marginal basins of India. Tectonophysics 135, 307–327.

Biswas, S.K., 1996. Mesozoic volcanism in the east coast basins of India. Ind. J. Geol. 68, 237–254.

Biswas, S.K., Deshpande, S.V., 1983. Geology and hydrocarbon prospects of Kutch, Saurashtra and Narmada basin. Petro. Asia. J. 6, 111–126.

Bjarnason, I., Menke, W., Flovenz, O., Caress, D., 1993. Tomographic image of the spreading center in southern Iceland. J. Geophys. Res. 98, 6607–6622.

Bohlen, S.R., 1987. Pressure-temperature-time paths and a tectonic model for the evolution of granulites. J. Geol. 95, 617–632.

Boland, A.V., Ellis, R.M., 1989. Velocity structure of the Kapuskasing uplift, northern Ontario, from seismic refraction studies. J. Geophys. Res. 94, 7189–7204.

Borah, K., Rai, S.S., Gupta, S., Prakasam, K.S., Kumar, S., Sivaram, K., 2014. Preserved and modified mid-Archean crustal blocks in Dharwar craton: seismological evidence. Precambrian Res. 246, 16–34.

Bose, R.N., Arora, C.L., 1969. Longitudinal wave velocities in various rock formations in India. J. Eng. Geol. 4, 78–87.

Bossart, P., Dietrich, D., Greco, A., Ottiger, R., Ramsay, J.G., 1988. The tectonic structure of the Hazara-Kashmir Syntexis, southern Himalayas, Pakistan. Tectonics 7, 273–297.

Bowring, S.A., Williams, I.S., Compston, W., 1989. 3.96 Ga gneisses from the Slave Province, Northwest Territories, Canada. Geology 17, 971–975.

Boyd, N.K., Clement, W.P., Smithson, S.B., 1992. The nature of the Moho in the Archean Minnesota Gneiss Terrane (abstract). Eos. Trans. AGU 73 (43), 354–355. Fall Meeting suppl.

Bratt, S.R., Purdy, G.M., 1984. Structure and variability of oceanic crust and flanks of the East Pacific rise between 11° and 13°N. J. Geophys. Res. 89, 6111–6125.

Brown, G.C., Mussett, A.E., 1993. The Inaccessible Earth. An Integrated View to Its Structure and Composition. Springer, The Netherlands. ISBN: 978-0-412-48160-4. (Print). https://doi.org/10.1007/978-94-011-1516-2.

Brown, R.E., Koeberl, C.K., Montanari, A., Bice, D.M., 2009. Evidence for a change in Milankovitch forcing caused by extraterrestrial events at Massignano, Italy, Eocene-Oligocene boundary GSSP. Geology 452, 119–137. https://doi.org/10.1130/2009.2452(08).

Burke, K., Dewey, J.F., 1973. Plume generated triple junctions: key indicators in applying plate tectonics to old rocks. J. Geol. 81, 406–433.

Campbell, I.H., Griffiths, R.W., 1990. Implications of mantle plume structure for the evolution of flood volcanics. Earth Planet. Sci. Lett. 99, 79–93.

Carbonell, R., Josep, G., Jordi, D., Alba, G., Mimoun, H., Fadila, O., Puy, A., Antonio, T., Luisa, A.M., Imma, P., Alan, L., 2013. A 700 km long crustal transect across northern *Morocco:* Vienna. EGU Gen. Assem. 15. EGU2013-8313.

Cermak, V., Rybach, L. (Eds.), 1991. Terrestrial Heat Flow and the Lithosphere Structure. Springer-Verlag, Berlin and Heidelberg. 507 pp. (Proceeding s volume of international conference at Bechyne, Czechoslovakia, June 1–6, 1987 giving a good overview of the interpretation of heat flow data in terms of the continental lithosphere structure.).

Chakravorty, K.C., Ghosh, D.P., 1960. In: Proc. World. Conf. Earthquake Eng., Tokyo, pp. 1633–1642.

Chamberlain, C.P., Koons, P.O., Meltzer, A.S., Park, S.K., Draw, D., Zeitler, P., Poage, M.A., 2002. Overview of hydrothermal activity associated with active orogenesis and metamorphism: Nanga Parbat, Pakistan Himalaya. Am. J. Sci. 302, 726–748.

Chandra, R., 1999. Geochemistry and petrogenesis of the layered sequence in Grinner ijolite series, India. In: Srivastava, R.K., Hall, R.P. (Eds.), The Role of Differentiation and Allied Factors in Magmatism in Diverse Tectonic Settings. Oxford & IBH Publishing Co., New Delhi, India, pp. 155–194

Chandrakala, K., Pandey, O.P., Mall, D.M., Sarkar, D., 2010. Seismic signatures of a Proterozoic thermal plume below southwestern part of the Cuddapah basin, Dharwar craton. J. Geol. Soc. India 76, 565–572.

Chandrasekhar, D.V., Mishra, D.C., Poornachandra Rao, G.V.S., Mallikharjuna Rao, J., 2002. Gravity and magnetic signatures of volcanic plugs related to Deccan volcanism in Saurashtra, India and their physical and geochemical properties. Earth Planet. Sci. Lett. 201, 277–292.

Chatterjee, S., 1986. The Palaeo-position of India. J. SE Asia Earth Sci. 1, 145–189.

Chatterjee, G.C., Ghosh, P.K., 1970. Tectonics framework of peninsular Gondwanas in India. Rec. Geol. Surv. India 98 (2), 1–15.

Chekunov, A.V., Sollogub, V.B., Starostenko, V.l., Kharechko, G.E., Rusakov, A.M., Kozlenko, V.G., Kostyukevich, A.S., 1984. Structure of the Earth's crust and upper mantle below Hindustan and the northern part of the Indian Ocean from geophysical data. Tectonophysics 101, 63–73.

Chen, W.-P., Molnar, P., 1981. Constraints on the seismic wave velocity structure beneath the Tibetan Plateau and their tectonic implications. J. Geophys. Res. 86, 5937–5962.

Cherepanova, Yu., Artemieva, I.M., Thybo, H., Chemia, Z., 2013. Crustal structure of the Siberian Craton and the West Siberian Basin: an appraisal of existing seismic data. Tectonophysics 609, 154–183.

Chetty, T.R.K., Bhaskar Rao, Y.J., Narayana, B.L., 2003. A structural cross section along Krishnagiri-Palani corridor, Southern Granulite Terrene of India. In: Ramakrishnan, M. (Ed.), Tectonics of Southern Granulite Terrene, Kuppam-Palani Geotransect. Geol. Soc. India, Mem., vol. 50, 255–277.

Chian, D.P., Louden, K., 1992. The structure of Archean–Ketilidian crust along the continental shelf of southwestern Greenland from a seismic refraction profile. Can. J. Earth Sci. 29, 301–313.

Choudhary, A.K., Gopalan, K., Anjaneya Sastry, C., 1984. Present status of the geochronology of the Precambrian rocks of Rajasthan. Tectonophysics 105, 131–140.

Chauhan, R.K.S., Singh, R.N., 1965. Crustal studies in Himalayan region. J. Indian Geophys. Union 2, 51–57.

Choukroune, P., Ludden, J.N., Chardon, D., Calvert, A.J., Bouhallier, H., 1997. Archean crustal growth and tectonic processes: a comparison of the Superior Province, Canada and the Dharwar craton, India. In: Burg, J.P., Ford, M. (Eds.), Orogeny Through Time. Geol. Soc. Lond., Spec. Publ. 121, 63–98.

Christensen, N.I., 1996. Poisson's ratio and crustal seismology. J. Geophys. Res. 101, 3139–3156.

Christensen, N.I., Mooney, W.D., 1995. Seismic velocity structure and composition of the continental crust: a review. J. Geophys. Res. 100 (B7), 9761–9788.

Christeson, G.L., Purdy, G.M., Fryer, G.J., 1994. Seismic constraints on shallow crustal emplacement processes at the fast spreading East Pacific Rise. J. Geophys. Res. 99, 17957–17974.

Collins, A.S., Razakamanana, T., Windely, B.F., 2000. Neoproterozoic extensional detachment in central Madagascar: implications for the collapse of the East African orogeny. Geol. Mag. 137, 39–51.

Compston, W., Pidgeon, R.T., 1986. Jack Hills, evidence of more very old detrital zircons in Western Australia. Nature 321, 766–769.

Condie, K.C., 2005. Earth as an Evolving Planetary System. Elsevier, Amsterdam/Boston, ISBN: 0120883929.

Cook, F.A., 2002. Fine structure of the continental reflection Moho. Geol. Soc. Am. Bull., 64–79. https://doi.org/10.1130/0016-7606(2002)114b0064.

Cook, F.A., White, D.J., Jones, A.G., Eaton, D., Hall, J., Clowes, R.M., 2010. How the crust meets the mantle: lithoprobe perspectives on the Mohorovic discontinuity and crust–mantle transition. Can. J. Earth Sci. 351, 315–351. https://doi.org/10.1139/E09-076.

Courtillot, V., Feraud, G., Malusk, H., Vandamme, D., Moreau, M.G., Besse, J., 1988. Deccan flood volcanics and the Cretaceous-Tertiary boundary. Nature 333, 843–846.

Curray, J.R., Shor, G.G., Raitt, R.W., Henry, M., 1977. Seismic refraction and reflection studies of crustal structure of the Eastern Sunda and Western Banda Arcs. J. Geophys. Res. 82, 2479–2489.

Curray, J.R., Emmel, F.J., Moore, D.G., Raitt, R.W., 1982. Structure tectonics and geological history of the northeastern Indian Ocean. In: Narin, A.E.M., Stehyl, F.G. (Eds.), The Ocean Basins and Margins. Plenum Press, New York, pp. 399–450.

Dahl-Jensen, T., Larsen, T.B., Woelbern, I., et al., 2003. Depth to Moho in Greenland: receiver-function analysis suggests two Proterozoic blocks in Greenland. EPSL 205, 379–393.

Damodara, N., Vijaya Rao, V., Kalachand, Sain, Prasad, A.S.S.S.R.S., Murty, A.S.N., 2017. Basement configuration of the West Bengal sedimentary basin, India as revealed by seismic refraction tomography: its tectonic implications. Geophys. J. Int. 208, 1490–1507.

Das, D., Saha, T., Bhattacharya, R., 1993. Sedimentological studies of subsurface Gondwana sediments of Bengal basin. In: Gondwana Geological Magazine. Spl. Birbal Sahani Centenary, National Symposium on Gondwana of India. Gondwana Geological Society, Nagpur, pp. 19–33.

Dassarma, D.C., 1988. Post-orogenic deformation of the Precambrian crust in northeast Rajasthan. In: Roy, A.B. (Ed.), Precambrian of the Aravalli Mountain, Rajasthan, India. Geol. Soc. India, Mem. 7, 109–120.

Datta, N.R., Mitra, N.D., Bandopadhyay, S.K., 1983. Recent trends in the study of Gondwana basins of Peninsular and extra Peninsular India, Petroliferous basins of India. Pet. Asia J. 6, 159–169.

Davey, F.J., Stern, T.A., 1990. Crustal seismic observations across the convergent plate boundary, North Island, New Zealand. In: Leven, J.H., Finlayson, D.M., Wright, C., Dooley, J.C., Kennett, B.L.N. (Eds.), Seismic Probing of Continents and Their Margins. Tectonophysics 173, 283–296.

Davidson, J.P., Arculus, R.J., 2006. The significance of Phanerozoic arc magmatism in generating continental crust. In: Brown, M., Rushmer, T. (Eds.), Evolution and Differentiation of the Continental Crust. Cambridge Univ. Press, New York, pp. 135–172.

De Wit, M.J., Roering, C., Hart, R.J., Armstrong, R.A., de Ronde, C.E.J., Green, R.W.E., Tredoux, M., Peberdy, E., Hart, R.A., 1992. Formation of an Archaean continent. Nature 357, 553–562.

Deb, M., Thorpe, R.I., Cumming, G., Wagner, P.A., 1989. Age, source and stratigraphic implications of Pb isotope data for conformable, sediment hosted base metal deposits in the Proterozoic Aravalli-Delhi orogenic belt, N-W India. Precambrian Res. 43, 1–22.

Deb, M., Talwar, A.K., Tewari, A., Banerjee, A.K., 1995. Bimodal volcanism in South Delhi Fold Belt: a suite of differentiated felsic lava at Jharivav, north Gujarat. In: Sinha-Roy, S., Gupta, K.R. (Eds.), Continental Crust of N-W and Central India. Geol. Soc. India, Mem. 31, 259–278.

DeBari, S.M., Sleep, N.H., 1991. High-Mg, low-Al bulk composition of the Talkeetna island arc, Alaska: implications for primary magmas and the nature of arc crust. Geol. Soc. Am. Bull. 103, 37–47.

Dubey, R.K., Bayana, J.C., Chaudhury, H.M., 1973. Crustal structure of the peninsular India. Pure Appl. Geophys. 109, 1717–1727.

DeCelles, P.G., Robinson, D.M., Zandt, G., 2002. Implications of shortening in the Himalayan fold-thrust belt for uplift of the Tibetan plateau. Tectonics 21, 12-1–12-25.

Dewey, J.F., Cande, S., Pitman, W.C., 1989. Tectonic evolution of the Indian/Eurasia Collision Zone. Eclogae Geol. Helv. 82, 717–734.

Dhuime, B., Hawkesworth, C.J., Cawood, P.A., Storey, C.D., 2012. A change in the geodynamics of continental growth 3 billion years ago. Science 355, 1334–1336.

Divakara Rao, V., Gupta, H.K., Gupta, S.B., Mall, D.M., Mishra, D.C., Murty, P.R.K., Narayana, B.L., Reddy, P.R., Tewari, H.C., 1998. A Geotransect in the Central Indian

Shield, across the Narmada-Son Lineament and the Central Indian Suture. Int. Geol. Rev. 40, 1021–1037.
Dixit, M.M., Satyavani, N., Sarkar, D., Khare, P., Reddy, P.R., 2000. Velocity inversion in the Lodhika area, Saurashtra peninsula, western India. First Break 18, 499–504.
Dixit, M.M., Tewari, H.C., Visweswara Rao, V., 2010. Two dimensional velocity model of the crust beneath south Cambay basin, India from refraction and wide-angle reflection data. Geophys. J. Int. 181, 635–652.
Drummond, B.J., 1979. A crustal profile across the Archaen Pilbara and northern Yilgarn cratons, northwest Australia. BMR J. Aust. Geol. Geophys. 4, 171–180.
Drummond, B.J., Collins, C.D.N., 1986. Seismic evidence for the underplating of the lower continental crust of Australia. Earth Planet. Sci. Lett. 79, 361–372.
Dube, R.K., Bayana, J.C., Chaudhury, H.M., 1973. Crustal structure of the peninsular India. Pure Appl. Geophys. 109, 1717–1727.
Duncan, R.A., Pyle, D.G., 1988. Rapid eruption of the Deccan flood volcanics at the Cretaceous-Tertiary boundary. Nature 333, 841–843.
Durheim, R.J., Mooney, W.D., 1994. Evolution of the Precambrian Lithosphere: geological and geochemical constraints. J. Geophys. Res. 99, 15359–15374.
Durrheim, R.J., Mooney, W.D., 1991. Archcan and Proterozoic crustal evolution: evidence from crustal seismology. Geology 19, 606–609.
Durrheim, R.J., Barker, W.H., Green, R.W.E., 1992. Seismic studies in the Limpopo Belt. Precambrian Res. 55, 187–200.
Dziewonski, A.M., Anderson, D.L., 1981. Preliminary reference earth model (PREM). Phys. Earth Planet. Inter. 25, 297–356.
Eaton, J.P., 1966. Crustal structure in northern and central California from seismic evidence, geology of Northern California. In: Bailey, E.H. (Ed.), Bull. Calif. Div. Min. Geol. 190, 419–426.
Edwards, R.A., Minshull, T.A., Flueh, E.R., Kopp, C., 2008. Dalrymple Trough: an active oblique-slip ocean-continent boundary in the northwest Indian Ocean. Earth Planet. Sci. Lett. 272, 437–445.
Enderle, U., Tittgemeyer, M., Itzin, M., Prodehl, C., Fuchs, K., 1997. Scales of structure in the lithosphere; images of processes. Tectonophysics 275, 165–198.
EUGENO-S Working Group, 1988. Crustal structure and tectonic evolution of the transition between the Baltic Shield and the North German Caledonides (the EUGENO-S project). Tectonophysics 150 (3), 253–348.
Evangelidis, C.P., Minshull, T.A., Henstock, T.J., 2004. Three-dimensional crustal structure of Ascension Island from active source seismic tomography. Geophys. J. Int. 159, 311–325.
Fleitout, L., Froidevaux, C., 1982. Tectonics and topography for a lithosphere containing density heterogeneities. Tectonics 1, 21–56.
Fountain, D.M., Salisbury, M.H., Percival, J., 1990. Seismic structure of the continental crust based on rock velocity measurements from the Kapuskasing Uplift. J. Geophys. Res. 95, 1167–1186.
Fuchs, K., Vinnik, L.P., Prodehl, C., 1987. Exploring heterogeneities of the continental mantle by high-resolution seismic experiments. In: Composition, Structure and Dynamics of the Lithosphere-Asthenosphere System. Geodynamics Series, vol. 16. Am. Geophys. Union, Washington, DC, pp. 137–154.
Fuloria, R.C., 1993. Geology and hydrocarbon prospects of Mahanadi basin, India. In: Biswas, et al., (Ed.), Proc. Second Seminar on Petroliferous basins in India. Indian Petrol., Dehradun, India, vol. 1, pp. 355–369.
Gaherty, J.B., Kato, M., Jordan, T.H., 1999. Seismological structure of the upper mantle: a regional comparison of seismic layering. Phys. Earth Planet. Inter. 110, 21–41.
Galve, A., Jiang, M., Hirn, A., Sapin, M., Laigle, M., de Voogd, B., Gallart, J., Qian, C.H., 2006. Explosion seismic P and S velocity and attenuation constraints on the lower crust of

the North–Central Tibetan plateau, and comparison with the Tethyan Himalayas: implications on composition, mineralogy, temperature and tectonic evolution. Tectonophysics 412, 141–157.

Gao, S., Zhang, B.R., Jin, Z.M., Kern, H., Luo, T.H., Zhao, Z.D., 1998. How mafic of the lower continental crust? Earth Planet. Sci. Lett. 161, 101–117.

Gaur, V., Priestley, K., 1997. Shear wave velocity structure beneath the Achaean granites around Hyderabad, inferred from receiver function analysis. Proc. Indian Acad. Sci. 106, 1–8.

Ghosh, D.B., 1975. The nature of Narmada-Son lineament. Geol. Surv. India Misc. Publ. 34 (part 1), 119.

Ghosh, J.G., De Wit, M.J., Zartman, R.E., 2004. Age and tectonic evolution of Neoproterozoic ductile shear zones in the Southern Granulite Terrene of India, with implications for Gondwana studies. Tectonics, 23, TC3006.

Gibbs, A.K., 1986. A qualitative assessment: reflection seismology. In: Barazangi, M., Brown, L. (Eds.), Seismic Reflection Profile of Precambrian: The Continental Crust. In: Geodynamic Ser., vol. 14. A. G. U. Publ., Washington, DC, pp. 95–106.

Giese, P., Scheuber, E., Schilling, F., Schmitz, M., Wigger, P., 1999. Crustal thickening processes in the Central Andes and the different natures of the Moho-discontinuity. South Am. J. Geosci. 12, 201–220.

Glikson, A.Y., Lambert, L.B., 1976. Vertical zonation and petrogenesis of the early Precambrian crust in western Australia. Tectonophysics 30, 55–89.

Gokaran, S.G., Rao, C.K., Singh, B.P., 1995. Crustal structure in southeast Rajasthan using Magnetotelluric techniques. In: Sinha-Roy, S., Gupta, K.R. (Eds.), Continental Crust of NW and Central India. Geol. Soc. India, Mem. 31, 373–381.

Gokaran, S.G., Gupta, G., Rao, C.K., Selvaraj, C., 2002. Electrical structure across the Indus-Tsangpo suture and the Shyok suture zones in the NW Himalayas using magnetotelluric studies. Geophys. Res. Lett. 29, 92.1–92.4.

Grad, M., Guterch, A., Keller, G.R., Janik, T., Hegeds, E., Vozár, J., Ślczka, A., Tiira, T., Yliniemi, J., 2006. Lithospheric structure beneath trans-Carpathian transect from Precambrian platform to Pannonian basin: CELEBRATION 2000 seismic profile CEL05. J. Geophys. Res. 111, 1–23.

Green, R.W.E., 1983. Seismic refraction observations the Damara in Orogen flanking and craton and their bearing on deep seate processes the Orogen. Spec. Publ. Geol. Soc. S. Afr. 11, 355–367.

Green, R.W.E., Durrheim, R.J., 1990. A seismic refraction investigation the Namaqual and of Metamorphic Complex, South Africa. J. Geophys. Res. 95, 19927–19932.

Grevemeyer, I., Weigel, W., Whitmarsh, R.B., Avedik, F., Deghani, G.A., 1997. The Ageir Rist: crustal structure of an extinct spreading axis. Mar. Geophys. Res. 19, 1–23.

Griffin, W.L., O'Reilly, S.Y., 1986. Mantle-derived sapphirine. Min. Mag. 50, 635–640.

Griffin, W.L., O'Reilly, S.Y., 1987. Is the Moho the crust-mantle boundary? Geology 15, 241–244.

Griffin, W.L., Sutherland, F.L., Hollis, J.D., 1987. Geothermal profile and crust-mantle transition beneath east-central Queensland: vulcanology, xenolith petrology and seismic data. J. Volcanol. Geotherm. Res. 31, 177–203.

Guerra, I., Luongo, G., Maistrello, M., Scarascia, S., 1983. Deep Seismic Sounding along the profile Lawrencepur—Sango Sar (Nanga Parbat). Boll. Geofis. Teor. Appl. 25, 211–219.

Guillot, S., Garzanti, E., Baratoux, D., Marquer, D., Maheo, G., De Sigoyer, J., 2003. Reconstructing the total shortening history of the NW Himalayas. Geochem. Geophys. Geosyst. 4, 1046.

Gupta, M.L., 1981. Surface heat flow and igneous intrusion in the Cambay basin, India. J. Volcanol. Geotherm. Res. 10, 279–292.

Gupta, H.K., Dwivedy, K.K., 1996. Drilling at Latur earthquake region exposes a Peninsular Gneiss basement. J. Geol. Soc. India 47, 129–131.

Gupta, H.K., Naraian, H., 1967. Crustal structure in Himalayas and Tibet plateau region from surface wave dispersion. Bull. Seismol. Soc. Am. 57, 235–248.

Gupta, S., Rai, S.S., 2005. Structure and evolution of south Indian crust using teleseismic waveform modeling. Himal. Geol. 26, 109–123.

Gupta, H.K., Rastogi, B.K., Chadha, R.K., Mandal, P., Sarma, C.S.P., 1997. Enhanced reservoir-induced earthquakes in Koyna region, India, during 1993–95. J. Seismol. 1, 47–53.

Gupta, S., Rai, S.S., Prakasam, K.S., Srinagesh, D., Bansal, B.K., Chadha, R.K., Priestley, K., Gaur, V.K., 2003a. The nature of the crust in southern India: implications for Precambrian crustal evolution. Geophys. Res. Lett. 30. https://doi.org/10.1029/2002gl016770.

Gupta, S., Rai, S.S., Prakasam, K.S., Srinagesh, D., Chadha, R.K., Prisetley, K., Gaur, V.K., 2003b. First evidence for anomalous thick crust beneath mid-Archean western Dharwar craton. Curr. Sci. 84, 1219–1226.

Guru Rajesh, K., Chetty, T.R.K., 2006. Structure and tectonics of the Achankovil Shear Zone, South India. Gondwana Res. 10, 86–98.

Haines, S.S., Klemperer, S.L., Brown, L.D., Jingru, G., Mechie, J., Meissner, R., Ross, A., Wenjin, Z., 2003. INDEPTH-III seismic data: from surface observations to deep crustal processes in Tibet. Tectonics 22, 1–20.

Hammer, P.T.C., Clowes, R.M., 1997. Moho reflectivity patterns—a comparison of Canadian lithoprobe transects. Tectonophysics 269, 179–198.

Harding, A.J., Orcutt, J., Kappus, M., et al., 1989. The structure of young oceanic crust at 13° N on the East Pacific Rise from expanding spread profiles. J. Geophys. Res. 94, 12163–12196.

Harinarayana, T., Naganjaneyulu, K., Manoj, C., Patro, B.P.K., Kareemunnisa Begum, S., Murthy, D.N., Rao, M., Kumaraswamy, V.T.C., Virupakshi, G., 2003. Magnetotelluric investigations along the Kuppam-Palani Geotransect, South India—2-D modelling results. In: Ramakrishnan, M. (Ed.), Tectonics of Southern Granulite Terrene, Kuppam-Palani Geotransect. Geol. Soc. India, Mem. 50, 107–124.

Harris, N.B.K., Santosh, M., Taylor, P.N., 1994. The crustal evolution in South India: constraints from Nd isotopes. J. Geol. 102, 139–150.

Harrison, T.M., Ryerson, F.J., Le Fort, P., Yin, A., Lovera, O.M., Catlos, E.J., 1997. A late Miocene-Pliocene origin for Central Himalayan inverted metamorphism. Earth Plan. Sci. Lett. 146, E1–E7.

Hauck, M.L., Nelson, K.D., Brown, L.D., Zhao, W., Ross, A.R., 1998. Crustal structure of the Himalayan Orogen from project INDEPTH deep seismic reflection profiles at ~90° E. Tectonics 17, 481–500.

Hawkesworth, C.J., Kemp, A.I.S., 2006. Evolution of the continental crust. Nature 443, 811–817.

Hawkesworth, C.J., Kempton, P.D., Rogers, N.W., Ellam, R.M., van Calsteren, P.W., 1990. Continental mantle lithosphere, and shallow level enrichment processes in the Earth's mantle. Earth Planet. Sci. Lett. 96, 256–268.

Helmstaedt, H., Gurney, J.J., 1995. Geotectonic controls of primary diamond deposits: implications for area selection. J. Geochem. Explor. 53, 125–144.

Henstock, T.J., et al., 1993. The accretion of oceanic crust by episodic sill intrusion. J. Geophys. Res. 98, 4143–4161.

Hernandez-Pardo, O., von Frese, R.R.B., Kim, J.W., 2007. Crustal thickness variations and seismicity of northwestern South America. Earth Sci. Res. J. 11 (1), 81–94.

Heron, A.M., 1953. The geology of the central Rajputana. Mem. Geol. Surv. India 79, 389.

Hirn, A., 1988. Features of the crust-mantle structure of Himalaya-Tibet: a comparison with seismic traverses of Alpine, Pyrenean and Veriscan orogenic belts. Philos. Trans. R. Soc. Lond. A326, 17–32.

Hirn, A., Lepine, J.C., Jobart, G., Sapin, M., Wittlinger, G., Xin, X.Z., Yuan, G.E., Jing, W.X., Wen, T.J., Bai, X.S., Pandey, M.R., Tater, J.M., 1984a. Crustal structure and variability of the Himalayan border of Tibet. Nature 307, 23–25.

Hirn, A., Nercessian, A., Sapin, M., Jobart, G., Xin, X.Z., Yuan, G.E., Yuan, L.D., Wen, T.J., 1984b. Lhasa block and bordering sutures—a continuation of 500 km Moho traverse through Tibet. Nature 307, 25–27.

Hirn, A., Jiang, M., Diaz, J., Nercessain, A., Lu, Q.T., Lepine, J.C., Shi, D.N., Sachpazi, M., Pandey, M.R., Ma, K., Gallart, J., 1995. Seismic anisotropy as an indicator of mantle flow beneath the Himalayas and Tibet. Nature 375, 571–574.

Hirn, A., Sapin, M., Lepine, J.C., Diaz, J., Mei, J., 1997. Increase in melt fraction along a north-south traverse below the Tibetan plateau: evidence from seismology. Tectonophysics 273, 17–30.

Hoffman, P.F., 1988. United Plates of America, the birth of a craton: early Proterozoic assembly and growth of Laurentia. Annu. Rev. Earth Planet. Sci. 16, 543–603.

Hooper, P.R., 1990. The timing of crustal extension and the eruption of continental flood volcanics. Nature 345, 246–249.

Hurtado, O., Leuro, E., 2000. Modelo gravimétrico del espesor de la corteza terrestre en Colombia. Geofis. Colomb. 4, 11–22.

Iyer, H.M., Gaur, V.K., Rai, S.S., Ramesh, D.S., Rao, C.V.R., Srinagesh, D., Suryaprakasam, K., 1989. High velocity anomaly beneath the Deccan Volcanic Province: evidence from seismic tomography. Proc. Indian Acad. Sci. (Earth Planet Sci.) 98, 31–60.

Jagadeesh, S., Rai, S.S., 2008. Thickness, composition, and evolution of the Indian Precambrian crust inferred from broadband seismological measurements. Precambrian Res. 162, 4–15. https://doi.org/10.1016/j.precamres.2007.07.009.

Jagannathan, C.R., Ratnam, C., Baishya, N.C., Das Gupta, U., 1983. Geology of the offshore Mahanadi basin. Petrol. Asia J. 4, 101–104.

Jain, S.C., Nair, K.K.K., Yedekar, B.D., 1995. Geology of the Son-Narmada-Tapti Lineament Zone in Central India. Geoscientific studies of the Son-Narmada-Tapti Lineament Zone. Project CRUMSONATA. Geol. Surv. India Spec. Publ. 10, 1–154.

James, D.E., Niu, F., Rokosky, J., 2003. Crustal structure of the Kaapvaal Craton and its significance for early crustal evolution. Lithos 71, 413–429.

Jarchow, C.M., Thompson, G.A., 1989. The nature of Mohorovicic discontinuity. Annu. Rev. Earth Planet. Sci. 17, 475–506.

Jiang, M., Galve, A., Hirn, A., de Voogd, B., Laigle, M., Su, H.P., Diaz, J., Lépine, J.C., Wang, Y.X., 2006. A crustal thickening and variations in architecture from the Qaidam basin to the Qang Tang (North-Central Tibetan plateau) from wide-angle reflection seismology. Tectonophysics 412, 121–140.

Jones, K.A., Warner, M.R., Morgan, R.P.L., Morgan, J.V., Barton, P.J., Price, C.E., 1996. Coincident normal-incidence and wide-angle reflections from the Moho: evidence for crustal seismic anisotropy. Tectonophysics 264, 205–217.

Jordan, T.H., 1978. Composition and development of the continental tectosphere. Nature 274, 544–548.

Julia, J., Jagadeesh, S., Rai, S.S., Owens, T.J., 2009. Deep crustal structure of the Indian shield from joint inversion of P wave receiver functions and Rayleigh wave group velocities implications for Precambrian crustal evolution. J. Geophys. Res. 114. https://doi.org/10.1029/2008JB006261.

Kaila, K.L., Bhatia, S.C., 1981. Gravity study along the Kavali-Udipi deep seismic sounding profile in the Indian peninsular shield: some inferences about the origin of Anorthosites and Eastern Ghat Orogeny. Tectonophysics 79, 129–143.

Kaila, K.L., Sain, K., 1997. Variation of crustal velocity structure in India as determined from DSS studies and their implications on regional tectonics. J. Geol. Soc. India 49, 395–407.

Kaila, K.L., Tewari, H.C., 1985. Structural trends in the Cuddapah basin from Deep Seismic Soundings (DSS) and their tectonic implications. Tectonophysics 115, 69–86.

Kaila, K.L., Reddy, P.R., Narain, H., 1968. Crustal structure in the Himalayan foothills area of north India from P-wave data of shallow earthquakes. Bull. Seismol. Soc. Am. 58, 597–612.

Kaila, K.L., Roy Choudhury, K., Reddy, P.R., Krishna, V.G., Narain, Hari, Subbotin, S.I., Sollogub, V.B., Chekunov, A.V., Kharetchko, G.E., Lazarenko, M.A., Ilchenko, T.V., 1979. Crustal structure along Kavali-Udipi profile in the Indian Peninsular Shield from the deep seismic soundings. J. Geol. Soc. India 20, 307–333.

Kaila, K.L., Reddy, P.R., Dixit, M.M., Lazarenko, M.A., 1979b. Deep crustal structure at Koyna, Maharashtra, indicated by deep seismic soundings. J. Geol. Soc. India 22, 1–16.

Kaila, K.L., Tewari, H.C., Sarma, P.L.N., 1980. Crustal structure from deep seismic sounding studies along Navibander-Amreli profile in Saurashtra, India. Mem. Geol. Soc. India 3, 218–232.

Kaila, K.L., Murty, P.R.K., Rao, V.K., Kharetchko, G.E., 1981a. Crustal structure from deep seismic sounding along the Koyna II (Kelsi-Loni) profile in the Deccan Trap area, India. Tectonophysics 73, 365–384.

Kaila, K.L., Krishna, V.G., Mall, D.M., 1981b. Crustal structure along Mehmadabad-Billimora profile in the Cambay basin, India from deep Seismic soundings. Tectonophysics 76, 99–130.

Kaila, K.L., Tripathi, K.M., Dixit, M.M., 1984. Crustal structure along Wular Lake-Gulmarg-Naoshera profile across Pir Panjal range of the Himalayas from deep seismic sounding. J. Geol. Soc. India 25, 706–719.

Kaila, K.L., Tewari, H.C., Roy Choudhury, K., Rao, V.K., Sridhar, A.R., Mall, D.M., 1987. Crustal structure of the northern part of the Proterozoic Cuddapah basin of India from deep seismic soundings and gravity data. Tectonophysics 140, 1–12.

Kaila, K.L., Murthy, P.R.K., Rao, V.K., Venkateswarlu, N., 1990a. Deep seismic sounding in the Godavari Graben and Godavari (costal) basins, India. Tectonophysics 173, 307–317.

Kaila, K.L., Tewari, H.C., Krishna, V.G., Dixit, M.M., Sarkar, D., Reddy, M.S., 1990b. Deep Seismic sounding studies in the North Cambay and Sanchor basins, India. Geophys. J. Int. 103, 621–637.

Kaila, K.L., Reddy, P.R., Mall, D.M., Venkateswarlu, N., Krishna, V.G., Prasad, A.S.S.S.R.S., 1992. Crustal structure of the West Bengal basin, India, from deep seismic sounding investigations. Geophys. J. Int. 111, 45–66.

Kaila, K.L., Murty, P.R.K., Madhava Rao, N., Rao, I.B.P., Koteswar Rao, P., Sridhar, A.R., Murthy, A.S.N., Vijaya Rao, V., Rajendra Prasad, B., 1996. Structure of the crystalline basement in West Bengal basin, India, as determined by DSS studies. Geophys. J. Int. 124, 175–188.

Kailasam, L.N., 1976. Geophysical studies of the major sedimentary basins of the Indian craton, their deep structural features and evolution. Tectonophysics 36, 225–245.

Karanth, R.V., Sant, D.A., 1995. Lineaments and dyke swarms of the lower Narmada valley and Saurashtra, western India. Geol. Soc. India Mem. 33, 425–434.

Kawakatsu, H., Kumar, P., Takei, Y., Shinohara, M., Kanazawa, T., Araki, E., Suyehiro, K., 2009. Seismic evidence for sharp lithosphere-asthenosphere boundaries of oceanic plates. Science 324, 499–502.

Kayal, J.R., 1996. Precursor seismicity, foreshocks and aftershocks of the Uttarkashi earthquake of October 20, 1991 at Garhwal Himalayas. Tectonophysics 263, 339–345.

Kelemen, P.B., Koga, K., Shimizu, N., 1997. Geochemistry of gabbro sills in the crust–mantle transition zone of the Oman ophiolite: implications for the origin of the oceanic lower crust. Earth Planet. Sci. Lett. 146, 475–488.

Kennett, B.L.N., Widiyantoro, S., 1999. A low seismic wave speed anomaly beneath northwestern India: a seismic signature of the Deccan hotspot? Earth Planet. Sci. Lett. 165, 145–155.

Khan, F.H., 1991. Geology of Bangladesh. Wiley Eastern Ltd., New Delhi, p. 207

Kind, R., Yuan, X., Saul, J., Nelson, D., Sobolev, S.V., Mechie, J., Zhao, W., Kosarev, G., Ni, J., Achauer, U., Jiang, M., 2002. Seismic images of crust and upper mantle beneath Tibet: evidence for Eurasian plate subduction. Science 298, 1219–1221.

Kiselev, S., Vinnik, L., Oreshin, S., Gupta, S., Rai, S.S., Singh, A., Kumar, M.R., Mohan, G., 2008. Lithosphere of the Dharwar craton by joint inversion of P and S receiver functions. Geophys. J. Int. 173, 1106–1118. https://doi.org/10.1111/j.1365-246X.2008.03777.

Kola-Ojo, O., Meissner, R., 2001. Southern Tibet: its deep structure and some tectonic implications. J. Asian Earth Sci. 19, 240–256.

Kosarev, G., Kind, R., Sobolev, S.V., Yuan, X., Hanka, W., Oreshin, S., 1999. Seismic evidence for a detached India in lithosphere mantle beneath Tibet. Science 283, 1306–1309.

Kostyuchenko, S.L., Romanyuk, T.V., 1997. Nature of the Mezen gravity maximum (translated from Fizika Zemli 1997, 12, 3–22). Izv. Phys. Solid Earth 33 (12), 961–980.

Koulakov, I., Tychkov, S., Bushenkova, N., Vasilevsky, A., 2002. Structure and dynamics of the upper mantle beneath the alpine–Himalayan orogenic belt, from teleseismic tomography. Tectonophysics 358, 77–96.

Krishna, V.G., Vijaya Rao, V., 2005. Processing and modelling of short-offset seismic refraction-coincident deep seismic reflection data sets in sedimentary basins: an approach for exploring the underlying deep crustal structures. Geophys. J. Int. 163, 1112–1122.

Krishna, V.G., Vijaya Rao, V., 2011. Velocity modelling of a complex deep crustal structure across the mesoproterozoic south Delhi Fold Belt, NW India, from joint interpretation of coincident seismic wide-angle and near-offset reflection data: an approach using unusual reflections in wide-angle records. J. Geophys. Res. 116, B01307. https://doi.org/10.1029/2009JB006660.

Krishna, V.G., Kaila, K.L., Reddy, P.R., 1989. Synthetic seismogram modelling of crustal seismic record sections from the Koyna DSS profiles in the Western India. In: Properties and Processes of Earth's lower crust. Am. Geophys. Union Geophys. Monogr. 51 (IUGG 6), 143–157.

Krishna, V.G., Kaila, K.L., Rao, G.S.P., Reddy, P.R., 1990. Reflection from low angle thrusts at mid-crustal depths: examples from the Proterozoic Cuddapah basin. In: 4th International Symposium On Deep Reflection Profiling of Continental Lithosphere, Bayreunth, FRG, Terra Abs. vol. 126, p. 50.

Krishna, V.G., Kaila, K.L., Reddy, P.R., 1991. Low velocity layers in the subcrustal lithosphere beneath the Deccan Trap region of western India. Phys. Earth and Planet. Int. 67, 288–302.

Kroner, A., 1984. Evolution, growth and stabilization of the Precambrian lithosphere. Phys. Chem. Earth 15, 69–106.

Kumar, P., 2002. Seismic Structure of the Continental Crust in the Central Indian Region and its Tectonic Implications, Thesis submitted to Osmania University, Hyderabad (unpublished).

Kumar, P., Kawakatsu, H., 2011. Imaging the seismic lithosphere-asthenosphere boundary of the oceanic plate. Geochem. Geophys. Geosyst. 12. https://doi.org/10.1029/2010GC003358.

Kumar, A., Padma Kumari, V.M., Dayal, A.M., Murty, D.S.N., Gopalan, K., 1993. Rb-Sr ages of Proterozoic kimberlites of India: evidence from contemporaneous emplacement. Precambrian Res. 62, 227–237.

Kumar, P., Tewari, H.C., Sain, K., 1999. Velocity-depth relationship in selected parts of Indian crust. J. Geol. Soc. India 54, 129–136.

Kumar, P., Tewari, H.C., Khandekar, G., 2000. A high velocity layer at shallow crustal depths in the Narmada region. Geophys. J. Int. 142, 95–107.

Kumar, M.R., Saul, J., Sarkar, D., Kind, R., Shukla, A.K., 2001. Crustal structure of the Indian shield: new constraints from teleseismic receiver functions. Geophys. Res. Lett. 28 (7), 1339–1342.

Kumar, M.R., Solomon Raju, P., Uma Devi, E., Saul, J., Ramesh, D.S., 2004. Crustal structure variations in northeast India from converted phases. Geophys. Res. Lett. 31. https://doi.org/10.1029/2004GL020576.

Kumar, P., Kind, R., Priestley, K., Dahl-Jensen, T., 2007. Crustal structure of Iceland and Greenland from receiver function studies. J. Geophys. Res. https://doi.org/10.1029/2005JB003991.

Kumar, N., Sharma, J., Arora, B.R., Mukhopadhaya, S., 2009. Seismotectonic model of the Kangra- Chamba sector of northwest Himalaya: constraints from joint hypocenter determination and focal mechanism. Bull. Seismo. Soc. Am. 99, 95–109.

Kumar, P., Yuan, X., Kind, R., Mechie, J., 2012. The lithosphere-asthenosphere boundary observed with USArray receiver functions. Solid Earth 4 (1), 1–31.

Kumar, P., Kumar, M.R., Srijayanthi, G., Arora, K., Srinagesh, D., Chadha, R.K., Sen, M.K., 2013. Imaging the lithosphere-asthenosphere boundary of the Indian plate using converted wave techniques. J. Geophys. Res. 118, 5307–5319.

Kumar, P., Sen, M.K., Haldar, C., 2014a. Estimation of shear velocity contrast from transmitted Ps amplitude variation with ray-parameter. Geophys. J. Int. https://doi.org/10.1093/gji/ggu213.

Kumar, P., Talukdar, K., Sen, M.K., 2014b. Lithospheric structure below transantarctic mountain using receiver function analysis of TAMSEIS data. J. Geol. Soc. India 83, 483–492.

Kusky, T.M., Polet, A., 1999. Growth of granite-greenstone terranes at convergent margins, and stabilization of Archean cratons. Tectonophysics 305, 43–73.

Lagabrielle, Y., Bideau, D., Cannat, M., Karson, J.A., Mevel, C., 1998. Ultramafic-mafic plutonic rock suites exposed along the Mid-Atlantic Ridge (10N-30N): symmetrical-asymmetrical distribution and implications for seafloor spreading processes. In: Buck, W.R. et al. (Eds.), Faulting and Magmatism at Mid-Ocean Ridges. Geophys. Monogr. Ser., AGU, Washington, DC.

Laxminarayana, G., 1995. Gondwana sedimentation in the Chinthalpudi sub-basin Godavari Valley Andhra Pradesh. J. Geol. Soc. India 46, 375–383.

Le Fort, P., 1996. Evolution of the Himalayas. In: Yin, A., Harrison, T.M. (Eds.), The Tectonic Evolution of Asia. Cambridge University Press, Cambridge, pp. 95–109.

Leaver, D.S., Mooney, W.D., Kohler, W.M., 1984. A refraction study of the Oregon cascades. J. Geophys. Res. 89, 3121–3134.

Li, Y.H., Tian, X.B., Wu, Q.J., Zeng, R.S., Zhang, R.Q., 2006. The Poisson ratio and crustal structure of the central Tibetan plateau inferred from Indepth-III teleseismic waveforms: geological and geophysical implications. Chin. J. Geophys. 49, 924–931. https://doi.org/10.1002/cjg2.913.

Li, Y.H., Wu, Q.J., Tian, X.B., 2009. Crustal structure in the Yunnan region determined by modeling receiver functions. Chin. J. Geophys. 52, 67–80.

Lowe, D.R., 1994. Accretionary history of the Archean Barberton Greenstone Belt (3.55–3.22 Ga), southern Africa. Geology 22, 1099–1102.

Lundberg, N., Reed, D.U., Liu, C.S., Lieske Jr., J., 1997. Fore arc-basin closure and arc accretion in the submarine suture zone south of Taiwan. Tectonophysics 274, 5–23.

Luosto, U., Tiira, T., Korhonen, H., Azbel, I., Burrain, V., Buyanov, A., Kosminskaya, I., Ionkis, V., Sharov, N., 1990. Crust and upper mantle structure along the DSS Baltic profile in SE Finland. Geophys. J. Int. 101, 89–110.

MacDougall, J.D., Gopalan, K., Lugmair, G.W., Roy, A.B., 1983. The banded gneissic complex of Rajasthan. India: early crust from depleted mantle at 3.5 AE? Trans. Am. Geophys. Union 64, 351.

MacLeod, C.J., Yaouancq, G., 2000. A fossil melt lens in the Oman ophiolite: implications for magma chamber processes at fast spreading ridges. Earth Planet. Sci. Lett. 176 (3), 357–373.

Mahadevan, T.M., 1994. Deep continental structure of India: a review. Geol. Soc. India, Bangalore, pp. 1–569.

Mahalik, N.K., 2000. Mahanadi Delta, Geology, Resources and Biodiversity. AIT Alumni Association (India Chapter), New Delhi, pp. 1–169.

Makarov, V.I., Alekseev, D.V., Batalev, V.Y., Bataleva, E.A., Belyaev, I.V., Bragin, V.D., Dergunov, N.T., Efimova, N.N., Leonov, M.G., Munirova, L.M., Pavlenkin, A.D., Roecker, S., Roslov, Y.V., Rybinc, A.K., Shchelochkov, G.G., 2010. Underthrusting of Tarim beneath the Tien Shan and deep structure of their junction zone: main results of seismic experiment along MANAS profile Kashgar-Song Köl. Geotectonics 44, 102–126.

Makovsky, Y., Klemperer, S.L., Huang, L.Y., Lu, D.Y., Project INDEPTH Team, 1996a. Structure elements of the southern Tethyan Himalaya crust from wide-angle seismic data. Tectonics 15, 100–997.

Makovsky, Y., Klemperer, S.L., Ratschbacher, L., Brown, L.D., Li, M., Zhao, W., Meng, F., 1996b. INDEPTH wide-angle reflection observation of P-wave-to-S-wave conversion from crustal bright spots in Tibet. Science 274, 1690–1691.

Makowski, Y., Klemperer, S., 1999. Measuring the seismic properties of Tibetan bright spots: evidence for free aqueous fluids in the Tibetan middle crust. J. Geophys. Res. 104, 10795–10825.

Mall, D.M., Kaila, K.L., Rao, V.K., 1991. Magnetic body interpreted at mid crustal level between Jabalpur and Mandla as an indicator for the source regions for Deccan basalt. In: First International Seminar and Exhibition on Exploration Geophysics in Nineteen Nineties, Extended Abstracts. vol. 1.

Mall, D.M., Rao, V.K., Reddy, P.R., 1999. Deep sub-crustal features in the Bengal basin, Seismic signatures for plume activity. Geophys. Res. Lett. 26, 2545–2548.

Mall, D.M., Reddy, P.R., Mooney, W.D., 2008. Collision tectonics of the central Indian suture zone as inferred from a deep seismic sounding study. Tectonophysics 460, 116–123.

Mandal, P., Rastogi, B.K., Gupta Harsh, K., 2000. Recent Indian earthquakes. Curr. Sci. 79, 1334.

Mandal, B., Sen, M.K., Vaidya, V.V., Mann, J., 2014. Deep seismic image enhancement with the common reflection surface (CRS) stack method: evidence from the Aravalli–Delhi fold belt of northwestern India. Geophys. J. Int. 196, 902–917.

Mechie, J., Kind, R., 2013. A model of the crust and mantle structure down to 700 km depth beneath the Lhasa to Golmud transect across the Tibetan plateau as derived from seismological data. Tectonophysics 606, 187–197.

Meissner, R., 1986. The Continental Crust: A Geophysical Approach. vol. 426. Academic, San Diego, CA.

Meissner, R., Weaver, T.H., 1989. Continental crustal structure. In: James, D.E. (Ed.), The Encyclopaedia of Solid Earth Geophysics. Vam Nostrand Reinhold Co., New York, pp. 75–89.

Meissner, R., Tilmann, F., Haines, S., 2004. About the lithospheric structure of central Tibet, based on seismic data from the INDEPTH III profile. Tectonophysics 380, 1–25.

Mengel, K., Kern, H., 1992. Evolution of the petrological and seismic Moho—implications for the continental crust–mantle boundary. Terra Nova 4, 109–116.

Merh, S.S., 1995. Geology of Gujarat. Geological Society of India, Bangalore, India.

Mishra, D.C., 1992. Mid-continent gravity high of central India and Gondwana tectonics. Tectonophysics 212, 153–161.

Mishra, D.C., 2002. Crustal structure of India and its environs based on geophysical studies. Indian Miner. 56, 27–96.

Mishra, D.C., Kumar, R., 1998. Characteristics of faults associated with Narmada Son lineament and rock types in Jabalpur sector. Curr. Sci. 75, 308–310.

Mishra, D.C., Gupta, S.B., Rao, M.B.S.V., Venkataryudu, M., Laxman, G., 1987. Godavari basin—a geophysical study. J. Geol. Soc. India 30, 469–476.

Mishra, D.C., et al., 2001. Major lineaments and gravitymagnetic trends in Saurashtra, India. Curr. Sci. 80, 270–280.

Mitra, S., Priestley, K., Gaur, V.K., Rai, S.S., 2006. Shear-wave structure of the south Indian lithosphere from Rayleigh wave phase-velocity measurements. Bull. Seismol. Soc. Am. 96, 1551–1559. https://doi.org/10.1785/0120050116.

Miyashiro, A., Aki, K., Şengör, A.M. (Eds.), 1982. Orogeny. John Wiley, New York.

Mohan, G., Rai, S.S., Panzac, G.F., 1997. Shear velocity structure of the laterally heterogeneous crust and uppermost mantle beneath the Indian region. Tectonophysics 277, 259–270.

Mohinuddin, S.K., Satyanarayana, K., Rao, G.N., 1993. Cretaceous sedimentation in the subsurface of Krishna-Godavari basin. J. Geol. Soc. India 41, 533–539.

Molnar, P., 1988. A review of geophysical constraints on the deep structure of the Tibetan plateau, the Himalayan and the Karakoram and their tectonic implications. Philos. Trans. R. Soc. Lond. A 326, 33–88.

Molnar, P., Tapponnier, P., 1975. Cenozoic tectonics of Asia: effects of a continental collision. Science 189, 419–426.

Mooney, W.D., Braile, L.W., 1989. The seismic structure of the continental crust and upper mantle of North America. In: Bally, A.W., Palmer, A.R. (Eds.), The Geology of North America—An overview. In: The Geology of North America, A, Geological Society of America, Boulder, CO, pp. 39–52.

Mooney, W.D., Brocher, T.M., 1987. Coincident seismic reflection/refraction studies of the continental lithosphere: a global review. Rev. Geophys. 25, 723–742.

Mooney, W.D., Meissner, R., 1992. Multi-genetic origin of crustal reflectivity: a review of seismic reflection profiling of the continental lower crust and Moho. In: Fountain, D.M., Arculus, R., Kay, R.W. (Eds.), Continental Lower Crust. Elsevier, Amsterdam, pp. 45–79.

Mooney, W.D., Gettings, M.E., Blank, H.R., Healy, J.H., 1985. Saudi Arabian seismic refraction profile: a travel time interpretation of crustal and upper mantle structure. Tectonophysics 11, 173–246.

Mooney, W.D., Laske, G., Masters, G., 1998. CRUST 5.1: a global crustal model at 5 degrees × 5 degrees. J. Geophys. Res. 103, 727–747.

Morgan, W.J., 1981. Hot spot tracks and opening of the Atlantic and the Indian Oceans. In: Emiliani, C. (Ed.), The Sea, The Oceanic Lithosphere, vol. 7. Wiley-Interscience, Hoboken, NJ, pp. 443–487.

Morgan, P.J., Chen, Y.J., 1993. The dependence of ridge-axis morphology and geochemistry on spreading rate and crustal thickness. Nature 364, 706–708.

Mukherjee, S.M., 1942. Seismological features of the satpura earthquake of the 14th March 1938. Proc. Indian Acad. Sci., Sec. A 16 (3), 167–176.

Müller, R.D., Sdrolias, M., Gaina, C., Roest, W.R., 1997. Age, spreading rates, and spreading asymmetry of the world's ocean crust. Geochem. Geophys. Geosyst. https://doi.org/10.1029/2007GC001743.

Murty, A.S.N., Mall, D.M., Murty, P.R.K., Reddy, P.R., 1998. Two-dimensional crustal velocity structure along Hirapur-Mandla profile from seismic refraction and wide-angle reflection data. Pure Appl. Geophys. 152, 247–266.

Murty, A.S.N., Tewari, H.C., Reddy, P.R., 2004. 2-D crustal velocity structure along Hirapur-Mandla profile in Central India: an update. Pure Appl. Geophys. 161, 165–184.

Murty, A.S.N., Tewari, H.C., Kumar, P., Reddy, P.R., 2005. Sub-crustal low velocity layers in central India and their implications. PAGEOPH 162, 2409–2431.

Murty, A.S.N., Sain, K., Tewari, H.C., Rajendra Prasad, B., 2008. Crustal velocity inhomogeneities along the Hirapur-Mandla profile, central India and its tectonic implications. J. Asian Earth Sci. 31, 533–545.

Mutter, J.C., Hélène, D.C., 2013. The Mohorovicic discontinuity in ocean basins: some observations from seismic data. Tectonophysics 609, 314–330.

Naganjaneyulu, K., Harinarayana, T., 2004. Deep crustal electrical signatures of eastern Dharwar craton, India. Gondwana Res. 7, 951–960.

Nageswara Rao, B., Kumar, N., Singh, A.P., Prabhakar Rao, M.R.K., Mall, D.M., Singh, B., 2011. Crustal density structure across the Central Indian Shear Zone from gravity data. J. Asian Earth Sci. 42, 341–353.

Nair, K.K., Jain, S.C., Yedekar, D.B., 1995. Stratigraphy, structure and geochemistry of Mahakosal Greenstone Belt. Geol. Soc. India, Mem. 31, 403–432.

Nair, S.K., Gao, S.S., Liu, K.H., Silver, P.G., 2006. Southern African crustal evolution and composition: constraints from receiver function studies. J. Geophys. Res. 111, B02304.

Naqvi, S.M., Rogers, J.J.W., 1987. Precambrian Geology of India. Oxford University, New York, NY. p. 223.

Narayanaswamy, S., 1966. Tectonics of Cuddapah basin. J. Geol. Soc. India 7, 33–50.

Nayak, P.N., 1990. Deep crustal configuration of Central India. Geol. Surv. India Spec. Publ. 28, 67–98.

Negi, B.S., 1951. Gravity and magnetic surveys in subsurface structures in the Borsad area, Khera district. Geol Survey of India Report (unpublished).

Nelson, K.D., 1991. A unified view of craton evolution motivated by recent deep seismic reflection and refraction results. Geophys. J. R. Astron. Soc. 105, 25–35.

Nelson, K.D., McBride, J.H., Arnow, J.A., Oliver, J.E., Brown, L.D., Kaufman, S., 1985. New COCORP profiling in the southeastern United States. Part II: Brunswick and east coast magnetic anomalies, opening of the north-central Atlantic Ocean. Geology 13, 718–721.

Nelson, K., Zhao, W., Brown, L., Kuo, J., Che, J., Liu, X., Klemperer, S., Makovsky, Y., Meissner, R., Mechie, J., Kind, R., Wenzel, F., Ni, J., Nabelek, J., Leshou, C., Tan, H., Wei, W., Jones, A., Booker, J., Unsworth, M., Kidd, W., Hauck, M., Alsdorf, D., Ross, A., Cogan, M., Wu, C., Sandvol, E., Edwards, M., 1996. Partially molten middle crust beneath southern Tibet: synthesis of project INDEPTH results. Science (New York, N.Y.) 274, 1684–1688.

NGRI/GPH-2, 1975. Bouguer Gravity Anomaly Map of India, 1:5 Million Scale.

Nguuri, T., et al., 2001. Crustal structure beneath southern Africa and its implications for the formation and evolution of the Kaapvaal and Zimbabwe Cratons. Geophys. Res. Lett. 28 (13), 2501–2504.

Ni, J.F., Barazangi, M., 1984. Seismotectonics of the Himalayan collision zone: geometry of the underthrusting beneath the Himalaya. J. Geophys. Res. 89, 1147–1163.

Ni, J.F., Ibenbrahim, A., Roecker, S.W., 1991. Three-dimensional velocity structure and hypocenters of earthquakes beneath the Hazara Arc, Pakistan: geometry of the underthrusting Indian plate. J. Geophys. Res. 96, 19865–19877.

Niu, F., James, D.E., 2002. Fine structure of the lowermost crust beneath the Kaapvaal Craton and its implications for crustal formation and evolution. Earth Planet. Sci. Lett. 200, 121–130.

Ojeda, A., Havskov, J., 2001. Crustal structure and local seismicity in Colombia. J. Seismol. 5, 575–593.

Oliver, J., 1982. Changes at the crust—mantle boundary. Nature 299, 398–399.

Olson, P., 1989. Mantle convection and plumes. In: James, D.E. (Ed.), The Encyclopedia of Solid Earth Geophysics. Van Nostrand Reinhold, New York, pp. 788–802.

O'Reilly, Y., 1989. Xenolith types, distribution and transport. In: Johnson, R.W. (Ed.), Intraplate Volcanism in Eastern Australia and New Zealand. Cambridge University Press, Cambridge, pp. 249–253.

O'Reilly, S.Y., Griffin, W.L., 1985. A xenolith-derived geotherm for southeastern Australia and its geophysical implications. Tectonophysics 111, 41–63.
O'Reilly, S.Y., Griffin, W.L., 1991. Petrologic constraints on geophysical modelling, southeastern Australia. Geol. Soc. Aust. 17, 157–162.
Oreshin, S., Kishelev, S., Vinnik, L., Prakasam, K.S., Rai, S.S., Makeyeva, L., Savvin, Y., 2008. Crust and mantle beneath western Himalaya, Ladakh and western Tibet from integrated seismic data. Earth Planet. Sci. Lett. 271, 275–287.
Osler, J.C., Louden, K.E., 1995. Extinct spreading centre in the Labrador Sea: crustal structure from a two-dimensional seismic refraction velocity model. J. Geophys. Res. 100, 2261–2278.
Owens, T.J., Zandt, G., 1997. Implications of crustal property variations for models of Tibetan plateau evolution. Nature 387, 37–43. https://doi.org/10.1038/387037a0.
Oxburg, E.R., Parmentier, E.M., 1977. Compositional and density stratification in ocean lithosphere—causes and consequences. J. Geol. Soc. Lond. 133, 343–355.
Panda, P.K., 1985. Geothermal maps of India and their significance in the source assessment. Petrol. Asia J. 10, 145–150.
Pande, K., Venkatesan, T.R., Gopalan, K., Krishnamurty, P., Macdougall, J.D., 1988. 40Ar/39Ar ages of Alkali volcanics from Kutch, Deccan Volcanic province from India. In: Subbarao, K.V. (Ed.), Deccan Flood Volcanics. Geol. Soc. India Mem. 10, 145–150.
Pandey, A.P., Chadha, R.K., 2003. Surface loading and triggered earthquakes in the Koyna-Warna region, western India. Phys. Earth Planet. Int. 139, 207–223.
Pavlenkova, N.I., 1987. Properties of middle and lower crust in platform areas. Ann. Geophys. 5 (6), 651–656.
Pavlenkova, N.I., 1988. The nature of seismic boundaries in the continental lithosphere. Tectonophysics 154, 211–225.
Percival, J.A., 1989. Granulite terranes and the lower crust of the Superior Province. In: Mereu, R.F., Mueller, St, Fountain, D.M. (Eds.), Properties and Processes of Earth's Lower Crust. Am. Geophys. Union Monogr. 51, 301–310.
Percival, J.A., Card, K.D., 1985. Structure and evolution of the Archcan crust in Central Superior Province, Canada: evolution of Archean supracrustal sequences. Geol. Assoc. Can. Spec. Pap. 28, 179–192.
Peucat, J.J., Vidal, P., Bernard-Griffiths, J., Condie, K.C., 1989. Sm and Nd isotopic systematics in the archaean low to high-grade transition zone of southern India: syn-accretion vs post-accretion granulites. J. Geol. 97, 537–550.
Poveda, E., Monsalve, G., Vargas, C.A., 2015. Receiver functions and crustal structure of the northwestern Andean region, Colombia. J. Geophys. Res. https://doi.org/10.1002/2014JB011304.
Powell, C.M., Roots, S.R., Veevers, J.J., 1988. Pre-breakup continental extension in east Gondwana land and early opening of the eastern Indian ocean. Tectonophysics 155, 261–283.
Prasad, A.S.S.S.R.S., Sarkar, D., Reddy, P.R., 2002. Identification and usage of multiples in crustal seismics—an application in the Bengal basin, India. Curr. Sci. 82, 1033–1037.
Pratt, L., Costain, J.K., Glover, L., 1988. Rocks exposed/buried Mesozoic basins positive Bouguer gravity. J. Geophys. Res. 93, 6649–6667.
Priestley, K., Jackson, J., McKenzie, D., 2007. Lithospheric structure and deep earthquakes beneath India, the Himalaya and southern Tibet. Geophys. J. Int. 172, 345–362.
Prodehl, C., 1979. Crustal structure of the western United States. U. S. Geol. Surv. Prof. Pap. 1034, 74.
Prodehl, C., Kennett, B., Artemieva, I.M., Thybo, H., 2013. 100 years of seismic research on the Moho. Tectonophysics 609, 9–44.
Rabbel, W., Kaban, M., Tesauro, M., 2013. Contrasts of seismic velocity, density and strength across the Moho. Tectonophysics 609, 437–455.

Radhakirhsna, T., Ramakrishnan, M., 1993. Udipi-Kavali transect across south India. Geol. Soc. India, Mem. 25, 45–61.
Radhakrishna, B.P., 1989. Suspect tectono-stratigraphic terrane elements in the Indian subcontinent. J. Geol. Soc. India 34, 1–24.
Radhakrishna, B.P., Naqvi, S.M., 1986. Precambrian continental crust of India and its evolution. Geol. Soc. India, Mem. 25, 45–61.
Rai, S.S., Keith, P., Prakasam, K.S., Drinagesh, D., Gaur, V.K., Du, Z., 2003. Crustal shear velocity structure of the south Indian shield. J. Geophys. Res. 108 (B2), 10-1–10-12, 2088.
Rai, S.S., Vijay Kumar, T., Jagadeesh, S., 2005. Seismic evidence for significant crustal thickening beneath Jabalpur earthquake, 21 May 1997, source region in Narmada-Son lineament, central India. Geophys. Res. Lett. 32, L22306. https://doi.org/10.1029/2005GL023580.
Rai, S.S., Keith, P., Gaur, V.K., Mitra, S., Singh, M.P., Searle, S., 2006. Configuration of the Indian Moho beneath the NW Himalaya and Laddakh. Geophys. Res. Lett. 33, 5, L15308. https://doi.org/10.1029/2006GL026076.
Raith, M., Srikantappa, C., Buhl, D., Koehler, H., 1999. Nilgiri enderbites, South India: nature and age constraints on protolith formation, high-grade metamorphism and cooling history. Precambrian Res. 98, 129–150.
Raiverman, V., 2002. Foreland Sedimentation in Himalayan Tectonic Regime: A Relook at the Orogenic Processes. Bishen Singh Mahendra Pal Singh, Dehra Dun. p. 3785.
Rajendra Prasad, B., Vijaya Rao, V., 2006. Deep seismic reflection study over the Vindhyan of Rajasthan: Implications for geophysical setting of the basin. J. Earth Syst. Sci. 115, 135–147.
Rajendra Prasad, B., Tewari, H.C., Vijaya Rao, V., Dixit, M.M., Reddy, P.R., 1998. Structure and tectonics of the Proterozoic Aravalli-Delhi Fold Belt in the northwestern India from deep seismic reflection studies. Tectonophysics 288, 31–41.
Rajendra Prasad, B., Vijaya Rao, V., Reddy, P.R., 1999. Seismic and magnetotelluric studies over a crustal scale fault zone for imaging a metallogenic province of Aravalli-Delhi Fold Belt region. Curr. Sci. 76, 1027–1031.
Rajendra Prasad, B., Behera, L., Koteswara, Rao, P., 2006. A tomographic image of upper crustal structure using P and S wave seismic refraction data in the southern granulite terrain (SGT), India. Geophys. Res. Lett. 33, 14. https://doi.org/10.1029/2006GL026307.
Rajendra Prasad, B., Kesava Rao, G., Mall, D.M., Koteswara Rao, P., Raju, S., Reddy, M.S., Rao, G.S.P., Sridhar, V., Prasad, A.S.S.R.S., 2007. Tectonic implications of seismic reflectivity pattern observed over the Precambrian Southern Granulite Terrene, India. Precambrian Res. 153, 1–10.
Rajendra Prasad, B., Klemperer, S.L., Vijaya Rao, V., Tewari, H.C., Khare, P., 2011. Crustal structure beneath the Sub-Himalayan fold-thrust belt, Kangra recess, northwest India from seismic reflection profiling: implications for Late Palaeoproterozoic orogenesis and modern earthquake hazard. Earth Planet. Sci. Lett. 308, 218–228.
Raju, A.T.R., 1968. Geological evolution of Assam and Cambay Tertiary basins of India. Bull. Am. Assoc. Petrol. Geol. 52, 2422–2437.
Raju, A.T.R., Srinivasan, S., 1983. More hydrocarbon from well explored Cambay basin. Petrol. Asia J. 6, 25–35.
Raju, A.T.R., Chaube, A.N., Chowdhary, C.R., 1971. Deccan trap and the Geological framework of the Cambay basin. Bull. Volcanol. 35, 522–538.
Ramakrishnan, M., 2003. Craton–mobile belt relations in southern granulite terrane. In: Ramakrishnan, M. (Ed.), Tectonics of Southern Granulite Terrene, Kuppam–Palani Geotransect. Geol. Soc. India, Mem. 50, 1–24.
Ramakrishnan, M., Viswanantha, M.N., Swami Nath, J., 1976. Basement-cover relationship of peninsular gneiss with high-grade schist and greenstone belts of southern Karnataka. J. Geol. Soc. India 17, 97–111.

Ramesh, D.S., Kumar, M.R., Uma Devi, E., Raju, P.S., 2005. Moho geometry and upper mantle images of northeast India. Geophys. Res. Lett. 32. https://doi.org/10.1029/2005gl022789.

Rao, N.D.J., 1968. An interpretation of gravity data of Cambay basin. ONGC Bull. 4, 74–80.

Rao, G.N., 1993. Geology and hydrocarbon prospects of east coast sedimentary basins of India with special reference to Krishna-Godavari basin. J. Geol. Soc. India 41, 444–454.

Rao, G.S.P., Tewari, H.C., 2005. The seismic structure of the Saurashtra crust in northwestern India and its relationship with the Reunion plume. Geophys. J. Int. 160, 319–331.

Rao, C.K., Gokaran, S.G., Singh, B.P., 1995. Upper crustal structure in the Torni-Region, Central India using magnetotelluric studies. J. Geomagn. Geoelectr. 47, 411–420.

Rao, I.B.P., Murty, P.R.K., Rao, P.K., Murty, A.S.N., Rao, N.M., Kaila, K.L., 1999. Structure of the lower crust revealed by one- and two-dimensional modelling of wide-angle reflections, West Bengal basin, India. PAGEOPH 156, 701–718.

Rao, G.S.P., Tewari, H.C., Rao, V.K., 2000. Velocity structure in parts of the Gondwana Godavari Graben. J. Geol. Soc. India 56, 373–384.

Rao Rammohan, T., Sai Ram, K., Rao, Y.V., 1999. Sedimentological and geochemical characteristics of the lower Gondwana rocks of Chinthalpudi sub-basin of the Godavari Valley of Andhra Pradesh. Indian Mineralogist 33, 151–168.

Rastogi, B.K., 1992. Seismotectonics inferred from earthquakes and earthquake sequences in India during 1980s. Science 62, 101–108.

Raval, U., 1989. On hotspots, Meso-Cenozoic tectonics, and possible thermal networking beneath the Indian continent. Advances in Geophysical Research India. Ind. Geophys. Un., Hyderabad, pp. 314–330.

Raval, U., Veeraswamy, K., 2003. India-Madagascar separation: break-up along a pre-existing mobile belt and chipping of the craton. Gondwana Res. 6, 467–485.

Ravindra Kumar, G.R., Chacko, T., 1994. Geothermobarometry of mafic granulites and metapelites from the Palghat Gap, South India: petrological evidence for isothermal uplift and rapid cooling. J. Metamorph. Geol. 12, 479–492.

Ravishanker, 1988. Heat flow map of India and discussion on its geological and economic significance. Indian Miner. 42, 89–110.

Ray, L., Senthil Kumar, P., Reddy, G.K., Roy, S., Rao, G.V., Srinivasan, R., Rao, R.U.M., 2003. High mantle heat flow in a Precambrian granulite province: evidence from southern India. J. Geophys. Res. 108 (B2), 2084.

Reddi, A.G.B., Ramakrishna, T.S., 1988. Subsurface structure of the shield area of Rajasthan-Gujarat as inferred from gravity. In: Roy, A.B. (Ed.), Precambrian of the Aravalli Mountain, Rajasthan, India. Geol. Soc. India, Mem. 1, 279–284.

Reddy, P.R., 1971. Crust and Upper Mantle Structurein India from Body wave Travl Time Studies of Shallow Earthquakes (Ph.D. thesis). Andhra University, Waltir, pp. 1–257.

Reddy, P.R., Murty, P.R.K., Rao, I.B.P., Khare, P., Kesava Rao, G., Mall, D.M., Koteswara Rao, P., Raju, S., Sridhar, V., Reddy, M.S., 1995. Deep crustal seismic reflection fabric pattern in Central India-preliminary interpretation. In: Sinha-Roy, S., Gupta, K.R. (Eds.), Continental Crust of NW and Central India. Geol. Soc. India, Mem. 31, 537–544.

Reddy, P.R., Murty, P.R.K., Rao, I.B.P., Mall, D.M., Rao, P.K., 2000. Coincident deep seismic reflection and refraction profiling, central India. In: Verma, O.P., Mahadevan, T.M. (Eds.), Research Highlights in Earth System Science. Indian Geol. Cong., 49–53.

Reddy, P.R., Venkateswarlu, N., Prasad, A.S.S.R.S., Koteswara Rao, P., 2002. Basement structure below the Coastal belt of Krishna-Godavari basin: correlation between seismic structure and well information. Gondwana Res. 5, 513–518.

Reddy, P.R., Rajendra Prasad, B., Vijaya Rao, V., Sain, K., Prasada Rao, P., Khare, P., Reddy, M.S., 2003. Deep seismic reflection/WAR studies along Kuppam-Palani transect in Southern Granulite terrane of India. In: Ramakrishnan, M. (Ed.), Tectonics of southern granulite terrene, Kuppam–Palani Geotransect. Geol. Soc. India, Mem. 50, 79–106.

Reddy, P.R., Chandrakala, K., Prasad, A.S.S.S.R.S., Rama Rao, Ch., 2004. Lateral and vertical crustal velocity and density variation in the Cuddapah basin and adjoining Eastern Dharwar craton. Curr. Sci. 87, 1607–1614.

Renne, P.R., Basu, A.R., 1991. Rapid eruption of the Siberian traps flood volcanics at the Permo-Triassic boundary. Science 253, 176–179.

Richards, M.A., Duncan, R.A., 1989. Flood volcanics and plume initiation events: active vs. passive rifting. EOS. Trans. Am. Geophys. Union. 70, 1357.

Ritter Michael, E., 2006. The Physical Environment: An Introduction to Physical Geography. http://www.uwsp.edu/geo/faculty/ritter/geog101/textbook/title_page.html.

Rodgers, A., Harben, P., 1999. Modeling the conversion of hydroacoustic to seismic energy at islands and continental margins: preliminary analysis of ascension island data. In: 21st Seismic Research Symposium: Technologies for Monitoring the Comprehensive Nuclear-Test-Ban Treaty Las Vegas, Nevada.

Roy, S.C., 1939. Seismometric study, 4. Mem. Geol. Surv. India 73, 49–74.

Roy, T.K., 1991. Structural styles in southern Cambay basin, India and role of Narmada geofracture in formation of giant hydrocarbon accumulation. ONGC Bull. 27, 15–56.

Roy, A.B., 2000. Geophysical modelling of the crust of the Aravalli Mountain and its neighbourhood-evidence of Mesozoic-Cainozoic reconstitution. In: Deb, M. (Ed.), Crustal Evolution and Metallogeny in the North-Western Indian Shield. Norosa Publishing House, New Delhi, pp. 203–214.

Roy Choudhury, K., Hargraves, R.B., 1981. Deep seismic sounding in India and the origin of continental crust. Nature 291, 648–650.

Rudnick, R.L., Fountain, D.M., 1995. Nature and composition of the continental crust: a lower crustal perspective. Rev. Geophys. 33, 267–309.

Rudnick, R.L., Gao, S., 2003. Composition of the continental crust. In: Holland, H.D., Turekian, K.K. (Eds.), Treatise on Geochemistry 3.01. Elsevier, New York, pp. 1–63.

Rychert, C.A., Shearer, P.M., 2011. Imaging the lithosphere-asthenosphere boundary beneath the Pacific using SS waveform modeling. J. Geophys. Res 116 (B7), B07307. https://doi.org/10.1029/2010JB008070.

Saintot, A., Brunet, M.-F., Yakovlev, F., et al., 2006. The Mesozoic–Cenozoic tectonic evolution of the greater Caucasus. European lithosphere dynamics. In: Gee, D., Stephenson, R. (Eds.), Geol. Soc. Lond., Spec. Publ. 32, 277–289.

Sajeev, K., Osanai, Y., Santosh, M., 2004. Ultrahigh temperature metamorphism followed by two-stage decompression of garnet-orthopyroxene-sillimanite granulites from Ganguvarpatti, Madurai block, southern India. Contrib. Mineral. Petrol. 148, 29–46.

Salmon, M., Kennett, B.L.N., Stern, T., Aitken, A.R.A., 2013. The Moho in Australia and New Zealand. Tectonophysics 609, 288–298.

Santosh, M., Yoshida, M., 2001. Pan-African extensional collapse along the Gondwana Suture. Gondwana Res. 4, 188–191.

Santosh, M., Tanaka, K., Yokoyama, K., Collins, A.S., 2005. Late Neoproterozoic-Cambrian felsic magmatism along transcurrent shear zones in South India: U–Pb electron microprobe ages and implications for the amalgamation of the Gondwana Supercontinent. Gondwana Res. 8, 31–42.

Sapin, M., Hirn, A., 1997. Seismic structure and evidence for evolution of the crust in the Himalayan convergence. Tectonophysics 273, 1–16.

Sapin, M., Wang, J.X., Hirn, A., Xu, Z.X., 1985. A seismic sounding in the crust of Lhasa block, Tibet. Ann. Geophys. 3, 637–646.

Sarkar, S.N., Gopalan, K., Trivedi, J.R., 1981. New data on the geochronology of the Precambrians of Bhandara-Durg, Central India. Indian J. Earth Sci. 8, 131–151.

Sarkar, G., Ray Burman, T., Corfu, F., 1989. Timing of continental arc-type magmatism in NW India: evidence form U-Pb zircon geochronology. J. Geol. 97, 607–612.
Sarkar, D., Reddy, P.R., Kaila, K.L., Prasad, A.S.S.S.R.S., 1995. Multiple diving waves and high-velocity gradients in the Bengal sedimentary basin. Geophys. J. Int. 121, 969–974.
Sarkar, D., Chandrakala, K., Padmavathi Devi, P., Sridhar, A.R., Sain, K., Reddy, P.R., 2001. Crustal velocity of western Dharwar craton, south India. J. Geodyn. 31, 227–241.
Sarkar, D., Kumar, M.R., Saul, J., Kind, R., Raju, P.S., Chadha, R.K., Shukla, A.K., 2003. A receiver function perspective of the Dharwar craton (India) crustal structure. Geophys. J. Int. 154, 205–211.
Sastri, V.V., 1995. Deltas of east coast of India and their hydrocarbon potential. In: Technology Trends in Petroleum Industry, Proc. Petrotech. 95, B.R. Publishing Co., Delhi, pp. 91–97
Sastri, V.V., Raju, A.T.R., Sinha, R.N., Venkatachala, B.S., 1974. Evolution of mesozoic sedimentary basins on the east coast of India. APEA 14, 29–41.
Satyavani, N., Dixit, M.M., Reddy, P.R., 2001. Crustal velocity along the Nagaur-Rian sector of the Aravalli fold belt, India, using reflection data. J. Geodyn. 31, 429–443.
Saul, J., Kumar, M.R., Sarkar, D., 2000. Correction to "Lithospheric and upper mantle structure of the Indian Shield, from teleseismic receiver functions". Geophys. Res. Lett. 0094-8276. https://doi.org/10.1029/2000GL900012.
Schmerr, N., 2012. The Gutenberg discontinuity: melt at the lithosphere-asthenosphere boundary. Science 335, 1480–1483.
Seeber, L., Armbruster, J.G., 1981. Great detachment of earthquakes along the Himalayan arc and long-term forecasting. In: Maurice Ewing Series, vol. 4. AGU, Washington, DC, pp. 259–277.
Seeber, L., Armbruster, J.G., Quittmeyer, R.C., 1981. Seismicity and continental subduction in the Himalayan arc. In: Gupta, H.K., Delany, F.M. (Eds.), Zagros, Hindukush, Himalayan-Geodynamic Evolution. In: Geodyn. Ser., vol. 3. AGU, Washington, DC, pp. 215–242.
Sen, D., Sen, S., 1983. Post-Neocene tectonism along the Aravalli range, Rajasthan, India. Tectonophysics 93, 75–98.
Sen Gupta, S., 1996. Geological and geophysical studies in western part of Bengal basin, India. Am. Assoc. Petr. Geol. Bull. 50, 1001–1017.
Sen Gupta, S.N., 1967. Structure of the Gulf of Cambay. In: Proc. Symp. Upper Mantle Project, Hyderabad, pp. 334–341.
Shams, F.A., Ahmad, S., 1979. Petrochemistry of some of granitic rocks from the Nanga Parbat massif, NW Himalayas, Pakistan. Geol. Bull. Univ. Peshawar 11, 181–187.
Shapiro, N.M., Ritzwoller, M.H., Molnar, P., Levin, V., 2004. Thinning and flow of Tibet crust constrained by seismic anisotropy. Science 305, 223–236.
Sharma, R.S., 1988. Patterns of metamorphism in the Precambrian rocks of the Aravalli mountain belt. In: Roy, A.B. (Ed.), Precambrian of the Aravalli Mountain, Rajasthan, India. Geol. Soc. India, Mem. 7, 33–75.
Sharma, S.R., Mall, D.M., 1998. Geothermal and seismic evidence for the fluids in the crust beneath, Koyna, India. Curr. Sci. 75 (10), 1070–1074.
Sharma, S.R., Sunder, A., Rao, V.K., Ramana, D.V., 1991. Surface heat flow and P_n velocity distribution in Peninsular India. J. Geodyn. 13, 67–76.
Shinohara, M., Fukano, T., Kanazawa, T., Araki, E., Suyehiro, K., Mochizuki, M., Nakahigashi, K., Yamada, T., Mochizuki, K., 2008. Upper mantle and crustal seismic structure beneath the northwestern Pacific basin using seafloor borehole broadband seismometer and ocean bottom seismometers. Phys. Earth Planet. Inter. 170, 95–106.
Singh, A.P., Mall, D.M., 1998. Crustal Accretion beneath Koyna Coastal Region, (India) and late Cretaceous Geodynamics. Tectonophysics 290, 285–297.

Singh, A.P., Meissner, R., 1995. Crustal configuration of the Narmada-Tapti region (India) from gravity studies. J. Geodyn. 20, 111–127.

Singh, D., Alat, C.A., Singh, R.N., Gupta, V.P., 1997. Source rock characteristics and hydrocarbon generating potential of Mesozoic sediments in Lodhika area, Saurashtra basin, Gujarat, India. In: Proc. Second International Petroleum Conference and Exhibition, PETROTECH-97, New Delhi, Oil and Natural Gas Corporation Limited (ONGC), Dehradun, India, pp. 205–207.

Singh, S.K., Dattatreyam, R.S., Shapiro, N.M., Mandal, P., Pacheco, J.F., Midha, R.K., 1999. Crustal and upper mantle structure of peninsular India and some parameters of 21 may 1997, Jabalpur earthquake (Mw = 5.8): results from a new revised broadband network. Bull. Seismol. Soc. Am. 89, 1631–1641.

Singh, A.P., Mishra, D.C., Vijaya Kumar, V., Vyaghreswar Rao, M.B.S., 2003. Gravity-magnetic signatures and crustal architecture along Kuppam-Palani Geotransect, South India. In: Ramakrishnan, M. (Ed.), Tectonics of Southern Granulite Terrene, Kuppam-Palani Geotransect. Geol. Soc. India, Mem. 50, 139–163.

Singh, A., Singh, C., Kennett, B.L.N., 2015. A review of crust and upper mantle structure beneath the Indian subcontinent. Tectonophysics 644–645, 1–21.

Sinha-Roy, S., 1988. Proterozoic Wilson cycles in Rajasthan. In: Roy, A.B. (Ed.), Precambrian of the Aravalli Mountain, Rajasthan, India. Geol. Soc. India, Mem. 7, 95–108.

Sinha-Roy, S., 1999. Proterozoic sutures in Rajasthan. In: Proc. Seminar on Geology of Rajasthan: Status and Perspective. MLS University, Udaipur, pp. 87–100.

Sinha-Roy, S., 2000. Plate tectonic evolution of crustal structure in Rajasthan. In: Verma, O.P., Mahadevan, T.M. (Eds.), Research Highlights in Earth System Sciences. Ind. Geol. Cong, Roorkee, pp. 17–25.

Sinha-Roy, S., Malhotra, G., Guha, D.B., 1995. A transect across Rajasthan Precambrian terrane in relation to geology, tectonics and crustal evolution of south central Rajasthan. In: Sinha Roy, S., Gupta, K.R. (Eds.), Continental crust of NW and central India. Geol. Soc. India, Mem. 31, 63–89.

Sivaji, Ch., Agrawal, B.N.P., 1995. Application of the relaxation techniques in mapping of crustal discontinuities from Bouguer anomalies over central India. In: Sinha-Roy, S., Gupta, K.R. (Eds.), Continental crust of NW and Central India. Geol. Soc. India, Mem. 31, 425–578.

Sivraman, T.V., Raval, U., 1995. U-Pb isotopic study of Zircons from a few granitoids of Delhi-Aravalli belts. J. Geol. Soc. India 46, 461–475.

Snelsonl, C.M., Keller, G.R., Miller, K.C., Rumpel, H., Prodehl, C., 2005. Regional crustal structure derived from the CD-ROM 99 seismic refraction/wide-angle reflection profile: the lower crust and upper mantle. Am. Geophys. Union. https://doi.org/10.1029/154GM21.

Snyder, A., England, R.W., McBridge, J.H., 1997. Linkage between mantle and crustal structures bearing on inherited structures on NW Scotland. J. Geol. Soc. Lond. 154, 79–83.

Sollogub, V.B., Litvinenko, I.V., Chekunov, A.V., Ankudinov, S.A., Iranov, A.A., Kalyuzhnaya, L.K., Kokorina, L.K., Tripolsky, A.A., 1973. New D.S.S.-data on the crustal structure of the Baltic and Ukrainian shields. Tectonophysics 20, 67–84.

Sollogub, V.B., Chekunov, A.V., Kharechko, G.E., Tripolsky, A., Babinets, V.A., 1977. Structure of the Earth's crust in the region of old platforms. Publ. Inst. Geophys. Pol. Acad. Sci. A-4 (115), 457–466.

Sridhar, A.R., Tewari, H.C., 2001. Existence of a sedimentary basin under the Deccan volcanics in the Narmada region. J. Geodyn. 31, 19–31.

Sridhar, A.R., Tewari, H.C., Vijaya Rao, V., Satyavani, N., Thakur, N.K., 2007. Crustal velocity structure of the Narmada-Son Lineament along the Thuadara-Sendhwa-Sindad profile in the NW part of central India and its geodynamic implications. J. Geol. Soc. India 69, 1147–1160.

Steinhart, J., 1967. Mohorovičićic Discontinuity. International Dictionary of Geophysics. vol. 2. Pergamon, Oxford. pp. 991–994.

Stern, R.J., 2008. Modern-style plate tectonics began in neoproterozoic time: an alternative interpretation of Earth's tectonic history. In: Condie, K., Pease, V. (Eds.), When Did Plate Tectonics Begin? In: vol. 440. Geol. Soc. Am., Special Paper, Boulder, CO, pp. 265–280.

Stern, R.J., Johnson, P., 2010. Continental lithosphere of the Arabian plate: a geologic, petrologic and geophysical synthesis. Earth Sci. Rev. 101, 29–67.

Subbarao, K.V., 1988. Introduction to Deccan flood volcanics. In: Subbarao, K.V. (Ed.), Deccan Flood Volcanics. Geol. Soc. India Mem. 10, 5–13.

Sugden, T.J., Deb, M., Windely, B.F., 1990. The tectonic setting of mineralisation in Proterozoic Aravalli Delhi orogenic belt, NW India. In: Naqvi, S.M. (Ed.), Precambrian Continental Crust and its Economic Resources, Developments in Precambrian Geology. Elsevier, Amsterdam, pp. 367–390.

Sychanthavong, S.P.H., Desai, S.D., 1977. Protoplate tectonics controlling the Precambrian deformations and metallogenic epochs of NW Peninsular India. Miner. Sci. Eng. 9, 218–237.

Tamashiro, I., Santosh, M., Sajeev, K., Morimoto, T., Tsunogae, T., 2004. Multistage orthopyroxene formation in ultrahigh temperature granulites of Ganguvarpatti, southern India: implications for complex metamorphic evolution during Gondwana assembly. J. Mineral. Petrol. Sci. 99, 279–297.

Tan, Y., Helmberger, D.V., 2007. Trans-Pacific upper mantle shear velocity structure. J. Geophys. Res. 112, B08301. https://doi.org/10.1029/2006JB004853.

Tandon, A.N., 1954. Study of the great Assam earthquake of august 1950 and its aftershocks. Indian J. Meteorol. Geophys. 5, 95–137.

Tapponier, P., Zhiqin, X., Roger, F., Meyer, B., Arnaud, N., Wittlinger, G., Jingsui, Y., 2001. Oblique stepwise rise and growth of the Tibet plateau. Science 294, 1671–1677.

Taylor, S.R., 1967. The origin and growth of continents. Tectonophysics 4, 17–34.

Taylor, S.R., McLennan, S.M., 1985. The Continental Crust: Its Composition and Evolution. Blackwell, Oxford. 312 pp.

Teng, J., Zhang, Z., Zhang, X., Wang, C., Gao, R., Yang, B., Qiao, Y., Deng, Y., 2013. Investigation of the Moho discontinuity beneath the Chinese mainland using deep seismic sounding profiles. Tectonophysics 609, 202–216.

Tewari, H.C., 1998. The effect of thin high velocity layers on seismic refraction data: an example from Mahanadi basin, India. PAGEOPH 151, 63–79.

Tewari, H.C., Kumar, P., 2003. Deep seismic sounding studies in India and their tectonic implications. In: Rajaram, Mita (Ed.), Geophysics: Window to Indian Geology. J. Virtual Explor. 12, 30–54 (Web Journal).

Tewari, H.C., Rao, V.K., 1987. A high velocity intrusive body in the upper crust in southeastern Cuddapah basin as delineated by deep seismic sounding and gravity modelling. Geol. Soc. India, Mem. 6, 349–356.

Tewari, H.C., Rao, V.K., 1990. Crustal velocity model in the eastern part of the Indian peninsular shield. J. Geol. Soc. India 36, 475–483.

Tewari, H.C., Rao, V.K., 1995. Seismic studies in the Cuddapah basin: a review. In: Tirupati '95, Seminar on Cuddapah basin, Geol. Soc. of India. Sri Venkateswara University, Tirupati, pp. 67–79.

Tewari, H.C., Vijaya Rao, V., 2003. Structure and tectonics of the Proterozoic Aravalli-Delhi geological province, NW Indian peninsular shield. In: Mahadevan, T.M., Arora, B.R., Gupta, K.R. (Eds.), Indian Continental Lithosphere: Emerging Research Trends. Geol. Soc. India, Mem. 53, 57–78.

Tewari, H.C., Rao, V.K., Kaila, K.L., 1986. A model of the crust beneath the Cuddapah from deep seismic soundings and gravity. In: Kaila, K.L., Tewari, H.C. (Eds.), Deep Seismic Soundings and Crustal Tectonics. A.E.G. Publication, Hyderabad, pp. 11–16.

Tewari, H.C., Mall, D.M., Kaila, K.L., 1988. Interpretation of gravity features in the Mahanadi basin based on deep seismic sounding studies. In: Bhattacharya, B.B. (Ed.), Frontiers of Exploration Geophysics. Oxford and IBH Publ. Co. Ltd., New Delhi, pp. 336–348

Tewari, H.C., Dixit, M.M., Sarkar, D., Kaila, K.L., 1991. A crustal density model across the Cambay basin, India, and its relationship with the Aravallis. Tectonophysics 194, 123–130.

Tewari, H.C., Murty, P.R.K., Dixit, M.M., 1996. Extension of the Pranhita-Godavari Graben Towards Southeast Under the Coastal Godavari Basin: Evidence From Refraction Seismics. In: Gondwana Nine, vol. 2. Oxford & IBH Publishing Co. Pvt. Ltd, New Delhi/Calcutta, pp. 953–962.

Tewari, H.C., Dixit, M.M., Madhava Rao, N., Venkateswarlu, N., Vijaya Rao, V., 1997a. Crustal thickening under the Paleo-Mesoproterozoic Delhi fold belt in northwestern India evidence from deep reflection profiling. Geophys. J. Int. 129, 657–668.

Tewari, H.C., Rajendra Prasad, B., Vijaya Rao, V., Reddy, P.R., Dixit, M.M., Rao, M.N., 1997b. Crustal reflectivity parameter for deciphering the evolutionary processes across the Proterozoic Aravalli-Delhi fold belt. J. Geol. Soc. India 50, 779–785.

Tewari, H.C., Divakar Rao, V., Narayana, B.L., Dixit, M.M., Madhava Rao, N., Murthy, A.S.N., Rajendra Prasad, B., Reddy, P.R., Venkateswarlu, N., Vijaya Rao, V., Mishra, D.C., Gupta, S.B., 1998. Nagaur-Jhalawar Geotransect across the Delhi/Aravalli Fold Belt in Northwest India. J. Geol. Soc. India 52, 153–161.

Tewari, H.C., Vijaya Rao, V., Rajendra Prasad, B., 2000. Tectonic significance of seismic reflectivity patterns: a study from the NW Indian shield. In: Deb, M. (Ed.), Crustal Evolution and Metallogeny in the North-Western Indian Shield. Narosa Publishing House, New Delhi, pp. 189–202.

Tewari, H.C., Murty, A.S.N., Kumar, P., Sridhar, A.R., 2001. A tectonic model of the Narmada zone. Curr. Sci. 20, 273–877.

Tewari, H.C., Rao, G.S.P., Rajendra Prasad, B., 2009. Uplifted crust in parts of western India. J. Geol. Soc. India 73, 479–488.

Thybo, H., 2001. Crustal structure along the EGT profile across the Tornquist fan interpreted from seismic, gravity and magnetic data. Tectonophysics 334, 155–190.

Thybo, H., et al., 2003. Upper lithospheric seismic velocity structure across the Pripyat trough and the Ukrainian shield along the EUROBRIDGE'97 profile. Tectonophysics 371 (1–4), 41–79.

Tiira, T., Hyvonen, T., Komminaho, K., Korja, A., Heikkinen, P., 2006. 3-D inversion of Moho discontinuity using wide-angle reflections in the Baltic Shield. In: EGU Abstracts, Vienna, April 2006. EGU06-A-06893.

Tiwari, S., 1983. The problem at Calcutta gravity high and its solution. Bull. Oil Nat. Gas Commission 20, 48–61.

Trofimov, V.A., 2006. Deep CMP seismic surveying along the Tatseis-2003 geotraverse across the Volga–Ural petroliferous province. Geotektonika 4, 3–20 (English translation in Geotectonics 40, 249–262).

Turcotte, D.L., 1989. Geophysical processes affecting the lower continental crust. In: Mereu, R.F., Mueller, St, Fountain, D.M. (Eds.), Properties and Processes of Earth's Lower Crust. Am. Geophys. Union Monogr. 51, 321–330.

Valdiya, K.S., 1984. Aspects of Tectonics: Focus on South-Central Asia. Tata McGraw-Hill, New Delhi, p. 319.

Valdiya, K.S., 2001. Tectonic resurgence of the Mysore plateau and surrounding regions in cratonic Southern India. Curr. Sci. 81, 1068–1089.

Vandamme, D., Courtillot, V., Besse, J., Montigny, R., 1991. Palaeomagnetism and age determinations of the Deccan traps (India): results of a Nagpur – Bombay traverse and review of earlier works. Rev. Geophys. 29, 159–190.

Van Kranendonk, M.J., Ivanic, T.J., Wingate, M.T.D., Kirkland, C.L., Wyche, S., 2011. Crustal evolution of the Murchison domain and implications for whole-of-Yilgarn tectonic models. In: GSWA 2011 Extended Abstracts: Promoting the Prospectivity of Western Australia. Geological Survey of Western Australia, Crawley, pp. 31–33.

Van Kranendonk, M.J., Ivanic, T.J., Wingate, M.T.D., Kirkland, C.L., Wyche, S., 2014. Long-lived, autochthonous development of the Archean Murchison Domain, and implications for Yilgarn Craton tectonics. Precambrian Res. 229, 49–92.

Venkatesan, T.R., Pande, K., Gopalan, K., 1986. 40Ar/39Ar dating of Deccan volcanics. J. Geol. Soc. India 27, 102–109.

Venkat Rao, K., Srirama, B.V., Ramasastry, P., 1990. A geophysical appraisal of Mahakoshal Group of upper Narmada valley. Precambrian of Central India. Geol. Surv. India Spec. Publ. 28, 99–117.

Venkatakrishnan, R., Dotiwalla, F., 1987. The Cuddapah Salient: a tectonic model for the Cuddapah basin. India, based on Landsat image interpretation. Tectonophysics 1136, 237–253.

Verma, R.K., Gupta, M.L., Hamza, V.M., Venkateshwar Rao, G., Rao, R.U.M., 1968. Heat flow and crustal structure near Cambay, Gujarat, India. Bull. Natl. Geophys. Res. 6, 49–59.

Vera, E.E., Diebold, J.B., 1994. Seismic imaging of oceanic layer 2A between 9°30′N and 10°N on the east Pacific rise from two-ship wide-aperture profiles. J. Geophys. Res. 99, 3031–3041.

Verma, R.K., Banerjee, P., 1992. Nature of continent al crust along the Narmada-Son lineament, inferred gravity and deep seismic sounding data. Tectonophysics 202, 375–397.

Vijaya Rao, V., Rajendra Prasad, B., 2006. Structure and evolution of the Cauvery Shear Zone system, Southern Granulite Terrene, India: evidence from deep seismic and other geophysical studies. Gondwana Res. 10, 29–40.

Vijaya Rao, V., Reddy, P.R., 2002. A mesoproterozoic supercontinent: evidence from the Indian Shield. In: Rogers, J.J.W., Santosh, M. (Eds.), Special Volume on Mesoproterozoic Supercontinent. Gondwana Res. 5, 63–74.

Vijaya Rao, V., Rajendra Prasad, B., Reddy, P.R., Tewari, H.C., 2000. Evolution of Proterozoic Aravalli-Delhi fold belt in the northwestern Indian Shield from seismic studies. Tectonophysics 327, 109–130.

Vijaya Rao, V., Sain, K., Reddy, P.R., Mooney, W.D., 2006. Crustal structure and tectonics of the northern part of the Southern Granulite terrane, India. Earth Planet. Sci. Lett. 251, 90–103.

Vijaya Rao, V., Sain, K., Krishna, V.G., 2007a. Modelling and inversion of single sided refraction data from the shot gathers of multifold deep seismic reflection profiling—an approach for deriving the shallow velocity structure. Geophys. J. Int. 169, 507–514.

Vijaya Rao, V., Sain, K., Rajendra Prasad, B., 2007b. Dipping Moho in the southern part of the Eastern Dharwar craton, India, as revealed by the coincident seismic reflection and refraction study. Curr. Sci. 93, 330–336.

Volpe, A.M., Macdougall, J.D., 1990. Geochemistry and isotope characteristics of mafic (Phulad Ophiolite) and related rocks in the Delhi super group, Rajasthan, India: implications for rifting in the Proterozoic. Precambrian Res. 48, 167–191.

Wang, Q., Bagdassarov, N., Shaocheng, J., 2013. The Moho as a transition zone: a revisit from seismic and electrical properties of minerals and rocks. Tectonophysics 609, 395–422.

West, W.D., 1962. The line of the Narbada and Son Valleys. Curr. Sci. 31, 143–144.

White, R.S., McKenzie, D.P., 1989. Magmatism at rift zones: the generation of volcanic continental margins and flood basalts. J. Geophys. Res. 94, 7685–7729.

White, D., Forsyth, D., Asudeh, I., Carr, S., Wu, H., Easton, R., Mereu, R., 2000. A seismic based cross-section of the Grenville Orogen in southern Ontario and western Quebec. Can. J. Earth Sci. 37, 183–192.

Willet, S.D., Beaumont, C., 1995. Subduction of Indian lithosphere mantle beneath Tibet inferred from models of continental collision. Nature 369, 642–645.

Xiao, W., Han, C., Yuan, C., Sun, M., Zhao, G., Shan, Y., 2010. Transitions among Mariana-, Japan-, Cordillera- and Alaska-type arc systems and their final juxtapositions leading to accretionary and collisional orogenesis. Geol. Soc. Lond., Spec. Publ. 338, 35–53.

Yang, Y., Li, A., Ritzwoller, M.H., 2008. Crustal and uppermost mantle structure in southern Africa revealed from ambient noise and teleseismic tomography. Geophys. J. Int. 174, 235–248.

Yedekar, D.B., Jain, S.C., Nair, K.K.K., Dutta, K.K., 1990. The central Indian collision suture. Precambrian of Central India. Geol. Surv. India Spec. Publ. 28, 1–43. Nagpur.

Zhao, W., Nelson, K.D., Project INDEPTH team, 1993. Deep seismic reflection evidence for continental underthrusting beneath southern Tibet. Nature 366, 557–559.

Zhao, W., et al., 2001. Crustal structure of central Tibet as derived from project INDEPTH-III wide-angle data. Geophys. J. Int. 145, 486–498.

Zhou, L., Chen, W.-P., Ozalaybey, S., 2000. Seismic properties of the central Indian shield Limei Zhou. Bull. Seismol. Soc. Am. 90, 1295–1304.

Zorin, Y.A., 1999. Geodynamics of the western part of the Mongolia-Okhotsk collisional belt, Trans-Baikal region (Russia) and Mongolia. Tectonophysics 306, 33–56.

INDEX

Note: Page numbers followed by *f* indicate figures, and *t* indicate tables.

F

G

H

O

P

Q

R

S

Printed in the United States
By Bookmasters